A Comprehensive Study of Volcanic Phenomena

Edited by Károly Németh

Published in London, United Kingdom

A Comprehensive Study of Volcanic Phenomena
http://dx.doi.org/10.5772/intechopen.1002554
Edited by Károly Németh

Contributors
Abdulrahman Sowaigh, Brice Molo, Daniel Alberto Gómez-Calderón, Emmanuel Chapron, Fawaz Moqeem, Francisco Abraham Paz-Moreno, Gülin Gencoglu Korkmaz, Hüseyin Kurt, Joan Martí, Károly Németh, Leonid Chernogor, Mahmoud Ashor, María del Sol Hernández-Bernal, Michel Meybeck, Mostafa Toni, Paolo Favali, Pedro Corona-Chávez, Pierre Lavina, Shigekazu Kusumoto, Tiziana Sgroi, Vladimir Sokolov, Yiyang Luo, Özlem Hacıoğlu

First published in London, United Kingdom, 2025 by IntechOpen
IntechOpen is the global imprint of INTECHOPEN LIMITED, registered in England and Wales, registration number: 11086078, 167-169 Great Portland Street, London, W1W 5PF, United Kingdom

For EU product safety concerns: IN TECH d.o.o., Prolaz Marije Krucifikse Kozulić 3, 51000 Rijeka, Croatia, info@intechopen.com or visit our website at intechopen.com.

British Library Cataloguing-in-Publication Data
A catalogue record for this book is available from the British Library

A Comprehensive Study of Volcanic Phenomena
Edited by Károly Németh
p. cm.

This title is part of the Earth Sciences Book Series, Volume 2
Series Editor: Maurizio Lazzari

Print ISBN 978-0-85466-626-3
Online ISBN 978-0-85466-625-6
eBook (PDF) ISBN 978-0-85466-627-0
ISSN 3049-8848

IntechOpen Book Series

Earth Sciences

Volume 2

Aims and Scope of the Series

The world of Earth Sciences, considering the interactions within the geosphere and between the geosphere–biosphere, is a place in which a large number of scientists find and have found over time their own relationship or sector of application precisely because it is absolutely transversal to many disciplines and subdisciplines, which do not necessarily fall within the Geosciences. The objective of this book series is to welcome original scientific contributions both in consolidated contexts and in new frontiers of research, as well as review papers included in the various disciplines of Earth Sciences, but above all, those that show a modern and transversal vision of applications and impacts on the community in a particular historical context, which, following the COVID-19 pandemic, has shifted global attention to sectors that were previously more neglected. In particular, those of mining research and fossil and renewable energy sources, environmental geology and the sustainable use of natural resources and impacts on the built environment, land consumption, geoarchaeology, forensic geology, geotourism/geoheritage, georisks and climate and environmental changes, considered at different scales, up to new applications of geostatistical and geospatial analysis, GIS and artificial intelligence for the definition of forecasting models and scenarios in various sectors of basic and applied research.

Meet the Series Editor

Dr. Maurizio Lazzari has a Ph.D. in Earth Science and is a researcher at the Italian National Research Council, Institute of Cultural Heritage Sciences. Since 2001, Dr. Lazzari has been a Professor of Pedology at the University of Basilicata (Italy) and a geoarchaeologist at the University of Salento (Italy). His research activities are focused on natural and anthropic hazards and risk factors, aimed at safeguarding and conserving settlements and the historical–monumental heritage of the Mediterranean, with particular attention to landslide processes, susceptibility maps, monitoring, and modelling. Since 2004, he has been working as a scientific coordinator for several national research projects studying landslides and triggering factors, natural and anthropogenic risks, geological and geomorphological mapping, soil erosion, preservation of historical and archaeological sites, enhancement of degraded areas, geo-touristic use, and the protection of the landscapes. He is the author of about 150 scientific publications in national and international journals, monographs, book chapters, and conference proceedings concerning applied geology, geomorphology, dynamics of artificial reservoirs, soil erosion, landslides, geoarchaeology, hydrogeological instability, natural hazards, monitoring, cultural landscape, UNESCO Heritage, geoarchaeology, and geo-tourism.

Meet the Volume Editor

Károly Németh is a Hungarian Academy of Sciences (DSc) doctor and an adjunct professor at Massey University in New Zealand. He holds dual Ph.D.s in geology and environmental sciences. He was a member of the executive committee of the International Association of Volcanology and Chemistry of Earth Interior (2011–2015) and vice president of the International Association of Sedimentologists (2022–2026). He is currently the chair of the IUGS Commission on Geological Heritage, Sub-commission on (Geo)Sites (2024–2028). He works for the Saudi Geological Survey and Hungary's Institute of Earth Physics and Space Science. His work focuses on monogenetic volcanism, volcano geology, and volcanic geoheritage.

Contents

Preface

Volcanoes are unique elements of the Earth system. They appear in many geotectonic situations and operate in various spatial and temporal scales. Small volcanoes are commonly referred to as monogenetic, emphasizing the nature of a "one-shot" activity style where a small-volume magma batch feeds an eruption that lasts a very short time. New eruptions then occur in different locations above a broad melting anomaly in the Earth's deep interior. Such volcanism often forms groups of volcanoes that are clustered or aligned and together form a volcanic field that may last over millions of years. Another type of volcano that characteristically evolves over a stable conduit with diverse eruption styles is defined as a polygenetic volcano. In such volcanoes, the magmatic feeding system is relatively stable in a time scale of tens to hundreds of thousand years. This type of volcanism can build volcanoes that can influence the host sedimentary basins, terrestrial or marine, over a long time and can contribute primary and reworked volcaniclastic material to the evolving basin. This type of volcano normally adds new landscape elements to the existing systems but can also act through impactful eruptions as landscape-forming systems that can significantly change large areas' geological and geomorphological appearance. In recent years, volcanology has rediscovered the importance of understanding the volcanic architecture and succession of eruption styles associated with small or large volcanoes. Volcano geology has advanced significantly in recent years by putting volcanism in the Earth's system perspective and providing key information to unfold the eruption mechanism locked in the volcanic geological record. Currently, volcano geology research is claiming its share within volcano science and influencing the application of advanced technologies to develop more realistic volcano models.

Alongside this new development to understand volcanic phenomena by going back to the fundamentals of geology, volcanology is incorporating new methods and models from the fields of geophysics, remote sensing, and numerical modelling. The combination of various levels of advanced geophysical methods to visualize the subsurface architecture of a magmatic–volcanic system—applying new geochemical systematics to trace the physicochemical processes leading to a volcanic eruption, using remote sensing data to map volcanic bodies and applying numerical modelling from process simulations to volcanic hazard map generation—can be seen as a major new development within volcano science. Volcanology is also experiencing an expansion to investigate areas normally left out of the mainstream research and incorporate other geological disciplines to solve volcanic eruption and process reconstruction from the magma rise to fragmentation, transportation, and deposition. More and more regions are paying attention to creating a unified stratigraphy framework within which volcanic rocks are handled similarly to normal sedimentary rocks.

As population growth increases dramatically, more regions where volcanic hazards are evident are occupied by human society and its infrastructures. This situation generates an ever-increasing risk to the human society that needs to be mitigated. In many cases, the benefit of volcanism to human society overrides the potential risk the inhabitants face. In this situation, the concept of living with volcanic hazards becomes a critical approach to managing daily life in an active volcanic terrain. Volcanic hazard mapping and probabilistic hazard estimates and predictions become a major subject of geohazard management in many Holocene volcanic regions. New mapping and visualization techniques and specific communication avenues associated with volcanic hazards have opened new research areas. Recognizing the geoheritage values of volcanic terrains and incorporating those in geoeducation and geotourism through an effective geoconservation strategy is also a new conceptual framework for understanding volcanic systems. In this book, a snapshot of these advancements is showcased.

Károly Németh
Saudi Geological Survey,
National Program of Earthquakes and Volcanoes,
Jeddah, Kingdom of Saudi Arabia,

Institute of Earth Physics and Space Science,
Sopron, Hungary

Section 1

Volcano Geology Approaches

Chapter 1

Volcano Geology: A Fundamental Tool to Quantify Volcanic Hazards

Joan Martí

Abstract

Volcanic hazard assessment relies on analyzing past eruptions documented in historical and geological records to identify potential future eruptive scenarios and assign probabilities to their occurrence. This process involves quantifying geological data, such as the composition, distribution, and textures of volcanic products, to understand magma properties, eruption dynamics, and depositional mechanisms. Such data are crucial for evaluating the potential impact, exposure, and vulnerability of areas at risk from future eruptions. Both probabilistic and deterministic approaches may be used, depending on the purpose of the hazard assessment, but geological data remain the foundation for models that predict volcanic hazards. However, volcanic systems are inherently complex, making it difficult to precisely quantify these data. As a result, only thresholds or ranges are often provided, contributing to significant uncertainty in the assessment. This uncertainty must be clearly communicated for accurate interpretation of the results. Additionally, many active volcanoes lack sufficient data, especially those with long intervals between eruptions, further complicating hazard assessments. Despite these challenges, volcanic hazard assessment is critical for reducing risks and preparing for future volcanic events.

Keywords: volcano geology, hazard assessment, simulation models, statistical models, monogenetic volcanism, polygenetic volcanism, uncertainty

1. Introduction

Volcanology has made great advances in the last decades, becoming a modern, interdisciplinary science aiming at quantifying volcanic processes and their associated hazards and impacts on society and the environment [1, 2]. The explosion of new knowledge and techniques (e.g., ion microprobe, numerical modeling, high-performance computing) has reduced the prominence and perceived value of basic geology, despite it remains as the main source of data for volcanic systems, processes, and hazard modeling. One of the most important changes has occurred with the emergence of experimental and mathematical models and through the application of statistical methods, including spatial statistics with the aid of GIS applications, which have allowed significant progress in the understanding of volcanic phenomena and in the prediction of future eruptions (e.g., [3–5]). Such models have helped us to visualize and quantify, for example, the changes occurring at the interior of volcanic systems due to magma movement or the eruption behavior of some volcanoes, so we can say that the irruption of simple and sophisticated physical and mathematical modeling has forced a true revolution inside volcanology during last years.

However, as it occasionally occurs with new exciting and rapid changes, not all contributions are good enough or fully satisfy the created expectations, such as the unrealistic ones to be able to predict and tell the location and style of eruptions in the human lifespan, or human investment scale typical in years to decades. Some of the current models explaining volcanological processes do not fully consider the physical constraints that the geological record imposes on some required assumptions and simplifications used to develop such models [6]. In fact, many of these models, in particular, those aimed at forecasting future volcanic events, require a detailed understanding of past events and of the behavior trends shown by volcanoes during their past evolution. Because models can describe only a simplified version of complex natural processes, geologic and field data are critical for constraining model input, necessary for model validation, and will ultimately improve model accuracy [7, 8].

Of particular, importance is the quantification of volcanic hazards and their potential impact, to estimate the risk they may represent (e.g., [9, 10]). This analysis requires the application of statistical models to deduce possible behavior trends in volcanic activity and of sophisticated simulation models to reproduce the occurrence of the different hazards in the current conditions to assess exposure and possible impacts. In both cases, a precise knowledge of the past eruptive history of the volcano is essential to obtain the input parameters necessary in both types of models. Moreover, when volcanoes have a long eruption recurrence, it is necessary to appeal to the study of the geological record, which significantly increases the uncertainty in the knowledge of the volcano's past and, consequently, the uncertainty of the results that can be obtained.

Volcanoes are among the most complex geological systems, as they are characterized by variable time scales and a high range of impacts and intensity of the related processes. Timing has a much wider range than normal geological processes, while their impact can be almost unrecognizable to landscape forming. So, if we go down in time and enter the realm of volcanic geology, we have to deal with and offer a model to be able to handle this huge scale variation in time and space. Moreover, volcanic processes are nonlinear and mostly hidden from the direct observation (e.g., volcano plumbing system, magma generation), which makes them difficult to model as the number of unknowns usually results too high [11]. Despite this, theoretical modeling is essential to constraint geological conceptual models, which sometimes tend to be too imaginative and not according to the laws of physics. At the same time, theoretical models need to be constrained with geological data; otherwise, they may be speculative and unrealistic, leading to misunderstanding the behavior of volcanic systems, particularly when we try to forecast how they will behave in the near future. Although the use of geophysical approaches and theoretical modeling is complementary and very convenient, we always must check their validity in the field; otherwise, huge interpretative mistakes can be made. In order to highlight and strengthen the role of volcano geology as a critical foundation for modern volcanology and, in particular, for the quantification of volcanic hazards, this contribution is a reflection on the significance of geology to volcanological modeling.

2. Advantages and restrictions of volcanological modeling

Before presenting the other sections included in this contribution, it is necessary to introduce certain concepts and constraints about volcanic modeling in general and volcanic hazard modeling, in particular, in order to avoid confusion on is being developed below.

Scientific modeling is broadly understood as the creation of physical, conceptual, or mathematical representations of real-world phenomena that are challenging to observe directly (e.g., [12–14]). These models play a crucial role in explaining and predicting the behavior of real objects or systems across various scientific fields, including physics, chemistry, ecology, and Earth sciences. While modeling is a fundamental aspect of modern science, it is important to note that scientific models are only approximations of the objects or systems they represent—not exact replicas. Consequently, scientists are continually working to improve and refine these models (e.g., [15, 16]).

Scientific models can take various forms, including conceptual, physical, mathematical, computational, statistical, theoretical, or analog (e.g., [17–22]). In volcanology, as in geology more broadly, classical models were primarily conceptual, aiming to illustrate the relationships between different components of a volcanic system through abstract representations that attempted to visualize complex processes. However, the widespread adoption of mathematical models in recent decades has allowed these conceptual models to be more rigorously defined using mathematical equations. These equations can represent the relationships between various variables and account for their potential changes over time. This advancement has made it possible to predict the behavior of a volcanic system or its components under different conditions. Additionally, statistical models have become essential for analyzing and interpreting data, often to make predictions or identify trends—crucial elements in hazard assessment and eruption forecasting. Finally, the increasing sophistication of computational models has enabled scientists to simulate the behavior of complex volcanic processes from magma formation to eruption.

However, scientific modeling has inherent limitations due to the fact that models are not complete representations of reality; they are based on simplified assumptions to create a more manageable version of complex systems [12, 16]. Essentially, any model—regardless of its type—comprises three key components: (i) an information input component, which supplies assumptions and data to the model; (ii) a processing component, which transforms these inputs into estimates; and (iii) a reporting component, which translates these estimates into useful actionable information [16].

In the context of volcanological models, geological and geophysical data—such as stratigraphic, lithological, structural, mineralogical, geochemical, and geochronological information—serve as the primary inputs. However, these data are often accompanied by very commonly significant uncertainty, both epistemic (related to knowledge limitations and the fact that the good observations available are low in numbers, so generalization will be impacted strongly by those available and known examples) and aleatoric (related to inherent randomness). As a result, the models produced must be interpreted with caution as the imprecision inherent in the data used to build these models inevitably carries over into the models' outcomes. This is particularly important when developing statistical and simulation models aimed at quantifying volcanic hazards.

The central challenge in quantifying volcanic hazards lies in how to manage uncertainty (e.g., [10, 11]). This challenge is at the heart of the ongoing debate over whether deterministic or probabilistic models are more effective (e.g., [7, 23–25]).

Deterministic models operate under the assumption that all phenomena are predictable—if all input parameters are known, the output can be precisely determined. The critical question here is: How many of the required input parameters can we actually know? On the other hand, probabilistic models incorporate randomness and uncertainty, meaning that even with full or partial knowledge of the inputs, the output can still vary due to inherent chance factors. As a result, deterministic models tend to offer a more definitive answer, while probabilistic models provide a range of possible outcomes, each with an associated probability. For example, deterministic lava flow modeling aims to predict the exact path, extent, and behavior of a lava flow based on a set of known physical laws, such as heat transfer, lava viscosity, topography, and flow dynamics. The model assumes that if all relevant variables (e.g., eruption parameters, terrain, and magma properties) are known and accurately measured, the behavior of the lava flow can be forecasted with precision. Probabilistic lava flow modeling, on the other hand, recognizes the inherent uncertainties in volcanic processes (such as varying eruption intensities, unpredictable magma behavior, and complex terrain interactions) and therefore predicts lava flow behavior in terms of probabilities. Instead of giving a single flow path, it provides a range of possible flow scenarios, each with a likelihood or probability of occurrence. Both types of models have their place in volcanology. However, it seems evident that anticipating volcanic processes based on past evidence is more effectively addressed through a probabilistic approach. Conversely, predicting the evolution of an ongoing volcanic event over time may be better suited to a deterministic framework. This distinction will be explored in greater detail later in this contribution.

To develop the statistical and physical models necessary for quantifying volcanic hazards, it is crucial to determine the type of geological information required and the level of accuracy needed. In the following sections, we will explore these aspects in detail.

3. Volcanic hazards

When assessing volcanic hazards, it is crucial to recognize that volcanoes are complex, multi-hazard systems. Volcanic eruptions can produce a wide range of direct hazards, such as lava flows, pyroclastic density currents (PDCs), gas emissions, and fallout, as well as indirect hazards such as lahars, tsunamis, landslides, earthquakes, and floods. These hazards often have intricate cause-and-effect relationships and can trigger cascading (cascading hazard) impacts on exposed elements.

Even during periods of quiescence, volcanoes can pose significant risks. Hazards, such as slope instability, meteorologically triggered lahars, debris flows, and avalanches, can arise, influenced by factors such as the lithology of volcanic rocks, soil properties, topography, local tectonics, and background activity. These potential hazards must be considered in hazard assessments as they can significantly affect land use and infrastructure planning.

Therefore, to identify the hazards associated with a specific volcano, it is essential to examine not only its past eruptive history but also its geology and the environmental conditions surrounding it. The dynamics of an eruption—primarily influenced by the composition and rheology of magma—will dictate the type and characteristics (such as extent, volume, and temperature) of the various direct and some indirect hazards the volcano might generate. Additionally, the geological features of the volcanic edifice play a crucial role in determining its response to external triggers, like seismic activity or meteorological events, which can lead to secondary hazards. The impact of these secondary hazards can vary significantly depending on the unique circumstances of each volcano.

A volcanic hazard is defined as the likelihood of a destructive volcanic event impacting a specific area within a certain timeframe [26–31]. This concept focuses on the physical phenomena associated with volcanic activity, including their recurrence, extent, and potential consequences. Evaluating volcanic hazards involves identifying areas that are susceptible to volcanic events, which is crucial for the development and implementation of emergency response strategies and land-use planning. In other words, hazard is the physical event having an impact on a particular area within a specific timeframe and therefore contains implicit spatial (i.e., the probability that the effects of the event will extend over a certain distance or surface area) and temporal (i.e., the probability that it will occur) probabilities.

Hazard assessment is a critical and foundational step in estimating risk and implementing any risk reduction program [31]. It serves as the first step in understanding how a volcano has behaved in the past, identifying the types of hazards it has produced, their extent, potential impacts, and the frequency of eruptions. The primary goal of hazard assessment is to categorize key past eruption scenarios and predict, which are most likely to occur in the future. Ultimately, the purpose of volcanic hazard assessment is to anticipate the nature of future eruptions [27, 31, 32].

This information is vital for land-use planning and the development of emergency services—two essential strategies for reducing risk. Land-use planning helps prevent development in hazardous areas, while emergency services focus on identifying safe zones and establishing evacuation routes in case of an eruption [33, 34]. Additionally, hazard assessment supports vulnerability analysis by providing insights into the type and intensity of impacts on exposed elements [35, 36].

Moreover, hazard assessment plays a crucial role in decision-making during volcanic crises, offering a basis for evaluating potential eruption scenarios and their associated impacts [37]. It also serves as a guide for educational programs and public awareness campaigns, helping to inform local populations about the volcanic hazards they may face [38].

All these previously mentioned aspects necessitate the quantification of volcanic hazards, either by assigning absolute values to the variables that influence their dynamics or by estimating the probabilities of their occurrence. In both cases, most of the required information is obtained from the geological record, which allows for the examination of deposits from past eruptions. This analysis helps to determine the main characteristics of these deposits and to establish their relative position within the volcano's chronological evolution.

4. Geological characteristics of volcanic products

The characteristics of volcanic deposits, including lithology, geometry, internal stratigraphy, facies variations, and contacts with the substrate or other deposits, are determined by their nature and the dynamics of the eruption. Lava flows, products of effusive volcanism, can form stratigraphic units with varying extents and thicknesses depending on the composition of the erupting magma and the magnitude of the eruption. Beyond petrological and mineralogical characterization, which will be conducted through the analysis of rock samples, other critical features will be identified by observing the surface, extent, and cross sections of the flows. The texture,

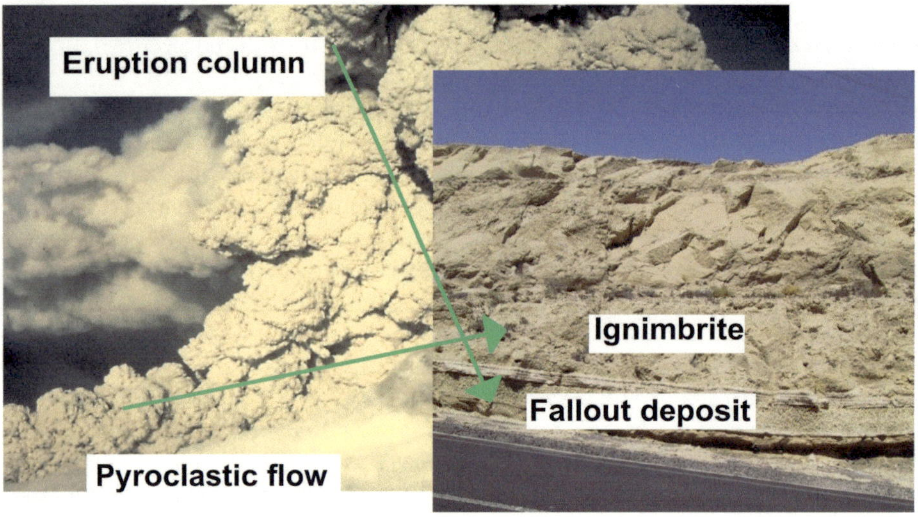

Figure 1.
Interpreting a volcanic succession in terms of eruptive processes, from the study of the field characteristics of volcanic deposits.

mineralogy, and composition of lava flow samples will provide essential constraints on magma rheology (e.g., [39, 40]), while field observations will offer insights into key aspects of lava flow emplacement, such as topographic influences, inflation processes, and total volume (e.g., [41–43]).

For pyroclastic and other volcaniclastic deposits, obtaining realistic input parameters for simulating the processes that control their emplacement requires careful consideration of the nature of the clastic components (**Figure 1**). This includes examining grain morphology, origin, grain-size distribution, degree of sorting, and fabric [44]. Additionally, as with lava flows, it is crucial to define the three-dimensional geometry of the deposit, which is influenced by the terrain's topography, the volume of the deposit, the transport and deposition mechanisms of volcanic materials, and the presence of non-depositional and post-depositional erosive processes. Identifying the sedimentological characteristics, such as stream-generated structures, is also important as it provides essential insights into the emplacement and depositional mechanisms [44]. Finally, the lateral and longitudinal variations in these characteristics (facies changes) offer valuable information on the processes that controlled the emplacement of volcaniclastic deposits (e.g., [45–48]). These variations are crucial for inferring the flow conditions necessary for simulating such volcanic hazards (e.g., [49–51]).

5. Volcanic stratigraphy: The time perspective

Volcanic stratigraphy is an essential tool for generating geological maps, reconstructing volcanic evolution and eruptive dynamics, studying physical volcanology and tephrostratigraphy, and assessing hazards [52]. It plays a critical role in tracing the chronological development of stratigraphic sequences that encompass all volcanic events and inter-eruptive periods within the geological record of a particular volcano or volcanic area. The precision and level of detail in volcanic stratigraphy vary, depending on the time scale of the study whether it is the complete history of a volcanic complex, the last few thousand years, the historical record, recent eruptions, or even the specifics of a single eruptive event. Consequently, different approaches and methodologies may be required tailored to the specific objectives of the research [52]. For example, the way we have to address a volcanic stratigraphy study referring to an ancient volcanic terrain, affected by tectonism and/or strong erosion, is very different than if we are dealing with a recent volcanic area, where all deposits are well-preserved and exposed. In the first case, it may be difficult to recognize the full stratigraphic succession and inside it making distinctions among deposits pertaining to the same or different eruptions. This will force us to concentrate on major discontinuities that may be associated with major geological changes and those subordinated ones that may be representative of changes in the volcano dynamics. However, by studying recent and well-preserved volcanic terrains, the stratigraphic detail may be much more precise, thus allowing recognition of the deposits from different eruptions and changes in the eruptive dynamics during the volcano evolution [52].

In volcanic hazard assessment, volcanic stratigraphy is crucial for interpreting volcanic successions in terms of eruptive sequences and eruptive scenarios, and consequently, to conduct temporal analysis in long-term hazard assessment (**Figure 2**) [31]. The primary goal of such studies is to determine the stratigraphic and chronological order in which the products of an eruption, a series of eruptions, or inter-eruptive periods (including epiclastic deposits or reworked volcanic materials) are preserved in the geological record [44, 52]. To achieve this, it is recommended to apply the same

Figure 2.
An example of a volcanic succession ("La Tarta") from Tenerife shows the emplacement of several mafic (dark) and felsic (light) airfall pyroclastic deposits at a single location. These deposits were generated by different eruptions originating from multiple volcanic vents, illustrating the complex layering of materials from varied sources over time.

principles used in classical stratigraphy [53]. This involves identifying and grouping the various units based on a hierarchy that enables the reconstruction of a temporal sequence of events or units of eruptive activity [52, 54].

As previously mentioned, volcanic deposits can be identified based on their lithology—such as mineralogy, petrology, color, degree of welding, and grain-size distribution—as well as their geometry and relative stratigraphic position. In complex stratigraphic successions, it is possible to distinguish between different deposits from the same eruption (referred to as a member) and different members that constitute a formation [52–55]. This distinction is often based on the presence of first-order discontinuities or time breaks, such as paleosols, erosional surfaces, or interbedded epiclastic deposits. However, it is important to note that erosional surfaces or interbedded epiclastic deposits do not always signify a significant interruption in eruptive activity. This distinction becomes particularly crucial when attempting to reconstruct an eruptive sequence from a poorly exposed succession of deposits, which may be affected by subsequent erosion or erosion occurring during the eruption itself.

Once volcanic stratigraphy is well-established through fieldwork, geochronological methods—such as radiometric isotope dating, paleomagnetism, palynology, and analysis of flora—can be applied to date the different units and determine the eruptive history of a volcanic system. This approach enables the calculation of eruption frequency, which is particularly important for volcanoes with long eruptive intervals and no historical records. Understanding eruption frequency is critical for conducting statistical temporal analyses in volcanic hazard assessments as it helps constrain predictions about when the next eruption may occur [31]. However, determining the "absolute" age of deposits is not always feasible, as it depends on factors, such as the quality and degree of alteration of the samples, their mineralogy, and the limitations of the dating methods. Therefore, it is crucial to establish at least the relative chronology, or volcanic stratigraphy, of the deposits under study.

6. Mapping volcanic areas

Mapping the geology of active volcanic regions is crucial for quantifying volcanic hazards and conducting long-term hazard assessments. This process should be integrated with volcanic stratigraphy and the detailed characterization of individual volcanic units, including facies analysis. Unlike conventional geological mapping, volcanological mapping lacks standardized criteria, leading to variability in how different volcanic zones are represented based on the specific focus (e.g., petrography, geochemistry, radiometry, volcanology, stratigraphy) of each researcher. However, for hazard assessment purposes, it is strongly recommended to delineate volcanic activity units using a combination of stratigraphic and volcanological criteria (**Figure 3**) [52, 54]. This approach, which treats all the deposits from an individual eruption (Member) as the primary unit, has the advantage of directly linking volcanic activity to its corresponding stratigraphic record. Moreover, volcanic stratigraphic units are invaluable for understanding the eruptive dynamics of specific eruptions or sequences of eruptions, thereby aiding in the quantification of volcanic hazards. This type of mapping can also incorporate inter-eruptive periods and deposits, such as debris flows and lahars, which appear as unconformities separating successive eruptive sequences and contribute to estimating eruptive recurrence [55]. The volcanic stratigraphies of Madeira Island (Portugal), Las Cañadas Caldera (Tenerife), or Etna (Italy), as reported in Ref. [52], are good examples of the application of this methodology.

DOI: http://dx.doi.org/10.5772/intechopen.1007213

Figure 3.
Overview from a south-to-north perspective of the southern sector of Tenerife, showcasing the profiles of the Las Cañadas caldera and Teide volcano in the background. In the foreground, recent monogenetic cones can be seen resting on or surrounded by the ignimbrites from Las Cañadas. Volcanological mapping of this region would primarily involve identifying the stratigraphic contacts between these deposits and analyzing their geometric relationships. This includes recognizing unconformities and time gaps, such as paleosols and erosion surfaces, which indicate breaks between eruptions. Once the relative volcanic stratigraphy is well-defined, geochronological dating can be applied to determine the eruptive frequency. The accuracy of this frequency will depend on how complete the stratigraphic succession is.

7. Volcanic susceptibility, exposure, and impact: Monogenetic vs. central systems

In volcanology, as with other natural phenomena, predicting the future behavior of a volcano relies on scientific knowledge of the relevant processes and data from past events. However, this field presents challenges that make volcano forecasting less straightforward for the general public to understand, especially when compared to weather forecasting. A significant limitation is the scarcity of observational data and limited experience in predicting volcanic activity. Modern volcanology is a relatively young science, and unlike meteorology—where consistent methods have been refined over the past two centuries—volcano prediction is still in its early stages. Volcanologists are working to understand eruption precursors and anticipate related hazards [7, 11]. While analyzing past eruptions provides a foundation for hazard assessment and inferring potential future outcomes, accurately quantifying these hazards and their impacts remains a difficult task.

Quantifying future volcanic hazards primarily depends on two key factors: the precise location of the eruption vent and the size and intensity of the eruption. These factors determine the type and dynamics of both direct and indirect hazards, which are also influenced by the characteristics of the surrounding terrain [31]. Predicting the location of a future vent requires an understanding of the stress distribution within the volcanic system. This involves compiling a wide range of structural geology data, such as the positions of past vents, vent alignments, fractures, faults, dykes, and *in situ* stress measurements, all categorized by relative age. This data is essential for inferring the current stress field orientation and its temporal variations [56]. The size and intensity of the eruption can be inferred from the compositional and rheological properties of potential eruptive magmas, which necessitates analyzing products from previous eruptions [40, 57, 58]. However, this geological information, while valuable, only provides indirect evidence for predicting volcanic behavior and is inherently subject to significant uncertainty. Additionally, the morphology and topography of the volcano must be considered as they strongly influence the emplacement of most eruption products and are critical for accurate hazard assessment [49, 59].

Volcanoes exhibit a diverse range of forms, tectonic settings, compositions, eruption dynamics, and recurrence intervals, making each one unique in its characteristics. There are two primary types of volcanic systems: polygenetic and monogenetic (**Figure 4**). Polygenetic volcanic systems are known for constructing large volcanic edifices, such as shield volcanoes and stratovolcanoes [60, 61], through the accumulation of lava and pyroclastic materials over multiple eruptions. Polygenetic or central volcanoes have an established vent that is used in dominant cases for new magma to reach the surface, a process that may span hundreds of thousands to millions of years. The eruption frequency of polygenetic systems varies widely, from several years (e.g., Etna) to thousands of years (e.g., Teide). A defining feature of these systems is the presence of a subsurface magma chamber, which is periodically filled and emptied during eruptive episodes. This magma chamber creates a relatively stable stress distribution around it, which typically leads to eruptions originating from a central crater [62, 63]. However, eruptions can also occur from secondary vents on the volcano's flanks. Over time, large central volcanoes may experience significant gravity-induced instabilities, leading to sector collapses that alter the morphology of the slopes and

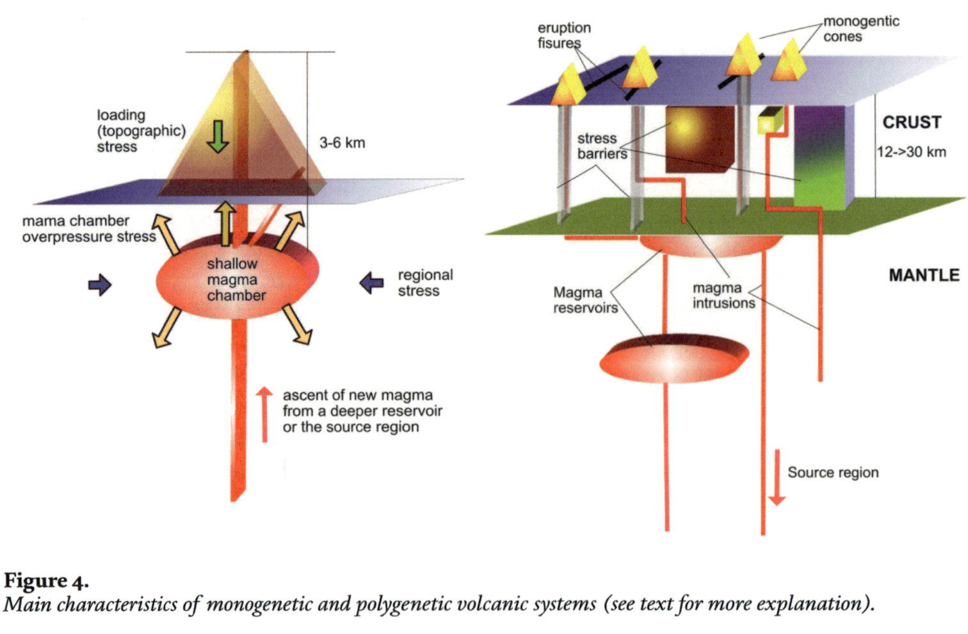

Figure 4.
Main characteristics of monogenetic and polygenetic volcanic systems (see text for more explanation).

basically resetting the original shallow magma plumbing system, shifting vents slightly offset from their original position [64].

Monogenetic volcanism represents the other end of the volcanic system spectrum, typically occurring in volcanic fields that contain anywhere from tens to thousands of small volcanoes, each formed by a single eruption. These eruptions are usually mafic in composition and involve relatively small volumes of magma, well below 1 km^3 DRE, typically in the range of 0.01 to 0.001 km^3. They are often low-intensity explosive eruptions (Strombolian), producing cinder/scoria cones and lava flows, with occasional phreatomagmatic phases due to the explosive interaction of magma with water. Basaltic monogenetic volcanic fields are the most common type of terrestrial volcanism. These fields can remain active for several million years, with eruption intervals ranging from several decades to several tens of thousands of years [60, 65–67]. The distribution of volcanic cones in these fields is largely governed by regional and local tectonic settings, but not by the presence of any shallow magma chamber [68, 69]. The diverse eruptive styles, edifice morphologies, and deposits found in monogenetic volcanoes result from a complex interplay of internal factors (such as magma composition, gas content, rheology, and volume) and external factors (including regional and local stress fields, stratigraphic and rheological contrasts in the substrate rock, and hydrogeology) that characterize each volcanic system [67]. While monogenetic volcanoes or eruptions (i.e., eruptions that occur only once from a specific vent) are most commonly associated with these basaltic fields, they can also occur as flank eruptions in association with polygenetic volcanoes [70, 71].

Volcanic susceptibility [56], defined as the spatial probability of vent opening, is a crucial first step in assessing volcanic hazards. Accurately identifying the potential location of future eruption vents is essential for developing realistic eruption scenarios and predicting their consequences. The hazards and impacts of an eruption can vary significantly depending on where magma breaches the surface within a given area. These differences arise because the hazards, as well as the topographic and geomorphological factors that influence them, can change drastically based on the vent location, affecting exposure and potential impacts. The point where magma will emerge is largely determined by the distribution of stress within the volcano or volcanic zone. This stress distribution is influenced by three primary factors: gravitational stresses (related to lithostatic pressure variations due to topography), tectonic stresses, and magma overpressure. These factors collectively dictate the most likely pathways for magma ascent within a volcanic system [62, 63, 69]. Since direct observation of the interior of active volcanic systems is impossible, we rely on indirect methods. These include geophysical techniques, which often lack precision, and the analysis of surface structural elements to infer the subsurface stress field. Additionally, comparisons with ancient volcanic systems, now exposed at the surface due to tectonic activity or erosion, provide valuable insights. By synthesizing this information, we can create maps that depict the spatial probability of future eruptions based on objective criteria. This forms the foundation for developing comprehensive volcanic hazard maps [56].

Volcano susceptibility maps will vary significantly depending on whether the system in question is monogenetic or polygenetic. In monogenetic systems, the maps will typically show a wide distribution of past eruptive vents, often grouped into clusters based on age and aligned with structural features such as fractures, eruption fissures, and faults that are contemporaneous with the vents. In contrast, polygenetic or central volcanoes tend to concentrate most past vents in the summit area, typically near the main crater. However, susceptibility zones may also extend along the flanks, which must be accounted for when creating hazard maps. Once we identify areas with the highest probabilities of new vent formation, we can run simulation models for various volcanic hazards. These models will help us assess exposure levels and identify the elements at risk in each sector that may be impacted by the potential hazards.

8. Probabilistic vs. deterministic approaches: Objectives and requirements

Both probabilistic and deterministic approaches have been applied in quantifying geohazards (e.g., [23–25, 72–77]). Deterministic models produce outcomes that are entirely determined by the specific parameter values and initial conditions, leading to a single, predictable result. In contrast, probabilistic (or stochastic) models incorporate elements of randomness, meaning that the same set of parameter values and initial conditions can yield a range of different outcomes.

The analysis and quantification of geological data for volcanic hazard assessment is an inherently complex task. It rarely yields precise numbers; instead, it typically produces ranges with considerable uncertainty, making it challenging to use as input parameters for hazard quantification. As a result, hazard quantification must always account for the uncertainty inherent in the scientific data and their interpretation. This presents a significant challenge when conveying scientific information to decision makers and other stakeholders involved in volcanic risk reduction. Consequently, the debate over the use of probabilistic versus deterministic approaches in volcanic hazard quantification remains ongoing [7, 23–25, 72–74].

A probabilistic approach, by definition, provides an estimation of uncertainty and is generally considered more suitable for volcanic hazard assessment [10, 11, 72–74]. This method aligns well with the nature of geological data, which often involves significant uncertainties both, due to lack of knowledge or incomplete information, which can be reduced with more data or better models (epistemic uncertainty), and to the inherent randomness or variability in a system, which cannot be reduced, even with more data (aleatoric uncertainty). Probabilistic approaches allow for the identification and modeling of various scenarios while also accounting for their return periods. However, assessing volcanic hazards probabilistically remains challenging due to the multitude of factors involved and the dynamic nature of volcanic hazards, which are increasingly influenced by other factors, such as climate change. Despite these challenges, probabilistic volcanic hazard assessment is preferred because it offers a more comprehensive analysis, quantifying both the impact and likelihood of events while accounting for associated uncertainties [72–74].

On the other hand, the deterministic approach to volcanic hazard assessment has notable disadvantages. It offers limited consideration of uncertainty, operating on the assumption that future events can be predicted with certainty—an assumption that may not hold true when using imprecise geological data. Additionally, it lacks flexibility as it provides a single-point hazard assessment that may not capture the full spectrum of possible outcomes. However, the deterministic approach can be more appropriate for forecasting the progression of an ongoing eruption, where many of the controlling parameters can be precisely monitored in real time. Moreover, deterministic hazard models are more appropriate to assess the impact of specific events on exposure [36].

Quantifying volcanic hazards requires accurate geological and historical records to identify all possible eruptive scenarios associated with a particular volcano or volcanic region. However, these records may be incomplete, historical accounts might lack precision, and geological evidence may not be fully preserved. This means our understanding of past events, which is crucial for anticipating future hazards, could be insufficient, potentially overlooking scenarios that have occurred in the past.

A probabilistic approach helps mitigate these limitations by incorporating past geological and historical events, expert knowledge, and theoretical models to include scenarios that could physically occur but are not documented in the available records. This approach can generate a comprehensive catalog of all possible events and assign a probability of occurrence to each. Even when the scientific data and knowledge are incomplete, a probabilistic approach acknowledges the inherent uncertainty and provides valuable insights into the likely occurrence of volcanic hazards.

9. Experts' opinion: Introducing subjectivity

To conduct quantitative volcanic hazard assessment, many different historical, geological, and geographical datasets are used as input components. The level of uncertainty is directly linked to the quality of the input data. In addition, there is also random uncertainty that cannot be reduced, which is associated with the proper volcanic processes and our ignorance about them. On many occasions during model (statistical, physical, mathematical, etc.) development, expert judgment, and proxies are used in the absence of past data, and the results are very sensitive to most of these assumptions and variations in input data. As such, the outputs of these models should be considered indicators of the order of magnitude of the results obtained, not as exact values. Better data quality and advances in science and modeling methodologies reduce the level of uncertainty, but it is crucial to interpret the results of any risk assessment against the backdrop of unavoidable uncertainty [78].

When evaluating volcanic susceptibility, expert opinion is crucial for assigning relative significance to the various structural criteria used to identify potential pathways for magma ascent, as well as for assessing the quality and reliability of the available data [56]. As mentioned earlier, the geological principle guiding this evaluation is that magma will ascend along the path of least energy consumption, which is determined by the distribution of the stress field within the volcano [69]. Since the stress field cannot be directly observed, we rely on surface structural features as indicators of the stress field at depth and therefore as potential pathways for magma to reach the surface. However, not all structural elements carry the same significance or degree of confidence. For instance, the age of these elements may reflect either a more recent or an older stress field, influencing their relevance in current assessments. To appropriately weigh each structural element when calculating the probability density function (PDF) necessary for creating a volcanic susceptibility map [56], expert judgment is required as this process cannot be fully automated. There are various methods available to incorporate expert opinion into the process of quantifying volcanic hazard (expert elicitation judgment, expected relative frequency, equal weight, analytic hierarchy process, etc.) [10, 79–81], and the choice of method will depend on the specific context of each case (i.e., amount of information available, quality of data, previous knowledge).

Similarly, when conducting temporal analyses and constructing probabilistic event trees to represent all possible eruptive scenarios for a particular volcano and assign probabilities of occurrence to them [82–84], expert opinion is essential as this process cannot be fully automated. The challenge lies in the potential incompleteness of the historical and geological records, especially for volcanoes with long periods of quiescence. These gaps in the record make it necessary to rely on expert knowledge to properly constrain the analysis and ensure that all plausible scenarios are considered.

The introduction of an expert's opinion may be regarded as a way to introduce certain subjectivity in the quantification of volcanic hazards, but it is the only way to identify the relative significance and reliability of the geological data considered. Each group of data (structural, stratigraphic, petrological, geochronological, etc.) will have a different relevance, depending on the aspect considered (eruption dynamics, volcanic hazards, etc.), and can only be evaluated by experts. Similarly, not all data will have the same degree of confidence or quality. While some data can be obtained directly by field measurements (tectonic lineations, vent locations, etc.), others are obtained by indirect methods, such as theoretical models of stress fields, structural geophysical data, and monitoring data, which will depend on several aspects related to data acquisition methods and data processing. Again, it will be the criteria of the expert(s) who have collected/computed the data, which will have to be taken into account to validate correctly such data.

10. Final considerations

Hazard assessment is primarily focused on achieving three key objectives: creating hazard maps, identifying all possible and most likely eruptive scenarios, and developing simulation models for various hazards. The creation of hazard maps is crucial for land-use planning, emergency planning, and crisis management. These maps are typically probabilistic, representing spatial probabilities, and can be designed for a single hazard, groups of hazards, or all hazards predicted for a specific area [85–87]. They may be qualitative or quantitative and can cover either specific restricted areas or entire volcanic fields. Constructed using a GIS platform, hazard maps are essential tools for illustrating and visualizing how a territory can be classified based on the degree of hazard exposure, making them highly valuable in territorial management.

A topic still debated in volcanological literature is whether hazard maps should be qualitative or quantitative [88, 89]. Hazard maps essentially represent probabilities, but the concept of probability is not always well understood. Qualitative hazard maps, which depict hazard levels as high, medium, low, or very low, may be sufficient for land-use planning and emergency preparedness [85–87]. However, when relying on geology as the primary source of information for volcanic hazard assessment, it is important to acknowledge that precise and detailed quantification is not always possible. This limitation depends on the quality and depth of knowledge of the volcanological record, which may sometimes be incomplete or insufficiently informative, leading to a hazard assessment with significant uncertainty. Using simulation models with limited or poor-quality input data can result in either overestimating or underestimating the hazard, providing a flawed basis for risk analysis and decision-making. Although achieving quantitative hazard maps is theoretically ideal, it is not always feasible. Therefore, it is crucial to assess the potential limitations in each case and deliver the best possible outcome based on the available information.

A similar challenge arises when identifying all possible and most probable eruptive scenarios for a specific volcano. This task is crucial for eruption forecasting as it is not only important to determine when and where a new eruption might occur, but also to understand what type of eruption to expect, its potential size and intensity, and the range of likely eruptive scenarios. However, the incompleteness of many geological and, at times, historical records can introduce significant uncertainty when identifying possible eruption scenarios and assigning probabilities to them [7]. In such cases, we might compare the volcano in question with similar, better-documented volcanoes, but this approach also requires caution due to the inherent unpredictability of volcanic activity.

The development of simulation models for various hazards is a crucial aspect of hazard assessment, aimed at calculating exposure, impact, and vulnerability parameters. In many cases, the input parameters for these simulations are derived from geological sources, requiring careful consideration. It is essential to calibrate model results against real-world cases where the controlling parameters are well understood. The accuracy of these models is heavily dependent on the quality of the input data. Using generalized models based on broad assumptions can be inappropriate, particularly when applied to specific volcanic sites. Therefore, it is imperative that the assumptions within these models are constrained by volcano geology, necessitating close collaboration between modelers and volcano geology experts. This partnership helps to prevent the development of models that may contain conceptual or physical errors, inconsistencies, or logical gaps. The consequences of using incorrect simulation models in the quantification of volcanic hazards can be disastrous, especially if these models are intended for forecasting potential volcanic activity. While technological advancements have significantly enhanced computational capabilities in hazard assessment, unrealistic models or those built on fabricated input parameters can be more dangerous than the eruptions themselves. It is always preferable to rely on geology, even if it carries some uncertainty, rather than inventing data and creating models that clearly contradict geological evidence [7, 90, 91].

A final aspect that has to be considered when trying to quantify volcanic hazards is to whom the results of such scientific assessment are going to be addressed. In most cases, the end users of such complex products will not be other scientists trained in volcanology, but decision makers or policymakers the background of whom will be probably not aligned with volcanology. The question here is: Do they understand geology, volcanology, hazard modeling, probabilities, or uncertainties? It is well-known that there are difficulties in approaching the science-policy encounter in the context of risk, especially under high uncertainty and involving multiple disciplines [92]. Hazard assessments containing an excess of information are hard to understand and may fail to inform.

To effectively communicate volcanic hazard assessments to policymakers, it is essential to simplify complex geological information and emphasize practical implications. Avoiding technical jargon and using clear, straightforward language helps convey the risks more effectively. Focus on the real-world impact of volcanic hazards rather than the scientific specifics. When possible, it is recommended to use relatable analogies or comparisons to make these concepts easier to understand, such as describing ashfall like a snowstorm, lava flows like slow-moving mud rivers, or volcanic gases as invisible, toxic fog. It is also important to explain the geological factors used to assess volcanic hazards in simple terms. For instance, talk about fractured rocks instead of tectonic controls on vent locations or describe magma as a viscous liquid affected by gas expansion rather than detailing complex eruption dynamics. Additionally, leveraging modern technology can greatly enhance the understanding of volcanic hazards. Virtual reality (VR) and augmented reality (AR) offer immersive, visual experiences that help policymakers grasp the potential threats more clearly [93, 94]. These tools not only improve preparedness and response capabilities but also serve as valuable resources for education and public outreach, making complex scientific information more accessible and engaging.

One of the most challenging aspects of communicating the results of volcanic hazard assessments is explaining the timescales involved in volcanic processes, particularly the frequency of eruptions. While it is relatively straightforward to describe the hazards that occur during an eruption—since these events often last only a matter of hours, days, or weeks—conveying the duration of post-eruptive processes, such as lava flow cooling and gas emissions, is more complex. These processes can take significantly longer, often lasting orders of magnitude longer than the eruption itself. This extended timeline can be difficult for policymakers and affected communities to grasp, especially those eager to restore land and property damaged by volcanic activity.

A recent example of this challenge is the eruption of La Palma, where lava flow cooling and CO_2 degassing may continue for decades. However, the greater difficulty arises when attempting to explain long-term hazard assessments, particularly those involving forecasts of future eruptions and eruption frequency [7, 11]. When eruption intervals are relatively short—on the scale of months or years—this is easier to comprehend. But when return periods extend into decades, centuries, or even millennia, risk perception shifts dramatically.

This shift is partly due to the limitations of human memory; people tend to quickly forget past disasters, especially if recovery was swift and effective [95, 96]. Additionally, politicians often focus on achieving results within a short time frame, which clashes with the much longer timescales that geological and volcanic processes typically follow. Consequently, experts and non-experts often have very different perceptions of time when discussing volcanic hazards.

To make the concept of eruption frequencies more understandable to the general public, it is helpful to draw comparisons with everyday situations. One effective analogy is weather forecasting: We know that storms will occur and can estimate their intensity, but we cannot predict the exact timing. Similarly, volcanologists can estimate how often a volcano is likely to erupt and the nature of the eruption, but they cannot predict precisely when it will happen. Additionally, comparing future eruptions to past ones from the same volcano can be valuable as it helps illustrate forecasts by showing the potential effects and outcomes based on historical events.

In summary, the quantification of volcanic hazards must be grounded in a thorough understanding of past eruptive activity, which is preserved in the geological record. This data must be carefully collected to ensure completeness and avoid losing key stratigraphic information. It forms the basis for developing statistical models that estimate eruption recurrence and help forecast future events. Additionally, geological data provides the range of values needed to simulate eruptive scenarios and specific volcanic hazards, which are crucial for conducting hazard assessments. However, volcanic geology rarely provides precise numbers, meaning that a certain level of uncertainty will always be present in the results. The critical challenge in quantifying volcanic hazards lies in calculating this uncertainty and effectively communicating it to non-experts, such as policymakers and affected communities.

Acknowledgements

This work has been partially funded by the European Commission (FP7 Theme: ENV.2011.1.3.3-1; Grant 282759: VUELCO; and the EC ECHO Grants SI2.695524 (VeTOOLS) and 826292 (EVE)).

Volcano Geology: A Fundamental Tool to Quantify Volcanic Hazards
DOI: http://dx.doi.org/10.5772/intechopen.1007213

Author details

Joan Martí
Department of Geology, Natural Risk Assessment and Management Survey (NRAMS), IDAEA-CSIC, Barcelona, Spain

*Address all correspondence to: j.marti@csic.es

References

[1] Heiken G. From Kīlauea Iki 1959 to Eyjafjallajökull 2010: How volcanology has changed! In: Bickford ME, editor. The Web of Geological Sciences: Advances, Impacts, and Interactions. Special Paper 500. Washington: Geological Society of America; 2013. pp. 33-63. DOI: 10.1130/2013.2500(02)

[2] Acocella V. Great challenges in volcanology: How does the volcano factory work? Frontiers in Earth Science. 2014;**2**:1-10. DOI: 10.3389/feart.2014.00004

[3] Mader HM, Coles SG, Connor CB, Connor LJ. Statistics in volcanology. Geological Society of London. 2006;**1**:285. DOI: 10.1144/IAVCEI001.3

[4] Fagents SA, Gregg TPG, Lopes RMC, editors. Modeling Volcanic Processes: The Physics and Mathematics of Volcanism. Cambridge, UK: Cambridge University Press; 2013. 421 p. ISBN 978-0-521-89543-9

[5] Kavanagh JL, Engwell SL, Martin SA. A review of laboratory and numerical modelling in volcanology. Solid Earth. 2018;**9**:531-571. DOI: 10.5194/se-9-531-2018

[6] Carniel R, Guzmán S. Chapter Machine Learning in Volcanology: A Review. London, UK: IntechOpen; 2022. DOI: 10.5772/intechopen.94217

[7] Poland MP, Anderson KR. Partly cloudy with a chance of lava flows: Forecasting volcanic eruptions in the twenty first century. Journal of Geophysical Research: Solid Earth. 2020;**125**:e2018JB016974. DOI: 10.1029/2018JB016974

[8] Bachèlery P. Understanding the geological history of volcanoes: An essential prerequisite to their monitoring. In: Lénat J-F, editor. Hazards and Monitoring of Volcanic Activity 1. Hoboken, New Jersey, U.S: Wiley Online Library; 2022. DOI: 10.1002/9781394163359.ch1

[9] Cagnoli B. Fuzzy logic in volcanology. Episodes Journal of International Geoscience. 1998;**21**(2):94-96

[10] Aspinall WP. Structured elicitation of expert judgment for probabilistic hazard and risk assessment in volcanic eruptions. In: Mader HM et al., editors. Statistics in Volcanology. Special Publication of IAVCEI # 1. London, UK: Geological Society of London; 2006. pp. 15-30

[11] Sparks RSJ. Forecasting volcanic eruptions. Earth and Planetary Science Letters. 2003;**210**(1-2):1-15. DOI: 10.1016/S0012-821X(03)00124-9

[12] Wartofsky MW. Models: Representation and the scientific understanding. In: Boston Studies in the Philosophy of Science. Vol. 48. 39 p. Dordrecht, Holland: Reidel Publishing Company; 1979. DOI: 10.1007/978-94-009-9357-0

[13] Börner K, Boyack KW, Milojević S, Morris S. An introduction to Modeling science: Basic model types, key definitions, and a general framework for the comparison of process models. In: Scharnhorst A, Börner K, van den Besselaar P, editors. Models of Science Dynamics. Understanding Complex Systems. Berlin, Heidelberg: Springer; 2012. DOI: 10.1007/978-3-642-23068-4_1

[14] Rogers K. Scientific modeling. Encyclopedia Britannica. Britannica.com;

2023. Available from: https://www.britannica.com/science/scientific-modeling [Accessed: August 22, 2024]

[15] Board of Governors of the Federal Reserve System. SR 11-7: Guidance on model risk management. 2011. Available from: http://www.federalreserve.gov/bankinforeg/srletters/sr1107a1.pdf

[16] Aggarwal A, Beck MB, Cann M, et al. Model Risk: Daring to Open the Black Box Technical Report in British Actuarial Journal March 2015. Cambridge, UK: Cambridge University Press; 2016. DOI: 10.1017/S1357321715000276

[17] Hesse MB. Models and Analogies in Science. Notre Dame, Indiana, USA: University of Notre Dame Press; 1966. 184 p

[18] Lin CC, Segel LA. Mathematics Applied to Deterministic Problems in the Natural Sciences. Philadelphia, USA: SIAM; 1988. 630 p. ISBN 1611971349, 9781611971347

[19] Morgan MS, Morrison M, editors. Models as Mediators: Perspectives on Natural and Social Science. Cambridge, UK: Cambridge University Press; 1999. DOI: 10.1017/CBO9780511660108.003

[20] Giere RN. Science without Laws: Model Systems, Cases, Exemplary Narratives. Chicago, USA: University of Chicago Press; 1999. DOI: 10.1093/mind/111.441.111

[21] Del Re G. Models and analogies in science. Hyle. 2000;**6**(1):5-15

[22] Gelman A, Carlin JB, Stern HS, Dunson DB, Vehtari A, Rubin DB. Bayesian Data Analysis. Columbia, USA: CRC Press; 2015. 675 p. DOI: 10.1201/b16018

[23] Donovan A, Oppenheimer C, Bravo M. The use of belief-based probabilistic methods in volcanology: Scientists' views and implications for risk assessments. Journal of Volcanology and Geothermal Research. 2012;**247-248**:168-180. DOI: 10.1016/j.jvolgeores.2012.08.011

[24] Rouwet D, Constantinescu R, Sandri L. Deterministic versus probabilistic volcano monitoring: Not "or" but "and". In: Gottsmann J, Newberg J, Scheu B, editors. Volcanic Unrest,. Advs in Volcanology. Berlin, Germany: Springer; 2019. pp. 35-46. DOI: 10.1007/11157_2017_8

[25] Marzocchi W, Selva J, Jordan TH. A unified probabilistic framework for volcanic hazard and eruption forecasting. Natural Hazards and Earth System Sciences. 2021;**21**:3509-3517. DOI: 10.5194/nhess-21-3509-2021

[26] Blong RJ. Volcanic Hazards: A Sourcebook on the Effects of Volcanic Eruptions. Sydney: Academic Press; 1984

[27] Blong RJ. Volcanic hazards and risk management. In: Sigurdsson H, editor. Encyclopedia of Volcanoes. San Diego: Academic; 2000. pp. 1215-1227

[28] Tilling RI. Volcanic hazards and their mitigation: Progress and problems. Reviews of Geophysics. 1989;**27**:237-269

[29] Tilling RI. Volcano hazards. In: Martí J, Ernst G, editors. Volcanoes and the Environment. Cambridge: Cambridge University Press; 2005. pp. 55-89

[30] Peterson DW, Tilling RI. Lava flow hazards. In: Sigurdsson H et al., editors. Encyclopedia of Volcanoes. San Diego: Academic; 2000. pp. 957-971

[31] Martí J. Assessing Volcanic Hazard: A Review. Oxford, UK: Oxford Handbooks

Online; 2017. Available from: https://academic.oup.com/edited-volume/41328/chapter/352327859 [Accessed: October 14, 2017]

[32] Sparks RSJ, Aspinall WP, Crosweller HS, Hincks TK. Risk and uncertainly assessment of volcanic hazards. In: Rougier J, Sparks RSJ, Hill L, editors. Risk and Uncertainty Assessment for Natural Hazards. Cambridge: Cambridge University Press; 2013. pp. 365-397

[33] Sharma A, Miyazaki H. Multi-hazard risk assessment in urban planning and development using AHP. Int. Arch. Photogramm. Remote Sens. Spatial Inf. Sci. 2019;**XLII-3/W8**:363-371. DOI: 10.5194/isprs-archives-XLII-3-W8-363-2019

[34] Alberico I, Petrosino P, Lirer L. Volcanic hazard and risk assessment in a multi-source volcanic area: The example of Napoli city (southern Italy). Natural Hazards and Earth System Sciences. 2011;**11**:1057-1070. DOI: 10.5194/nhess-11-1057-2011

[35] Scaini C, Felpeto A, Martí J, Carniel R. A GIS-based methodology for the estimation of potential volcanic damage and its application to Tenerife Island, Spain. Journal of Volcanology and Geothermal Research. 2014;**278-279**:40-58

[36] Zuccaro G et al. Economic impact of explosive volcanic eruptions: A simulation based ssessment model applied to Campania region volcanoes. Journal of Volcanology and Geothermal Research. 2013;**266**:1-15

[37] Woo G. Chapter 11 - cost–benefit analysis in volcanic risk. In: Shroder JF, Papale P, editors. Hazards and Disasters Series, Volcanic Hazards, Risks and Disasters. Amsterdam, The Netherlands: Elsevier; 2015. pp. 289-300. DOI: 10.1016/B978-0-12-396453-3.00011-3

[38] Doyle EEH, McClure J, Johnston DM, Paton D. Communicating likelihoods and probabilities in forecasts of volcanic eruptions. Journal of Volcanology and Geothermal Research. 2014;**272**:1-15

[39] Castruccio A, Rust AC, Sparks RSJ. Rheology and flow of crystal-bearing lavas: Insights from analogue gravity currents. Earth and Planetary Science Letters. 2010;**297**(3-4):471-480. DOI: 10.1016/j.epsl.2010.06.051

[40] Chevrel MO, Platz T, Hauber E, Baratoux D, Lavallée Y, Dingwell DB. Lava flow rheology: A comparison of morphological and petrological methods. Earth and Planetary Science Letters. 2013;**384**:109-120. DOI: 10.1016/j.epsl.2013.09.022

[41] Aufaristama M, Höskuldsson A, Jónsdóttir I, Ólafsdóttir R. Mapping and assessing surface morphology of Holocene lava field in Krafla (NE Iceland) using hyperspectral remote sensing. In: IOP Conference Series: Earth and Environmental Science, 29, Padjadjaran Earth Dialogues: International Symposium on Geophysical Issues, Bandung, Indonesia, 012002. London: IOP Publishing; 2015. DOI: 10.1088/1755-1315/29/1/012002

[42] Glaze LS, Baloga SM. Simulation of inflated pahoehoe lava flows. Journal of Volcanology and Geothermal Research. 2013;**255**:108-123. DOI: 10.1016/j.jvolgeores.2013.01.018

[43] Favalli M, Mazzarini F, Pareschi MT, Boschi E. Topographic control on lava flow paths at Mount Etna, Italy: Implications for hazard assessment. Journal of Geophysical Research, Solid Earth. 2007;**114**:F01019. DOI: 10.1029/2007JF000918

[44] Cas RAF, Wright JV. Volcanic Successions. London: Allen & Unwin; 1987. p. 528

[45] Smith GA. Facies sequences and geometries in continental volcaniclastic sediments. In: Fisher RV, Smith GA, editors. Sedimentation in Volcanic Settings. SEPM Special Publication n° 45. Tulsa, Oklahoma, USA: Society for Sedimentary Geology; 1991. pp. 109-121

[46] Smith RL. Zones and zonal variations in ash-flows. U.S. Geological Survey Professional Paper. 1960;**354-F**:149-159

[47] Sohn C, Sohn YK. Distinguishing between primary and secondary volcaniclastic deposits. Scientific Reports. 2019;**9**:12425. DOI: 10.1038/s41598-019-48933-4

[48] Suthren RJ. Facies analysis of volcaniclastic sediments: A review. Geological Society, London, Special Publications. 2007;**18**(1):123. DOI: 10.1144/GSL.SP.1985.018.01.07

[49] Sulpizio R, De Rosa R, Donato P. The influence of variable topography on the depositional behaviour of pyroclastic density currents: The examples of the Upper Pollara eruption (Salina Island, Southern Italy). Journal of Volcanology and Geothermal Research. 2008;**175**:367-385

[50] Sulpizio R, Dellino P, Doronzo DM, Sarocchi D. Pyroclastic density currents: State of the art and perspectives. Journal of Volcanology and Geothermal Research. 2014;**283**:36-65. DOI: 10.1016/j.jvolgeores.2014.06.014

[51] Cowlyn J, Kennedy BM, Gravley DM, Cronin SJ. A confidence-based assessment method for distinguishing pyroclastic density current deposits from other volcaniclastic units. Frontiers in Earth Science. 2020;**8**:1-14. DOI: 10.3389/feart.2020.581195

[52] Martí J, Groppelli G, Brum da Silveira A. Volcanic stratigraphy: A review. Journal of Volcanology and Geothermal Research. 2018;**357**:68-91. DOI: 10.1016/j.jvolgeores.2018.04.006

[53] Salvador A, editor. International stratigraphic guide. In: A Guide to Stratigraphic Classification, Terminology, and Procedure. Washington DC, USA: The International Union of Geological Sciences and the Geological Society of America; 1994. 214 p

[54] Fisher RV, Schmincke HU. Pyroclastic Rocks. Berlin, Germany: Springer Verlag; 1984. 472 p

[55] Groppelli G, Martí J. Volcanic stratigraphy – State of the art. In: Proceedings of STRATI 2013, First International Congress on Stratigraphy, Ciencias da Terra. Vol. 18. Portugal: University of Lisbon; 2013. pp. 99-104

[56] Martí J, Felpeto A. Methodology for the computation of volcanic susceptibility. An example for mafic and felsic eruptions on Tenerife (Canary Islands). Journal of Volcanology and Geothermal Research. 2010;**195**:69-77

[57] McBirney AR, Murase T. Rheological properties of magmas. Annual Review of Earth and Planetary Sciences. 1984;**12**:337-357

[58] Mader HM, Llewellin EW, Mueller SR. The rheology of two-phase magmas: A review and analysis. Journal of Volcanology and Geothermal Research. 2013;**257**:135-158. DOI: 10.1016/j.jvolgeores.2013.02.014

[59] Thouret JC. Volcanic geomorphology—An overview. Earth-Science Reviews. 1999;**47**:95-131

[60] Walker GPL. Basaltic volcanoes and volcanic systems. In: Sigurdsson H, editor. Encyclopedia of Volcanoes. San Francisco: Academic Press; 2000. pp. 283-289

[61] Davidson J, De Silva S. Composite volcanoes. In: Sigurdsson H, editor. Encyclopedia of Volcanoes. San Diego: Academic Press; 2000. pp. 663-681

[62] Pinel V, Jaupart C. Magma storage and horizontal dyke injection beneath a volcanic edifice. Earth and Planetary Science Letters. 2004;**221**:245-262. DOI: 10.1016/S0012-821X(04)00076-7

[63] Gudmundsson A, Brenner SL. On the conditions of sheet injections and eruptions in stratovolcanoes. Bulletin of Volcanology. 2005;**67**:768-782. DOI: 10.1007/s00445-005-0433-7

[64] Siebert L. Hazards of large volcanic debris avalanches and associated eruptive phenomena. In: Scarpa R, Tilling RI, editors. Monitoring and Mitigation of Volcano Hazards. Heidelberg: Springer-Verlag; 1996. pp. 541-572

[65] Wood CA. Morphometric evolution of cinder cones. Journal of Volcanology and Geothermal Research. 1980;**7**:387-413

[66] Lorenz V. Syn- and posteruptive hazards of Maar–Diatreme volcanoes. Journal of Volcanology and Geothermal Research. 2007;**159**(1-3):285-312

[67] Németh K. Monogenetic volcanic fields: Origin, sedimentary record, and relationship with polygenetic volcanism. In: Cañón-Tapia E, Szakács A, editors. What Is a Volcano? Boulder, CO: Geological Society of America; 2010. pp. 43-66

[68] Le Corvec N, Menand T, Lindsay J. Spatial distribution and alignments of volcanic Centers: Clues to the formation of monogenetic volcanic fields. Earth Science Reviews. 2013;**124**:96-114

[69] Martí J, López C, Bartolini S, Becerril L. Stress controls of monogenetic volcanism: A review. Frontiers in Earth Science. 2016;**4**(106):1-17. DOI: 10.3389/feart.2016.00106

[70] Ablay GJ, Martí J. Stratigraphy, structure, and volcanic evolution of the Pico Teide–Pico Viejo formation, Tenerife, Canary Islands. Journal of Volcanology and Geothermal Research. 2000;**103**:175-208

[71] Paolillo A, Principe C, Bisson M, Gianardi R, Giordano D, La Felice S. Volcanology of the Southwestern sector of Vesuvius volcano, Italy. Journal of Maps. 2016;**12**(Sup 1):425-440. DOI: 10.1080/17445647.2016.1234982

[72] Marzocchi W, Sandri L, Furlan C. A quantitative model for volcanic hazard assessment. In: Mader HM, Coles SG, Connor CB, Connor LJ, editors. Statistics in Volcanology. London, UK: The Geological Society; 2006. DOI: 10.1144/IAVCEI001.3

[73] Jaquet O, Connor C, Connor L. Probabilistic methodology for long-term assessment of volcanic hazards. Nuclear Technology. 2008;**163**(1):180-189. DOI: 10.13182/NT08-A3980

[74] Sparks RSJ, Aspinall W. Volcanic activity: Frontiers and challenges in forecasting, prediction and risk assessment. In: The State of the Planet: Frontiers and Challenges in Geophysics Geophysical Monograph 150, IUGG Volume 19 Copyright 2004 by the International Union of Geodesy and Geophysics and the American Geophysical Union. Washington DC, USA: American Geophysical Union; 2004. DOI: 10.1029/150GM28

[75] Panza GF, Vaccari F, Cazzaro R. Deterministic seismic Hazard assessment. In: Wenzel F, Lungu D, Novak O, editors. Vrancea Earthquakes: Tectonics, Hazard and Risk Mitigation. Advances in Natural and

Technological Hazards Research. Vol. 11. Dordrecht: Springer; 1999. DOI: 10.1007/978-94-011-4748-4_25

[76] Terlien MTJ, Van Westen CJ, van Asch TWJ. Deterministic modelling in Gis-based landslide hazard assessment. In: Carrara A, Guzzetti F, editors. Geographical Information Systems in Assessing Natural Hazards. Advances in Natural and Technological Hazards Research. Vol. 5. Dordrecht: Springer; 1995. DOI: 10.1007/978-94-015-8404-3_4

[77] Mazzorana B, Comiti F, Volcan C, et al. Determining flood hazard patterns through a combined stochastic–deterministic approach. Natural Hazards. 2011;**59**:301-316. DOI: 10.1007/s11069-011-9755-2

[78] Sobradelo R, Martí J. Statistics, probabilities, uncertainty and their interpretation for decision-making during volcanic crises. In: Fearnley CJ, Bird DK, Haynes K, McGuire B, Jolly G, editors. Observing the Volcano World: Volcano Crisis Communication. Advances in Volcanology. Berlin, Germany: Springer; 2018. pp. 571-583. DOI: 10.1007/11157_2017_15

[79] Cooke R. Experts in Uncertainty: Opinion and Subjective Probability in Science. Oxford, UK: Oxford University Press; 1991

[80] Saaty TL. The Analytic Hierarchy Process: Planning, Priority Setting, Resource Allocation. New York; London: McGraw-Hill International Book Co; 1980

[81] Flandoli F, Giorgi E, Aspinall WP, Neri A. Comparison of a new expert elicitation model with the classical model, equal weights and single experts, using a crossvalidation technique. Reliability Engineering and System Safety. 2011;**96**:1292-1310. DOI: 10.1016/j.ress.2011.05.012

[82] Newhall CG, Hoblitt RP. Constructing event trees for volcanic crisis. Bulletin of Volcanology. 2002;**64**:3-20

[83] Neri A, Aspinall W, Cioni R, Bertagnini A, Baxter P, Zuccaro G, et al. Developing an event tree for probabilistic Hazard and risk assessment at Vesuvius. Journal of Volcanology and Geothermal Research. 2008;**178**(3):397-415

[84] Sobradelo R, Martí J. Bayesian event tree for long-term volcanic hazard assessment: Application to Teide-Pico Viejo stratovolcanoes, Tenerife, Canary Islands. Journal of Geophysical Research: Solid Earth. 2010;**115**:B05206. DOI: 10.1029/2009JB006566

[85] Calder ES, Wagner K, Ogburn SE. Volcanic Hazard Maps. In: Brown K, Loughlin SC, Sparks RSJ, Vye-Brown C, et al., editors. Global Volcanic Hazards and Risk: Technical Background Paper for the Global Assessment Report on Disaster Risk Reduction 2015. Global Volcano Model and IAVCEI. Cambridge, UK: Cambridge University Press; 2015. pp. 335-342

[86] Lindsay JM, Charlton D, Clive MAT, Bertin D, Ogburn S, Wright H, et al. The diversity of volcanic hazard maps around the world: Insights from map makers. Journal of Applied Volcanology. 2023;**12**(1):8. DOI: 10.1186/s13617-023-00134-5

[87] Ogburn SE, Charlton D, Norgaard D, Wright HM, Calder ES, Lindsay J, et al. The volcanic hazard maps database: An initiative of the IAVCEI commission on volcanic hazards and risk. Journal of Applied Volcanology. 2023;**12**(1):2. DOI: 10.1186/s13617-022-00128-9

[88] Marzocchi W, Newhall CG, Woo G. The scientific Management of Volcanic Crises. Journal of

Volcanology and Geothermal Research. 2012;**247-248**:181-189

[89] Selva J, Marzocchi W, Sandri L, Costa A. Operational short-term volcanic hazard analysis: Methods and perspectives. In: Papale P, Shroder JF, editors. Volcanic Hazards, Risks and Disasters. Amsterdam: Elsevier; 2014. pp. 233-259

[90] Szakács A, Cañón-Tapia E. Some challenging new perspectives of volcanology. In: Szakács A, Cañón-Tapia E, editors. What Is a Volcano, the Geological Society Spacial Paper, 140. London, UK: The Geological Society; 2010. 123-140 pp

[91] Renschler CS. Scales and uncertainties in using models and GIS for volcano hazard prediction. Journal of Volcanology and Geothermal Research. 2005;**139**:73-87

[92] Donovan A. Critical volcanology? Thinking holistically about risk and uncertainty. Bulletin of Volcanology. 2019;**81**:20. DOI: 10.1007/s00445-019-1279-8

[93] Havenith HB, Cerfontaine P, Mreyen AS. How virtual reality can help visualise and assess geohazards. International Journal of Digital Earth. 2017;**12**(2):173-189. DOI: 10.1080/17538947.2017.1365960

[94] Alene GS, Irshad S, Moraru A, Depina I, Bruland O, Perkis A, et al. Virtual reality visualization of geophysical flows: A framework. Environmental Modelling & Software. 2024;**177**:106063. DOI: 10.1016/j.envsoft.2024.106063

[95] Pisa PF. Understanding memory transmission in disaster risk reduction practices: A case study from Japan. International Journal of Disaster Risk Reduction. 2024;**100**:104112. DOI: 10.1016/j.ijdrr.2023.104112

[96] Navarrete-Valladares C, Sandoval-Díaz J, Sandoval-Obando E. Experience and local memory of older people in the face of disasters: A systematic review. Frontiers in Public Health. 2023;**11**:1163561. DOI: 10.3389/fpubh.2023.1163561

Chapter 2

Caldera Formation and Post-Caldera Activity Influencing Formation or Activities of Faults: A Case Study of the Hohi Volcanic Zone, Central Kyushu, Japan

Shigekazu Kusumoto

Abstract

This study, which is a case study of the Shishimuta caldera located in the Hohi volcanic zone in central Kyushu, Japan, estimated the surface stress fields caused by the activity of a magma chamber under the caldera and discussed its effects on fault formation or activity in this region. Before these discussions, analytical solutions giving the magma chamber volume change required for initial caldera formation and information on the depth of the magma chamber were derived. These solutions were applied to the Shishimuta caldera, and the magma chamber depth consistent with geological data, and the minimum volume change for caldera formation was obtained. It was found that the estimated surface stress field caused by caldera formation could not explain the fault distribution near the caldera. In contrast, it was found that the potential activities of faults could be explained by stress fields due to the inflation of the magma chamber. This indicates the possibility that the magma chamber under the Shishimuta caldera was activated by inflation after caldera formation. This possibility is supported by the morphological and geological data.

Keywords: collapsed caldera, stress field, magma chamber, fault distribution, Hohi volcanic zone, Shishimuta caldera, Haneyama fault zone

1. Introduction

The relationships between volcanic activity and seismicity have been confirmed by many observational results and numerical simulations [1–7]. Earthquake activity correlates with volcanic eruptions in time and space [8–10]. Manga et al. [11] demonstrated the relationship between magnitude and hypocentral distance of earthquakes that trigger volcanic eruptions, including mud volcanoes. In addition, regional earthquakes may trigger volcanic eruptions [12] and enhance ongoing volcanic activities [13, 14].

Although considerable research has been conducted on activation and reactivation of volcanoes by earthquakes, volcanic activity also promotes the reactivation of existing faults. In the Campi Flegrei caldera, reactivation of the caldera ring-faults, intra-caldera faults, and eruptive fissures have been observed using Permanent Scatterer Synthetic Aperture Radar Interferometry (PSInSAR) [6]. In addition, the mechanics of fault slip around a pressured magma chamber has been discussed using numerical simulations [4, 5].

The partial or total collapse of a magma chamber roof results in caldera formation [15–17], which dramatically changes the stress fields in the inner crust and on the surface near the magma chamber [5, 18] and forms new fracture areas with intrusions of dikes and sills [18, 19]. Activation and reactivation of faults have been observed and discussed in the field [20–23]. Geological structures and geophysical phenomena related to caldera formation are important for understanding past, present, and future volcanic activities and tectonics. Therefore, in addition to geological and geophysical field studies, analogue experiments, numerical simulations, and theoretical studies have been conducted to elucidate the key parameters and mechanical processes controlling caldera formation.

Careful analogue experiments have revealed that shape and depth of the magma chamber provide substantial insight into caldera formation [24–30]. In addition, analogue experiments that incorporate a regional stress field have expanded the range of conditions that can be replicated [31, 32]. Such analogue experiments provide qualitative information on the collapse of the magma chamber and subsequent propagation of the collapse and have significantly advanced the understanding of the processes underlying caldera formation.

In simulations and theoretical studies, the stress fields caused by pressure changes within the magma chamber were estimated and evaluated at the surface and in the medium surrounding the chamber, and the shape (dip) and formation conditions of the ring-fault bordering the caldera were discussed [33–37]. Ambient stresses near the magma chamber influencing the formation of ring-faults and dike injections have been shown to play an important role in caldera formation. Excellent reviews of these issues and modelling procedures were published by Marti et al. [38] and Geyer and Marti [39].

The overall process of collapse caldera, including first-order aspects, such as the final caldera morphology and eruption type, depends on many factors. For example, pre-existing topography (e.g., volcanic edifices), structures (e.g., faults), and regional stress field control the subsidence amount, its pattern, and caldera shape [32, 40–42]. However, the key element in caldera formation is the removal of magmatic support from the crust overlying the chamber [43, 44]. Kusumoto and Gudmundsson [45] evaluated the surface stress field caused by the contraction of a sphere in an elastic medium and derived an analytical solution for the volume change necessary for the initial ring-fault formation as a two-dimensional (2-D) problem. They showed that the solution was proportional to the cubic depth of the magma chamber.

In this study, I discuss whether the distribution of active faults near the caldera and their behaviour can be explained by the stress fields created by deflation or inflation of the magma chamber. As a case study, I selected the Shishimuta caldera [46] located in the Hohi volcanic zone [47], central Kyushu, and its neighbouring active fault zone, the Haneyama fault zone. Because the Haneyama fault zone distributes across the Shishimuta caldera at the W-E direction, we can understand the relationship between the activity history of the caldera and the fault zone by evaluating the forming process of the Haneyama fault zone by stress changes due to the activity of the magma chamber. In the calculations of the stress fields owing to the activation of the magma chamber under the Shishimuta caldera, I employed the maximum volume change permissible within the elasticity range and discussed them under each condition of deflation and inflation. Consequently, an equation was derived to calculate the magma chamber volume change required for forming fractures at the surface using the von Mises criteria, which considers all stress components.

2. Shishimuta caldera and Haneyama fault zone

The Shishimuta caldera and the Haneyama fault zone are located in the Hohi volcanic zone, and the Shishimuta caldera is a buried caldera formed in the andesitic upper crust at about 1 Ma. The Haneyama fault zone is an active fault zone consisting of mainly normal faults younger than 0.9 Ma and distributes across the Shishimuta caldera at the W-E direction. The Hohi volcanic zone is located in the eastern part of central Kyushu (**Figure 1**) and is characterised by a wedge-shaped low-gravity anomaly area that narrows eastward (**Figure 2**). Detailed geological surveys have shown that this volcanic zone consists of a half-graben with volcanic activity that began at 6 Ma [47] and pull-apart basins that formed after the formation of the half-graben [50]. It was shown that these structures could be restored by numerical simulations based on dislocation modelling [51], and that the gravity anomalies of Beppu Bay (BB), Shonai Basin (SB), and Kuju Basin (KB), as shown in **Figure 2**, could also be attributed to these structures [52]. The gravity anomaly related to Shishimuta caldera (SC) was not restored. These results indicate that the tectonics suggested by Itoh et al.

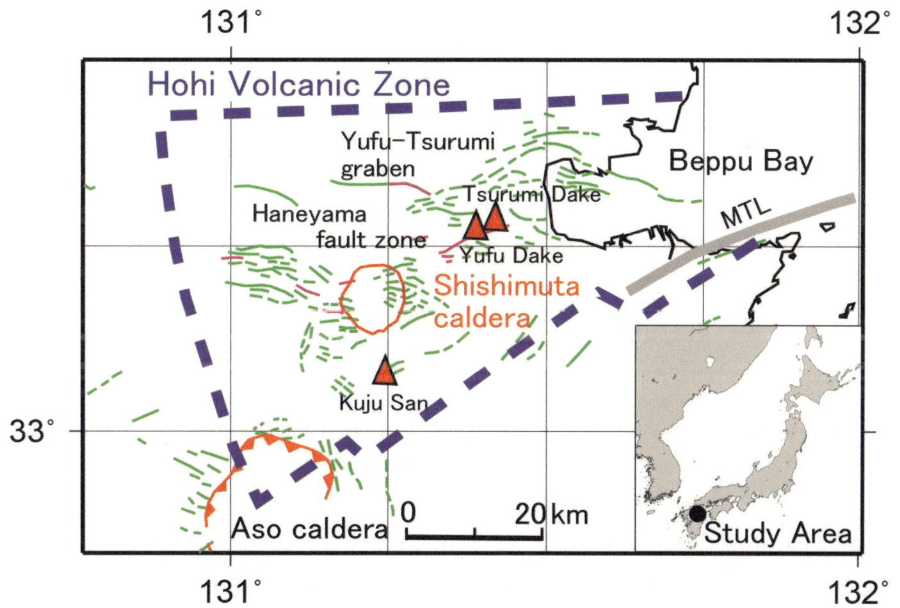

Figure 1.
Location map of the study area. The dashed blue line indicates the outline of the Hohi volcanic zone. Red lines indicate calderas. Green lines are active faults younger than 0.9 Ma, and purple lines are active faults which are older than 0.9 [48]. MTL: Median Tectonic Line.

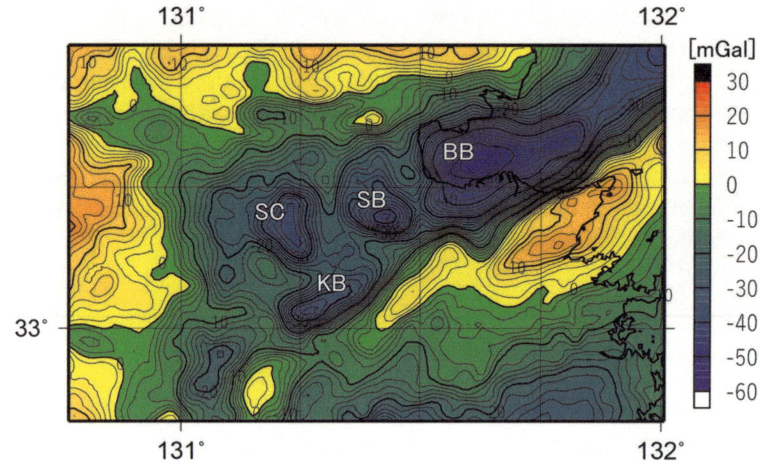

Figure 2.
Bouguer anomaly map. The Bouguer density was assumed to be 2300 kg/m³. The map is compiled from Komazawa's gravity database [49]. The contour interval is 0.2 mGal. BB: Beppu Bay. SB: Shonai Basin. KB: Kuju Basin. SC: Shishimuta caldera.

[50] can be proved using geophysical techniques. The transformation from half-graben formation to the pull-apart basin formation began 1.5 Ma; this tectonic phase change occurred through a counter-clockwise shift of the subducting Philippine Sea Plate and activated the Median Tectonic Line (MTL shown in **Figure 1**) as a right-lateral fault [50].

The Shishimuta caldera is a buried caldera, and the red circle shown in **Figure 1** indicates the caldera wall at a depth of 1 km, estimated from drilling core data and gravity anomalies [46]. It has a diameter of approximately 8–10 km. This caldera was identified as a caldera by the lack of Yabakei (1 Ma) and Imaichi (0.9 Ma) pyroclastic flows in drill core data at the underground of the conspicuous gravity-low area reaching −30 mGal. However, as shown in **Figure 2**, this caldera does not have the concentric circular low-gravity anomaly found in many calderas [53–55]; rather, it has a triangular low-gravity anomaly. In other areas, the calderas do not have concentric circular low-gravity anomalies. For example, the Cañas Dules caldera in Costa Rica is one of them, formed under strong structural control, dominated by two regional faults [56]. At present, it is unknown whether the formation of the Shishimuta caldera is controlled by existing faults. This caldera is the origin of the Yabakei and Imaichi pyroclastic flows distributed widely in central Kyushu [46, 57], with volumes estimated to be 40 km^3 [46] and 90 km^3 [57], respectively, based on detailed surface geological surveys. Based on drilling core data, gravity anomalies, active fault distribution, micro-earthquake activities, and geological surveys in and near the Shishimuta caldera, Kamata [46] estimated that this caldera is funnel-shaped. In addition, he estimated that a residual magma chamber for the Yabakei pyroclastic flow is assumed at a depth of 6–10 km, based on studies (hydrothermal experiments) at Aira caldera [58].

As shown in **Figure 1**, many active faults exist in the Hohi volcanic zone, most of which are short faults with lateral, normal, or both types of faults. Short normal fault swarms distributed near the Shishimuta caldera are collectively known as the Haneyama fault zone or Haneyama fault system. Morphological and geological surveys have confirmed that faults in the Haneyama fault zone were active after 0.9 Ma, mainly around 0.6 Ma [48]. These faults form small grabens, such as the Yufu-

Tsurumi graben [59]. These normal faults are considered to be related to volcanic activities [60]. However, normal faulting and related activities cannot be explained clearly by the stress fields caused by the activities of Shishimuta caldera or other volcanoes.

In general, observations and studies of stress or strain in the crust are indispensable for understanding the occurrence of earthquakes, and observations of crustal movements and seismic activity in the Kyushu district, including this area, have been conducted [61–63]. From a trigonometric survey, the strain field in Kyushu during the last hundred years indicates an N–S extension, and its magnitude in central Kyushu is larger than that in other areas [61]. Matsumoto et al. [64] reported that the stress conditions determined by focal mechanisms in a seismogenic layer shallower than 30 km have compressive and tensile axes in the WSW–ENE and NNW–SSE directions. Because the Haneyama fault zone consists of normal faults with strikes trending W–E (**Figure 1**), it is predicted that the accumulation of the present observed stress will generate normal fault-type earthquakes in this fault zone in the near future.

3. Simplified magma chamber model

3.1 Basic model

Here, an equation is derived for calculating the magma chamber volume change required to form a fracture area at the surface. The equation that provides the magma chamber volume change required for initial caldera formation has already been derived [45]. However, the equation represents a two-dimensional (2-D) solution because a finite sphere model calculating the surface stress field and the Coulomb failure criterion, considering the maximum and minimum principal stresses at the surface were employed to derive the equation as a magma chamber model and failure criterion. In this study, to approach the equation in a more practical situation, I employed the von Mises criterion to evaluate all principal stress fields and the point-source model (**Figure 3**), from which we can calculate three-dimensional (3-D) stress fields, and I rederived the equation giving the magma chamber volume change required for initial caldera formation.

Although actual magma chambers have complex configurations [65], in geophysical studies, an ellipsoidal magma chamber model with a sill-like chamber is widely

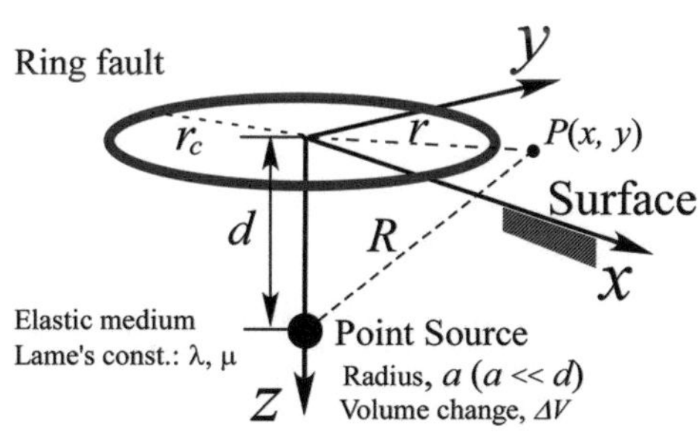

Figure 3.
Model of a small spherical magma chamber (point-source model) with depth to centre d*. Here,* r *and* r_c *are the horizontal distance from the surface point directly above the magma chamber to an arbitrary point and the periphery of the caldera (ring-fault) on the surface, respectively.*

regarded as appropriate, and analytical solutions expressing the stress and deformation fields caused by pressure changes within the ellipsoid have been presented by many researchers [66–68]. Ellipsoidal models have also been employed in numerical simulations [23, 33, 35, 37, 69]. In the present study, however, to find analytical solutions for the volume changes required for caldera formation, I employed a point-source model rather than an ellipsoid to facilitate mathematical treatment.

The horizontal and vertical surface deformation fields in the point-source model were analytically derived by Melan [70], Mindlin [71], Yamakawa [72], Mogi [73], Hagiwara [74], and Bonafede [75]. Bonafede [75] covered poro- and thermo-elastic effects in addition to pure elasticity. The key advantages of the point-source model are that the equations describing it are structurally simple, and there is no need to specify the source geometry. However, this approach has some disadvantages [33, 76–78]. For example, the differences between the solutions provided by point-source and finite sphere models for vertical deformation are substantial and significantly greater than the accuracy of modern geodetic surveys. Moreover, the point-source model does not provide information on the stress state of the magma chamber wall. However, the differences between the point-source and finite sphere model solutions were minor in terms of the surface pattern and extent of the rupture area [79].

In the present study, I employed Hagiwara's formula [74], which yields the same deformation field at the surface but involves simpler equations than other solutions [73]. The 3-D deformation components owing to the contraction of a small sphere are described as

$$u_r = \frac{a^3 r \Delta p}{4\mu}\left[\frac{1}{R_1^3} + \frac{\lambda + 3\mu}{\lambda + \mu}\frac{1}{R_2^3} - \frac{6z(z+d)}{R_2^5}\right], \tag{1}$$

$$u_z = \frac{a^3 \Delta p}{4\mu}\left[\frac{d-z}{R_1^3} + \frac{2d}{R_2^3} + \frac{\lambda - \mu}{\lambda + \mu}\frac{d+z}{R_2^3} + \frac{6z(z+d)^2}{R_2^5}\right], \tag{2}$$

where u_r and u_z are displacements in the r and z directions, respectively. The parameters a, d, and Δp are the radius, depth, and the internal pressure change of the small sphere, respectively (**Figure 3**). In addition, λ and μ are Lame's constants, and the parameters R_1 and R_2 are defined as.

$$R_1 = \sqrt{r^2 + (d-z)^2}, R_2 = \sqrt{r^2 + (d+z)^2}. \tag{3}$$

where $r^2 = x^2 + y^2$ is the radial distance from a point on the free surface above the point-source. In case of $z = 0$ and $\lambda = \mu$, Hagiwara's model is equivalent to the Mogi's formulations.

After transforming the cylindrical coordinates to Cartesian xyz coordinates, the 3-D strain fields were determined as follows:

$$\varepsilon_{xx} = \frac{\Delta V}{4\pi}\left[\frac{1}{R_1^3}\left(1 - \frac{3x^2}{R_1^2}\right) + \frac{\lambda + 3\mu}{\lambda + \mu}\frac{1}{R_2^3}\left(1 - \frac{3x^2}{R_2^2}\right) - \frac{6z(z+d)}{R_2^5}\left(1 - \frac{5x^2}{R_2^2}\right)\right], \tag{4}$$

$$\varepsilon_{yy} = \frac{\Delta V}{4\pi}\left[\frac{1}{R_1^3}\left(1 - \frac{3y^2}{R_1^2}\right) + \frac{\lambda + 3\mu}{\lambda + \mu}\frac{1}{R_2^3}\left(1 - \frac{3y^2}{R_2^2}\right) - \frac{6z(z+d)}{R_2^5}\left(1 - \frac{5y^2}{R_2^2}\right)\right], \tag{5}$$

$$\varepsilon_{zz} = \frac{-\Delta V}{4\pi}\left\{\frac{1}{R_1^3}\left[1 - \frac{3(d-z)^2}{R_1^2}\right] + \frac{\lambda-\mu}{\lambda+\mu}\frac{1}{R_2^3}\left[1 - \frac{3(d+z)^2}{R_2^2}\right] - \frac{6z(z+d)}{R_2^5}\left[3 - \frac{5(d+z)^2}{R_2^2}\right]\right\}, \tag{6}$$

$$\varepsilon_{xy} = \frac{-3\Delta V}{4\pi}xy\left[\frac{1}{R_1^5} + \frac{\lambda+3\mu}{\lambda+\mu}\frac{1}{R_2^5} - \frac{10z(z+d)}{R_2^7}\right], \tag{7}$$

$$\varepsilon_{yz} = \frac{-3\Delta V}{2\pi R_2^3}\frac{\lambda+2\mu}{\lambda+\mu}\frac{y(z+d)}{R_2^2}, \tag{8}$$

$$\varepsilon_{zx} = \frac{-3\Delta V}{2\pi R_2^3}\frac{\lambda+2\mu}{\lambda+\mu}\frac{x(d+z)}{R_2^2}. \tag{9}$$

These equations incorporate the relationship between the internal pressure change of the sphere and the volume change [80]:

$$\Delta p = \frac{\mu\Delta V}{\pi a^3}. \tag{10}$$

The stress fields are expressed by Hooke's law [81].

$$\sigma_{ij} = 2\mu\varepsilon_{ij} + \lambda\varepsilon_{ii}\delta_{ij}, \tag{11}$$

where δ_{ij} is the Kronecker delta. The surface stress fields were obtained using Eqs. (4)–(9), we set $z = 0$, as follows:

$$\sigma_{xx} = \frac{\Delta V}{\pi R^5}\frac{1}{\lambda+\mu}\left(k_1y^2 - k_2x^2 + k_3d^2\right), \tag{12}$$

$$\sigma_{yy} = \frac{\Delta V}{\pi R^5}\frac{1}{\lambda+\mu}\left(k_1x^2 - k_2y^2 + k_3d^2\right), \tag{13}$$

$$\sigma_{zz} = \frac{-\Delta V}{\pi R^5}\frac{\lambda(\lambda+2\mu)}{\lambda+\mu}\left(x^2 + y^2 - 2d^2\right), \tag{14}$$

$$\tau_{xy} = \frac{-3\Delta V}{\pi R^5}\frac{\mu(\lambda+2\mu)}{\lambda+\mu}xy, \tag{15}$$

$$\tau_{yz} = \frac{-3\Delta V}{\pi R^5}\frac{\mu(\lambda+2\mu)}{\lambda+\mu}yd, \tag{16}$$

$$\tau_{zx} = \frac{-3\Delta V}{\pi R^5}\frac{\mu(\lambda+2\mu)}{\lambda+\mu}xd, \tag{17}$$

where

$$R^2 = r^2 + d^2, \tag{18}$$

$$k_1 = -\lambda^2 + 2\mu^2, \tag{19}$$

$$k_2 = \lambda^2 + 3\lambda\mu + 4\mu^2, \tag{20}$$

$$k_3 = 2\lambda^2 + 3\lambda\mu + 2\mu^2. \tag{21}$$

3.2 Volume change required for forming fracture at the surface

The equation providing the volume change required to form a fracture or plastic area at the surface provides the upper limit for use within an elastic range. Here, the von Mises criterion was employed to evaluate the surface stress field caused by the volume change in the source because this criterion can consider all stress components.

The von Mises criterion is described [81] by

$$J_2 = \frac{\sigma_0^2}{3}, \tag{22}$$

where J_2 and σ_0 are the second invariants of the deviatoric stress tensor and the yield stress, respectively. J_2 is defined [81] as

$$J_2 = \frac{1}{2}\left(\sigma_{xx}^2 + \sigma_{yy}^2 + \sigma_{zz}^2\right) + \tau_{xy}^2 + \tau_{yz}^2 + \tau_{zx}^2. \tag{23}$$

Thus, from Eqs. (12)–(17), we obtain

$$J_2 = \frac{\mu^2 \Delta V^2}{3(\lambda+\mu)^2 \pi^2 R^{10}} \left(p_1 r^4 + p_2 r^2 d^2 + p_3 d^4\right), \tag{24}$$

where

$$p_1 = 7\lambda^2 + 26\lambda\mu + 28\mu^2, \tag{25}$$

$$p_2 = 26\lambda^2 + 112\lambda\mu + 104\mu^2, \tag{26}$$

$$p_3 = \lambda^2 - 4\lambda\mu + 4\mu^2, \tag{27}$$

The Eq. (24) can be rewritten using the Young's modulus (E) and Poisson's ratio (ν) as

$$J_2 = \frac{E^2 \Delta V^2}{3(1+\nu)^2 \pi^2 R^{10}} \left(q_1 r^4 + q_2 r^2 d^2 + q_3 d^4\right). \tag{28}$$

In Eq. (28), the relationships $\lambda = E\nu/[(1+\nu)(1-2\nu)]$ and $\mu = E/[2(1+\nu)]$ were used, and q_1, q_2, and q_3 are given as

$$q_1 = 9\nu^2 - 15\nu + 7, \tag{29}$$

$$q_2 = 18\nu^2 - 48\nu + 26, \tag{30}$$

$$q_3 = 9\nu^2 - 6\nu + 1. \tag{31}$$

As reported by Kusumoto and Takemura [79] and Kusumoto and Gudmundsson [45], a relationship between the source depth and surface fracture (ring-fault) radius can be derived by differentiating the fracture or elastic-plastic criterion and equating the derivative to zero. By differentiating Eq. (28) and equating the derivative to zero, $\partial J_2/\partial r = 0$, we obtain the following relationship:

$$r_c = \sqrt{\frac{-3(3\nu^2 - 9\nu + 5) + \sqrt{(2-3\nu)(135\nu^2 - 273\nu + 137)}}{9\nu^2 - 15\nu + 7}} \cdot d. \tag{32}$$

This is the relationship between the source depth and surface fracture or plastic zone radius, and the proportionality coefficient, α, is given in terms of Poisson's ratio as follows:

$$\alpha = \sqrt{\frac{-3(3\nu^2 - 9\nu + 5) + \sqrt{(2 - 3\nu)(135\nu^2 - 273\nu + 137)}}{9\nu^2 - 15\nu + 7}}. \quad (33)$$

By incorporating this relationship, we obtain the following equation expressing the volume change required for the fracture or plastic zone formation in the 3-D case from Eq. (28):

$$\Delta V = \frac{\pi\sigma_0(1+\nu)(\alpha^2+1)^{\frac{5}{2}}}{E\sqrt{q_1\alpha^4 + q_2\alpha^2 + q_3}} d^3. \quad (34)$$

As in the 2-D case [45], the volume change required for the fracture or plastic zone formation was proportional to the cube of the source depth.

As shown in Eq. (32), the relationship between the caldera radius and magma chamber depth is linear, and its proportionality coefficient is a function of Poisson's ratio. I derived the distribution of the proportionality coefficient in the theoretical range ($0 \leq v < 0.5$) of Poisson's ratio, in order to find the theoretical and general characteristics of the coefficient, although Poisson's ratio of rocks is 0.2–0.3 generally.

The results are shown in **Figure 4**. The coefficient α increases with increasing Poisson's ratio and decreases steeply when the ratio exceeds $v = 0.33$. If the source depth is constant, the fracture or plastic zone radius would be changed by a change in the coefficient α, and the radius reaches its maximum at $v = 0.33$.

The solutions shown here were determined by evaluating only the surface stresses caused by the contraction of the point-source, assuming an elastic-perfectly plastic material. However, actual magma chambers are not point-sources and have finite complex configurations [65]. In addition, the surface ring-fault is a result of fracture

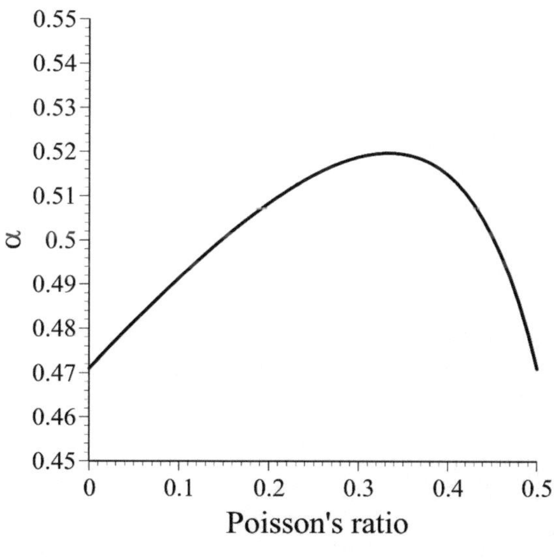

Figure 4.
Distribution of the proportionality coefficient α derived from the von Mises criterion for the theoretical range of Poisson's ratio (0 ≤ v < 0.5).

propagation from the chamber roof, as shown in many analogue and numerical simulations [24, 26, 27, 30, 36, 82–84]. Hence, as a first-order approximation, the solutions shown here provide information on the depth of the source and the volume change required for the fracture or plastic zone from a mechanical perspective.

4. Application to the Shishimuta caldera

4.1 Depth and volume change of the magma chamber

As previously mentioned, the Shishimuta caldera was the origin of the Yabakei pyroclastic flow at 1 Ma and the Imaichi pyroclastic flow at 0.9 Ma. It has been considered that the caldera was formed by these two large eruptions. In the modelling, because the Shishimuta caldera is the buried caldera and two pyroclastic flows had erupted in a short interval of 100,000 years, I assumed that the Shishimuta caldera was formed by a series of activities of one magma chamber, and I estimated the depth and volume change of the magma chamber that participated in a series of eruptions. I employed Eqs. (32) and (34), respectively, to estimate the depth and volume changes.

The Shishimuta caldera formed in the andesitic upper crust. The Poisson's ratio of the andesite was estimated to be 0.20 (v = 0.20) as an average from the database of Lama and Vutukuri [85]. Substituting this value into Eq. (33), we obtained α = 0.508, and the following was obtained from Eq. (32):

$$d = \frac{1}{0.508} \cdot r_c = 1.97 \cdot r_c. \tag{35}$$

From this equation, we can estimate a magma chamber depth of approximately 8.87 km, d = 8.87 km, because the average radius of the Shishimuta caldera is about 4.5 km, r_c = 4.5 km.

The estimated depth was within the magma depth range of 7–12 km suggested by Kamata [46]. Because the solutions presented in this study are provided by the point-source model, we should consider that the approximation of the solution is weak. However, the results obtained are roughly consistent with the magma chamber depth estimated from geological discussions, indicating that the solution is functional for providing information on the depth of the magma chamber as a first-order approximation.

When a magma chamber exists at a depth of d = 8.87 km, the necessary volume change for caldera formation can be estimated using Eq. (34). The Young's modulus of the andesite was estimated to be 45.3 GPa (E = 45.3 GPa) as an average from the database of Lama and Vutukuri [85]. Because the yield stress is not given in the database, I estimated the necessary volume change for the caldera such that the tensile strength (T_0) is employed as the yield stress (σ_0). When a tensile strength of T_0 = 8 MPa was employed from the database of Lama and Vutukuri [85] as the yield stress, the required volume change was estimated to be approximately 0.37 km^3.

4.2 Stress fields at the surface

By substituting this volume change ΔV of 0.37 km^3 in Eqs. (12)–(17), we can obtain the stress fields at the surface in the initial stage of Shishimuta caldera formation or during the deflation stage of the magma chamber. **Figure 5** shows the distributions of the three principal stress components caused by the deflation of the magma chamber. The principal stresses indicate that the dip angle of each principal stress axis was less than 40°.

Because subsidence caused by the contraction of the magma chamber occurred at the surface above the chamber, compressive stress fields appeared for each principal stress in the subsidence area. Although the maximum principal stress (σ_1) is compressive over the entire area, the minimum (σ_3) and intermediate (σ_2) principal stresses changed from the compressive field into the tensile field at an area slightly distant from the centre of the caldera. The principal axis of σ_3 is essentially distributed radially on the horizontal plane from the centre of the Shishimuta caldera, and the principal axes of σ_1 are essentially perpendicular to the axis of σ_3. Although the principal axes of σ_2 are perpendicular to the axis of σ_3 near the centre of the caldera, the axes are essentially vertical and are perpendicular to the axes of σ_1 and σ_3.

Figure 6 shows the distribution of the fracture or plastic zone evaluated using the von Mises criterion. The light-blue area shown in **Figure 6** indicates the area of $J_2 \geq \sigma_0^2/3$, namely the plastic or fracture area, which includes the Shishimuta caldera.

Figure 7 shows the surface distributions of the principal stress fields when the magma chamber was activated with an inflation of 0.37 km^3. The principal stresses shown also have dip angles which are less than 40° for each principal stress axis.

Tensile stress fields appeared for each principal stress in the uplift area because the uplift caused by the inflation of the magma chamber occurred at the surface just above the chamber. Although the minimum principal stress (σ_3) was tensile throughout the area, the maximum (σ_1) and intermediate (σ_2) principal stresses changed from the tensile to the compressive field in an area at a small distance from surface just above the chamber. The principal axes of σ_1 were distributed radially on the horizontal plane and were essentially parallel to the strike direction of each fault consisting of the Haneyama fault zone. The principal axes of σ_3 were essentially perpendicular to the axes of σ_1 and to the strike direction of the faults. Although the principal axes of σ_2 were perpendicular to the axis of σ_1 near the surface just above the chamber, they were essentially vertical and perpendicular to the axes of σ_1 and σ_3.

Figure 8 shows the distribution of the fracture or plastic zone evaluated using the von Mises criterion. The light-purple area shown in **Figure 8** indicates the area of $J_2 \geq \sigma_0^2/3$, that is, the plastic or fracture area.

5. Discussion

The distribution of each principal stress axis shown in **Figure 5** promotes ring-fault formation parallel to the intermediate or maximum principal stress axes. Although it is difficult to identify an initial ring-fault structure on the present surface because this area is covered by lava flows and ash, such structures might exist underground. Kamata [46] pointed out the existence of a caldera wall at a depth of 1 km based on drilling core data and gravity anomalies.

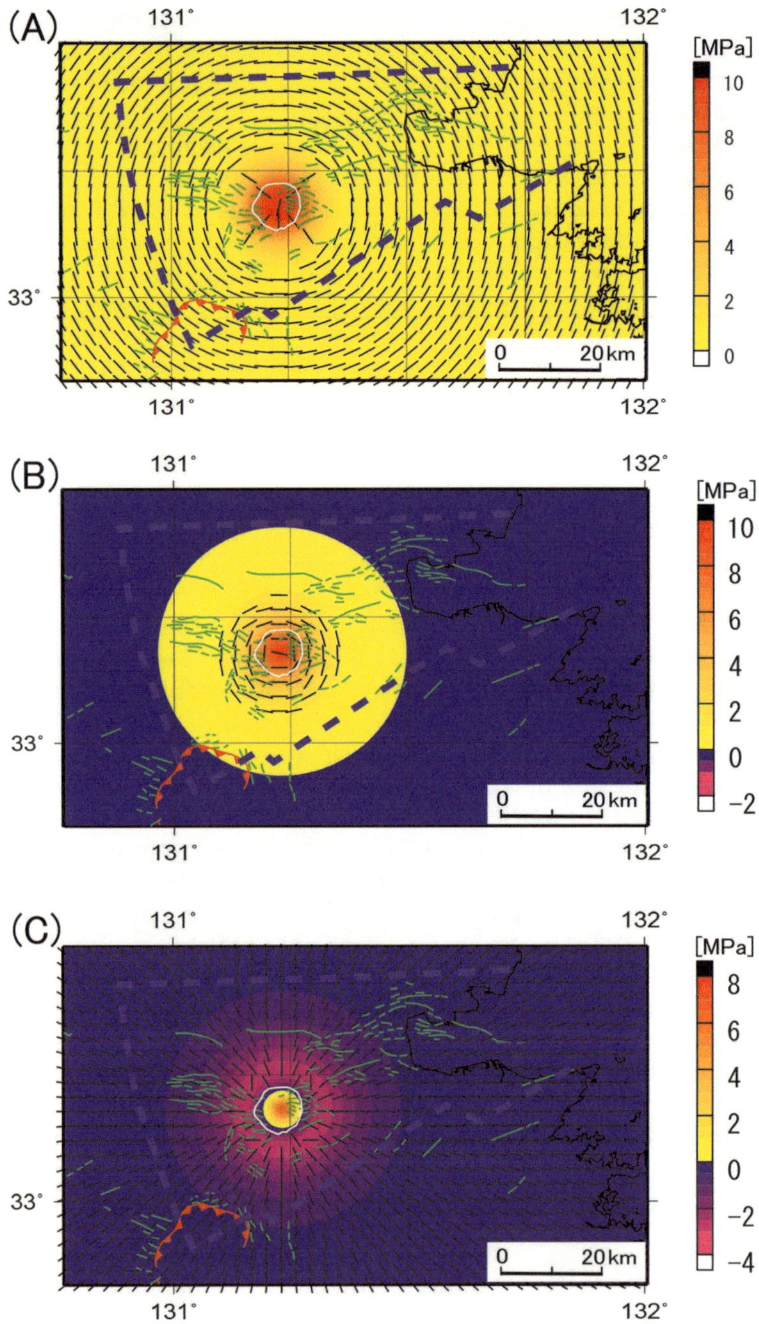

Figure 5.
Surface distributions of the principal stress fields (dip angle of each principal stress axis: less than 40°) and their principal axes caused by the formation of the Shishimuta caldera. Colour contours show the magnitude of each stress (MPa), and black ticks indicate the principal axis of each stress. Compression is expressed here positive. The tick length was drawn as a constant and did not indicate the magnitude of each principal stress. In each figure, the white, green, and red lines indicate the Shishimuta caldera, active faults, and Aso caldera, respectively. (A) Distributions of the maximum principal stress, σ_1, and its axis at the surface. (B) Distributions of the intermediate principal stress, σ_2, and its axis at the surface. (C) Distributions of the minimum principal stress, σ_3, and its axis at the surface.

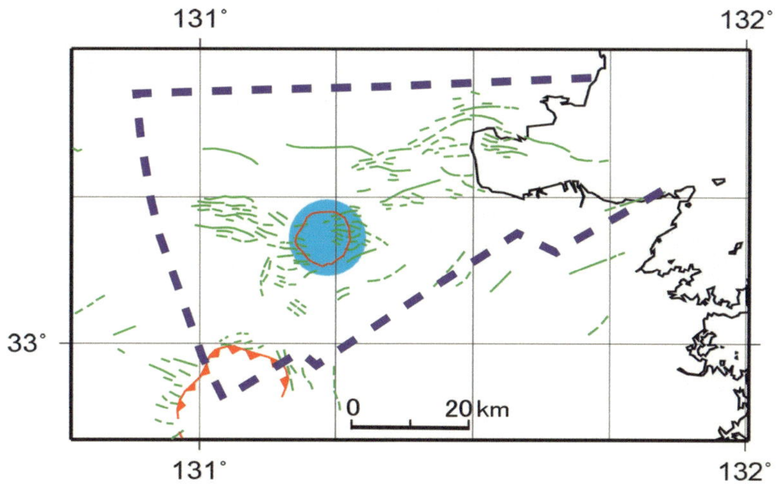

Figure 6.
The distribution of the fracture or plastic zone, assuming a deflation of 0.37 km³. The light-blue area indicates the fracture or plastic zone.

In the distribution of the principal stress axes shown in **Figure 5**, each normal fault constituting the Haneyama fault zone is parallel to the principal axis of σ_3 and perpendicular to σ_1, representing tensile and compressive stress, respectively. It is difficult to explain the normal faulting and related activities in these stress fields. Consequently, it was found that if the magma chamber under the Shishimuta caldera reactivates by deflation, the stress field due to the reactivation would not promote easy reactivation of the Haneyama fault zone as a normal fault zone.

The distributions of each principal stress axis shown in **Figure 7** make it difficult to promote ring-fault formation, although a circular fracture or plastic zone is formed. However, because each normal fault constituting the Haneyama fault zone is parallel to the principal axis of σ_1 and perpendicular to σ_3 (**Figure 7**), these stress fields can explain normal faulting and related activities. That is, if the magma chamber under the Shishimuta caldera reactivates with inflation, the stress field due to reactivation will easily promote reactivation of the Haneyama fault zone as a normal fault zone.

Results and discussion suggest that the stress field might have led to the formation and growth of the Haneyama fault zone due to the inflation of the magma chamber after the Shishimuta caldera formation. This suggests that the inflation of the chamber, rather than the contraction or collapse of the magma chamber including caldera formation, is required to reactivate normal faults in the fault zone. In fact, morphological and geological surveys have confirmed that faults in the Haneyama fault zone were active after 0.9 Ma, mainly around 0.6 Ma [48]. The results shown in **Figure 7** and the observations of faults in the Haneyama fault zone support the hypothesis that the normal faults distributed in the fault zone are related to volcanic activity, as suggested by Tsukuda [60]. However, because many volcanoes are situated in this region, a more detailed and practical discussion on factors such as the concrete volume change in reactivated faults and the location of the source (e.g. magma chamber) should be conducted by considering the active history of each volcano.

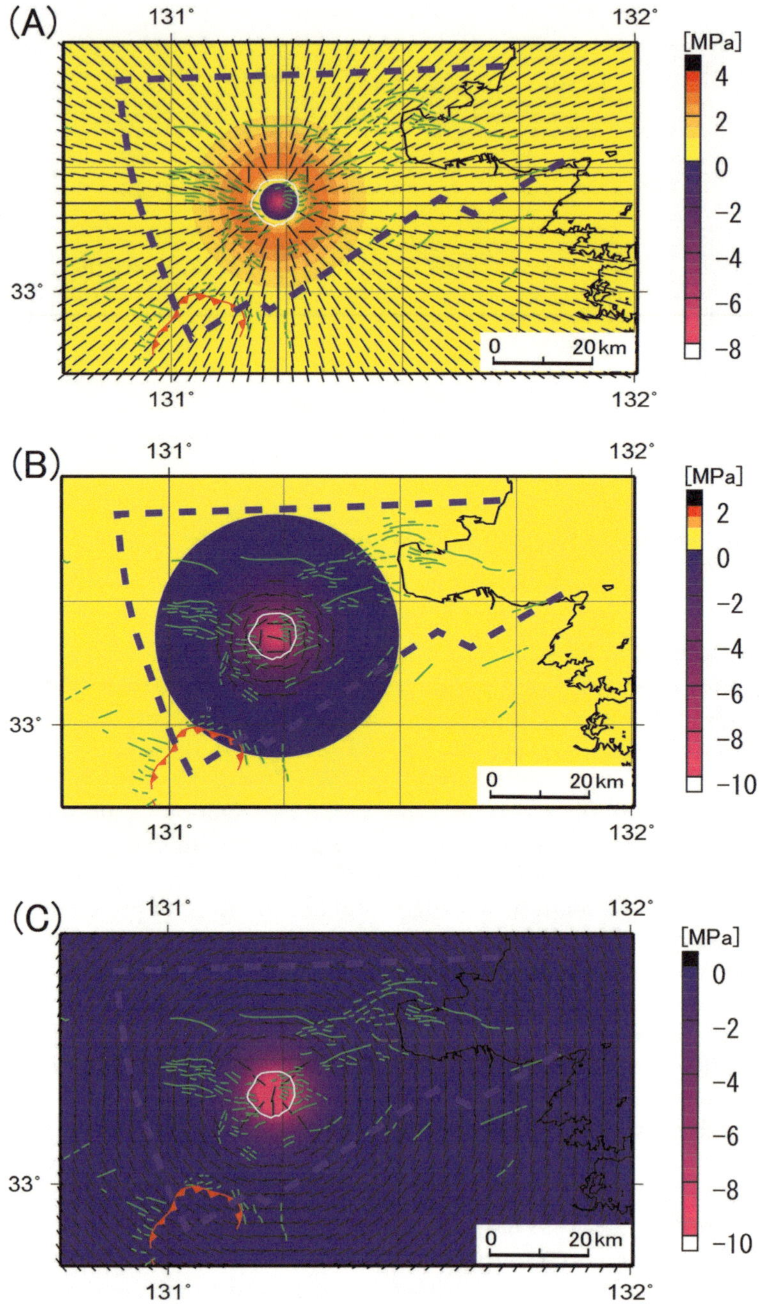

Figure 7.

The surface distributions of the principal stress fields (dip angle of each principal stress axis: less than 40°) and their principal axes caused by the inflation of the magma chamber, which is the origin of the Shishimuta caldera. Colour contours show the magnitude of each stress (MPa), and black ticks indicate the principal axis of each stress. Compression is expressed here positive. The tick length was drawn as a constant and did not indicate the magnitude of each principal stress. In each figure, the white, green, and red lines indicate the Shishimuta caldera, active faults, and Aso caldera, respectively. (A) Distributions of the maximum principal stress, σ_1, and its axis at the surface. (B) Distributions of the intermediate principal stress, σ_2, and its axis at the surface. (C) Distributions of the minimum principal stress, σ_3, and its axis at the surface.

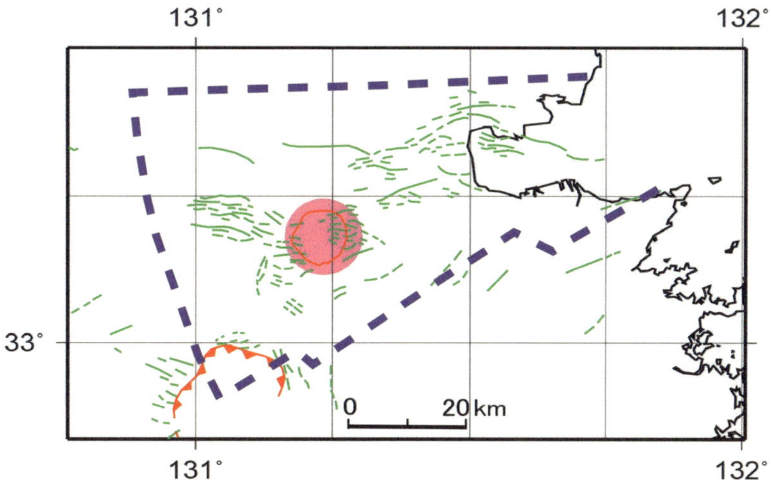

Figure 8.
Distribution of the fracture or plastic zone, assuming inflation of 0.37 km³. The light-purple area indicates a fracture or plastic zone.

6. Conclusion

In this study, I estimated the surface stress fields caused by the activity of the magma chamber of the Shishimuta caldera and discussed its effects on fault formation or activity in this region. Before this discussion, I derived analytical solutions providing the magma chamber volume change for caldera formation and connecting the radius of the caldera with the depth of the magma chamber. In this derivation, I employed a point-source model and the von Mises criterion as the magma chamber model and the elastic-fracture or elastic-plastic criterion, respectively. We found that the volume change was proportional to the cube of the magma chamber depth, and the radius of the caldera was dependent only on the Poisson's ratio and the depth of the magma chamber.

For field applications, the Shishimuta caldera was selected, and the depth of the magma chamber was estimated from the radius of the caldera. As a result, a magma chamber depth of approximately 9 km was obtained, which is comparable to the geological data, and the volume change for the initial caldera formation was estimated to be 0.37 km^3. I estimated the surface stress fields caused by deflation and inflation of 0.37 km^3 and found that if the magma chamber under the Shishimuta caldera reactivates by deflation, the Haneyama fault zone will not be formed or reactivated easily as a normal fault zone, although a ring-fault would be formed. However, if the magma chamber reactivates during inflation, the Haneyama fault zone will be activated or formed as a normal fault zone. In this case, a ring-fault will not be formed.

From these discussions and observations by morphological and geological surveys, I conclude that inflation of the magma chamber under the Shishimuta caldera promotes the formation and growth of the Haneyama fault zone after caldera formation. Currently, this region has a stress field which promotes normal faulting in the fault zone. In the future, to evaluate the possibility of earthquakes occurring in this fault zone, more realistic and detailed sources should be considered, and the effects of activities of other active volcanoes should be discussed with tectonic stress.

Acknowledgements

I am most grateful to the academic editor, Prof. Károly Németh, for his constructive review and comments on the manuscript. The manuscript was improved.

Conflict of interest

The author declares no conflict of interest.

Author details

Shigekazu Kusumoto
Graduate School of Science, Institute for Geothermal Sciences, Kyoto University, Japan

*Address all correspondence to: kusumoto.shigekazu.2m@kyoto-u.ac.jp

References

[1] Tilling RI, Dvorak JJ. Anatomy of a basaltic volcano. Nature. 1993;**363**: 125-132

[2] De Natale G, Zollo A, Ferraro A, Virieux J. Accurate fault mechanism determinations for a 1984 earthquake swarm at Campi Flegrei caldera (Italy) during an unrest episode: Implications for volcanological research. Journal of Geophysical Research. 1995;**100**: 24167-24185

[3] Toda S, Stein R, Sagiya T. Evidence from the AD 2000 Izu islands earthquake swarm that stressing rate governs seismicity. Nature. 2002;**419**: 58-61

[4] Gargani J, Geoffroy L, Gac S, Cravoisier S. Fault slip and coulomb stress variations around a pressured magma reservoir: Consequences on seismicity and magma intrusion. Terra Nova. 2006;**18**:403-411. DOI: 10.1111/j.1365-3121.2006.00705.x

[5] Walter TR. Facilitating dike intrusions into ring-faults. In: Gottsmann J, Martí J, editors. Caldera Volcanism: Analysis, Modelling and Response. Amsterdam: Elsevier; 2008. pp. 351-374

[6] Vilardo G, Isaia R, Ventura G, De Martino P, Terranova C. InSAR permanent Scatterer analysis reveals fault re-activation during inflation and deflation episodes at Campi Flegrei caldera. Remote Sensing of Environment. 2010;**114**: 2373-2383

[7] Gabrieli A, Wilson L, Lane S. Volcano–tectonic interactions as triggers of volcanic eruptions. Proceedings of the Geologists' Association. 2015;**126**: 675-682

[8] Linde AT, Sacks IS. Triggering of volcanic eruptions. Nature. 1998;**395**: 888-890

[9] Alam M, Kimura M. Statistical analysis of time-distance relationship between volcanic eruptions and great earthquakes in Japan. Earth, Planets and Space. 2004;**65**:179-192

[10] Lemarchand N, Grasso JR. Interactions between earthquakes and volcano activity. Geophysical Research Letters. 2007;**34**:L24303

[11] Manga M, Brumm M, Rudolph ML. Earthquake triggering of mud volcanoes. Marine and Petroleum Geology. 2009; **26**:1785-1798

[12] Lara LE, Naranjo JA, Moreno H. Rhyodacitic fissure eruption in southern Andes (cordon Caulle; 40.5 degrees S) after the 1960 (mw: 9.5) Chilean earthquake: A structural interpretation. Journal of Volcanology and Geothermal Research. 2004;**138**:127-138

[13] Harris AJL, Riepepe M. Regional earthquake as a trigger for enhanced volcanic activity: Evidence from MODIS thermal data. Geophysical Research Letters. 2007;**34**:L02304

[14] Takada Y, Fukushima Y. Volcanic subsidence triggered by the 2011 Tohoku earthquake in Japan. Nature Geoscience. 2013;**6**:637-641. DOI: 10.1038/ngeo1857

[15] Smith RL, Bailey RA. Resurgent calderas. Memoir of the Geological Society of America. 1968;**116**: 613-662

[16] Druitt TH, Sparks RS. On the formation of calderas during ignimbrite eruptions. Nature. 1984;**310**:679-681

[17] Lipman PW. Subsidence of ash-flow calderas: Relation to caldera size and magma-chamber geometry. Bulletin of Volcanology. 1997;**59**:198-218

[18] Gudmundsson A. How local stresses control magma-chamber ruptures, dyke injections, and eruptions in composite volcanoes. Earth-Science Reviews. 2006; **79**:1-31

[19] Gudmundsson A, Brenner SL. On the conditions of sheet injections and eruptions in stratovolcanoes. Bulletin of Volcanology. 2005;**67**:768-782

[20] Walter TR, Acocella V, Neri M, Amelung F. Feedback processes between magmatic events and flank movement at Mount Etna (Italy) during the 2002-2003 eruption. Journal of Geophysical Research. 2005;**110**(B10205):1-12

[21] Acocella V. Activating and reactivating pairs of nested collapses during calderaforming eruptions: Campi Flegrei (Italy). Geophysical Research Letters. 2008;**35**:L17304. DOI: 10.1029/2008GL035078

[22] Gudmundsson A, Friese N, Galindo I, Philipp SL. Dike-induced reverse faulting in a graben. Geology. 2008;**36**:123-126

[23] Holohan EP, Walter TR, Schöpfer MPJ, Walsh JJ, van Wyk de Vries B, Troll VR. Origins of oblique-slip faulting during caldera subsidence. Journal of Geophysical Research. 2013; **118**:1778-1794. DOI: 10.1002/jgrb.50057

[24] Komuro H. Experiments on caldron formation: A polygonal cauldron and ring fractures. Journal of Volcanology and Geothermal Research. 1987;**31**: 139-149

[25] Marti J, Ablay GJ, Redshaw LT, Sparks RSJ. Experimental studies of collapse calderas. Journal of the Geological Society of London. 1994;**151**: 919-929

[26] Acocella V, Cifelli F, Funiciello R. Analogue models of collapse calderas and resurgent domes. Journal of Volcanology and Geothermal Research. 2000;**104**:81-96

[27] Roche O, Druitt TH, Merle O. Experimental study of caldera formation. Journal of Geophysical Research. 2000;**105**:395-416

[28] Walter TR, Troll VR. Formation of caldera periphery faults: An experimental study. Bulletin of Volcanology. 2001;**63**:191-203

[29] Troll VR, Walter TR, Schmincke HU. Cycle caldera collapse: Piston or piecemeal subsidence? Field and experimental evidence. Geology. 2002;**30**:135-138

[30] Ruch J, Acocella V, Geshi N, Nobile A, Corbi F. Kinematic analysis of vertical collapse on volcanoes using experimental models time series. Journal of Geophysical Research. 2012;**117**: B07301. DOI: 10.1029/2012JB009229

[31] Acocella V, Funiciello R, Marotta E, Orsi G, de Vita S. The role of extensional structures on experimental calderas and resurgence. Journal of Volcanology and Geothermal Research. 2004;**129**:199-217

[32] Holohan EP, Troll VR, Walter TR, Münn S, McDonnell S, Shipton ZK. Elliptical calderas in active tectonic settings: An experimental approach. Journal of Volcanology and Geothermal Research. 2005;**144**:119-136

[33] Gudmundsson A. Formation and development of normal fault calderas and the initiation of large explosive eruptions. Bulletin of Volcanology. 1998; **60**:160-170

[34] Roche O, Druitt TH. Onset of caldera collapse during ignimbrite eruptions. Earth and Planetary Science Letters. 2001;**191**:191-202

[35] Folch A, Marti J. Geometrical and mechanical constraints on the formation of ring-fault calderas. Earth and Planetary Science Letters. 2004;**221**: 215-225

[36] Hardy S. Structural evolution of calderas: Insights from two-dimensional discrete element simulations. Geology. 2008;**36**:927-930. DOI: 10.1130/ G25133A.1

[37] Holohan EP, Schôpfer MPJ, Walsh JJ. Mechanical and geometric controls on the structural evolution of pit crater and caldera subsidence. Journal of Geophysical Research. 2011;**116**:B07202. DOI: 10.1029/2010JB008032

[38] Marti J, Geyer A, Folch A, Gottsmann J. A review on collapse caldera modelling. In: Gottsmann J, Martí J, editors. Caldera Volcanism: Analysis, Modelling and Response. Amsterdam: Elsevier; 2008. pp. 233-283

[39] Geyer A, Martí J. A short review of our current understanding of the development of ring faults during collapse caldera formation. Frontiers in Earth Science. 2014;**2**:22. DOI: 10.3389/ feart.2014.00022

[40] Acocella V, Korme AT, Salvini F, Funiciello R. Elliptic calderas in the Ethiopian rift: Control of pre-existing structures. Journal of Volcanology and Geothermal Research. 2002;**119**:189-203

[41] Lavallée Y, Stix J, Kennedy B, Richer M, Longpré M. Caldera subsidence in areas of variable topographic relief: Results from analogue modeling. Journal of Volcanology and Geothermal Research. 2004;**129**:219-236

[42] Kusumoto S, Nagao T. Relationship between the aspect ratio of a collapsed caldera and the regional stress field - their evaluations using a single small spherical magma chamber model in an anisotropic stress field. International Journal of Earth Science and Geophysics. 2016;**2**:006

[43] Anderson EM. The dynamics of formation of cone sheets, ring dykes and cauldron subsidences. Proceedings of the Royal Society of Edinburgh. 1936;**56**: 128-163

[44] Williams H. Calderas and their origin. University of California Publications. Bulletin of the Department of Geological Sciences. 1941;**25**:239-346

[45] Kusumoto S, Gudmundsson A. Magma-chamber volume changes associated with ring-fault initiation using a finite-sphere model: Application to the Aira caldera, Japan. Tectonophysics. 2009;**471**:58-66. DOI: 10.1016/jtecto. 2008.09.001

[46] Kamata H. Shishimuta caldera, the buried source of the Yabakei pyroclastic flow in the Hohi volcanic zone, Japan. Bulletin of Volcanology. 1989;**51**:41-50

[47] Kamata H. Volcanic and structural history of the Hohi volcanic zone, Central Kyushu, Japan. Bulletin of Volcanology. 1989;**51**:315-332

[48] Research Group for Active Faults. The Active Faults in Japan: Sheet Maps and Inventories. Rev. ed. Tokyo: University of Tokyo Press; 1991. 437 p

[49] Komazawa M. Gravity database of Japan. In: Digital Geoscience Map P-2. DVD ed. Tsukuba: Geological Survey of Japan; 2013. DVD

[50] Itoh Y, Takemura K, Kamata H. History of basin formation and tectonic

evolution at the termination of a large transcurrent fault system: Deformation mode of Central Kyushu, Japan. Tectonophysics. 1998;**284**:135-150

[51] Kusumoto S, Takemura K, Fukuda Y, Takemoto S. Restoration of the depression structure at the eastern part of Central Kyushu, Japan, by means of the dislocation modeling. Tectonophysics. 1999;**302**:287-296

[52] Kusumoto S, Fukuda Y, Takemura K. A distinction technique between volcanic and tectonic depression structures based on the restoration modeling of gravity anomaly: A case study of the Hohi volcanic zone, Central Kyushu, Japan. Journal of Volcanology and Geothermal Research. 1999;**90**: 183-189

[53] Aramaki S. Formation of the Aira caldera, Southern Kyushu, ~ 22,000 years ago. Journal of Geophysical Research. 1984;**89**:8485-8501

[54] Carle FS. Three-dimensional gravity modeling of the geologic structure of long valley caldera. Journal of Geophysical Research. 1988;**93**: 13237-13250

[55] Zurek J, Williams-Jones G. The shallow structure of Kīlauea caldera from high-resolution Bouguer gravity and total magnetic anomaly mapping: Insights into progressive magma reservoir growth. Journal of Geophysical Research. 2013;**118**:3742-3752

[56] Molina F, Martí J, Aguirre G, Vega E, Chavarría L. Stratigraphy and structure of the Cañas Dulces caldera (Costa Rica). Geological Society of America Bulletin. 2014;**126**:1465-1480

[57] Kamata H, Danhara T, Yamashita T, Hoshizumi H, Hayashida A, Takemura K. Correlation of the Azuki volcanic ash of the osaka group and the Ku6C volcanic ash of the kazusa group to the imaichi pyroclastic flow deposit in Central Kyushu, Japan – a co-ignimbrite ash erupted from Shishimuta caldera. Journal of Geological Society of Japan. 1994;**100**:848-866. (in Japanese with English abstract)

[58] Aramaki S. Hydrothermal determination of temperature and water pressure on the magma of Aira caldera, Japan. American Mineralogist. 1971;**56**: 1760-1768

[59] Hoshizumi H, Ono K, Mimura K, Noda T. Geology of the Beppu District, with Geological Sheet Map at 1: 50,000. Tsukuba: Geological Survey of Japan; 1988. 137 p. (in Japanese)

[60] Tsukuda E. Is the Central Kyushu really rifting north to south? Memory of Geological Society of Japan. 1992;**41**: 149-161. (in Japanese with English abstract)

[61] Tada T. Spreading of the okinawa trough and its relation to the crustal deformation in Kyusyu. Journal of Seismological Society of Japan, Series. 1984;**2**(37):407-415. (in Japanese with English abstract)

[62] Shimizu H, Umakoshi K, Matsuwo N. Seismic activity in middle and western Kyushu. Memory of Geological Society of Japan. 1993;**41**: 13-18. (in Japanese with English abstract)

[63] Sudo Y. Seismic activity and tectonic significance near the volcanic areas in Central Kyushu, Japan. Memory of Geological Society of Japan. 1993;**41**: 19-34. (in Japanese with English abstract)

[64] Matsumoto S, Nakao S, Ohkura T, Miyazaki M, Shimizu H, Abe Y, et al.

Spatial heterogeneities in tectonic stress in Kyushu, Japan and their relation to a major shear zone. Earth, Planets Space. 2015;**67**:172

[65] Cashman KV, Giordano G. Calderas and magma reservoirs. Journal of Volcanology and Geothermal Research. 2014;**288**:28-45

[66] Tsuchida E, Saito Y, Nakahara I, Kodama M. Stress concentration around a prolate spheroidal cavity in a semi-infinite elastic body under all-round tension. Bulletin of the Japan Society of Mechanical Engineers. 1982;**25**:493-500

[67] Davis PM. Surface deformation due to inflation of an arbitrarily oriented triaxial ellipsoidal cavity in an elastic half-space, with reference to Kilauea volcano, Hawaii. Journal of Geophysical Research. 1986;**91**:7429-7438

[68] Yang XM, Davis PM, Dieterich JS. Deformation from inflation of a dipping finite prolate spheroid in an elastic half-space as a model for volcanic stressing. Journal of Geophysical Research. 1988; **93**:4249-4257

[69] Gudmundsson A, Marti J, Turon E. Stress field generating ring faults in volcanoes. Geophysical Research Letters. 1997;**24**:1559-1562

[70] Melan E. Point force at internal pointing a semi-infinite plate. Journal of Applied Mathematics and Mechanics. 1932;**12**:343-346

[71] Mindlin RD. Force at a point in the interior of a semi-infinite solid. Physics. 1936;**7**:195-202

[72] Yamakawa N. On the strain produced in a semi-infinite elastic solid by an interior source of stress. Journal of Seismological Society of Japan, Series. 1955;**2**(8):84-98. (in Japanese with abstract and figure captions in English)

[73] Mogi K. Relations between eruptions of various volcanoes and the deformation of the ground surface around them. Bulletin of Earthquake Research Institute. 1958;**36**:99-134

[74] Hagiwara Y. Theory of Geogravity. Tokyo: Kyoritsu-Zensyo; 1978. 256 p. (In Japanese)

[75] Bonafede M. Axi-symmetric deformation of a thermo-poro-elastic half-space: Inflation of a magma chamber. Geophysical Journal International. 1990;**103**:289-299

[76] McTigue DF. Elastic stress and deformation near a finite spherical magma body: Resolution of the point source paradox. Journal of Geophysical Research. 1987;**92**:12931-12940

[77] Gudmundsson A. Magma chambers modeled as cavities explain the formation of rift zone central volcanoes and their eruption and intrusion statistics. Journal of Geophysical Research. 1998;**103**:7401-7412

[78] Trasatti E, Giunchi C, Bonafede M. Structural and rheological constants on source depth and overpressure estimates at the Campi Flegrei caldera, Italy. Journal of Volcanology and Geothermal Research. 2005;**144**:105-118

[79] Kusumoto S, Takemura K. Caldera geometry determined by the depth of the magma chamber. Earth, Planets and Space. 2005;**57**:e17-e20

[80] Williams CA, Wadge G. The effects of topography on magma chamber deformation models: Application to Mt. Etna and radar interferometry. Geophysical Research Letters. 1998;**25**: 1549-1552

[81] Jaeger JC, Cook NG, Zimmerman RW. Fundamentals of Rock Mechanics. 4th ed. London: Methuen; 2007. 488 p

[82] Brothelande E, Peltier A, Got JL, Merle O, Lardy M, Garaebiti E. Constraints on the source of resurgent doming inferred fromanalogue and numericalmodeling—Implications on the current feeding system of the Yenkahe dome–Yasur volcano complex (Vanuatu). Journal of Volcanology and Geothermal Research. 2016;**322**:225-240. DOI: 10.1016/j.jvolgeores.2015.11.023

[83] Liu YK, Ruch J, Vasyura-Bathke H, Jónsson S. Influence of ring faulting in localizing surface deformation at subsiding calderas. Earth and Planetary Science Letters. 2019;**526**:115784. DOI: 10.1016/j.epsl.2019.115784

[84] Maestrelli D, Bonini M, Corti G, Del Ventisette C, Moratti G, Montanari D. Exploring fault propagation and the role of inherited structures during caldera collapse through laboratory experiments. Journal of Volcanology and Geothermal Research. 2021;**414**:107232. DOI: 10.1016/j.jvolgeores.2021.107232

[85] Lama RD, Vutukuri VS. Handbook on Mechanical Properties of Rocks - Testing Techniques and Results – Vol. 2. Clausthal: Trans Tech Publications; 1978. 495 p

Chapter 3

Morphostructural Partitional Sites of Shoshonitic Volcanism in the Cahulote Graben, Michoacán Guanajuato Volcanic Field, México

Daniel Alberto Gómez-Calderón, Pedro Corona-Chávez, María del Sol Hernández-Bernal and Francisco Abraham Paz-Moreno

Abstract

The shoshonitic-type rocks are scarce in volcanic arcs (<3 vol.%), typically associated with mafic volcanism, and both suites are commonly connected to extensional tectonics and deep lithospheric magmatic source zones. The Michoacán Guanajuato Volcanic Field (MGVF), located at the central sector of the Transmexican Volcanic Belt (TMVB), has over 1400 volcanoes, mainly mafic-andesitic calc-alkaline type. Through a holistic approach, including morphostructural, volcano morphometry, petrography, geochemistry and geochronology data, this work presents a detailed study of Cahulote Graben (CG) in the Jorullo-Tacámbaro volcanic region. We have divided the CG into three morphostructural domains containing 116 volcanoes of several types. Morphostructural domains were developed by the interaction of four extensional and transtensional brittle deformation systems: NE-SW, NW-SE, E-W and NNW-SSE. CG is an extensional-transtensional NE-SW tectonic system. Three magmatic calc-alkaline, shoshonitic and intra-plate-like plumbing systems were arranged during the Quaternary (<2 Ma). Petrographic, major and trace elements support relatively fast ascent and different stages of magma stagnation from deep reservoirs. Comparison of Zr/Al and Ti/Al ratios and Nd–Sr isotopic signatures of CG shoshonites with others reported in the TMVB supports the hypothesis of the contrasting depth of source magmas and strongly suggests an extensive tectonic regime for these shoshonites. A volcanotectonic model is proposed to explain the emplacement mechanism of shoshonitic magmatism.

Keywords: shoshonitic volcanism, morphostructural analysis, extensional tectonics, Michoacán Guanajuato volcanic field, volcanotectonic

1. Introduction

In the central sector of the Transmexican Volcanic Belt (TMVB) is located the Michoacán Guanajuato volcanic field (MGVF; **Figure 1**), which is characterized by

IntechOpen

abundant small-volume volcanism, shield volcanoes and scarce composite volcanoes [1–5]. Within the MGVF has reported three locations with intraplate-like and shoshonitic mafic rocks: (i) Nueva Italia [6]; (ii) Zitácuaro [7, 8] and (iii) Jorullo-Tacámbaro [9, 10].

Since Hasenaka and Carmichael [1] outline the MGVF in 1985, several authors have worked in this monogenetic field as a laboratory from different approaches such as spatial vents distribution [2, 4, 11, 12], volcanic morphology/landform [1, 13, 14], and geochemical and petrological approach [9, 10, 15–20]. Nevertheless, the relationship between volcano morphology, tectonic structures and petrogenesis studies in the MGVF are uncommon [21, 22].

The different calc-alkaline, tholeiitic and alkaline magmatic series have been explained according to petrotectonic magmatic arc settings considering the distance of the trench to inland, as well as the geometry and kinematics of the underlying plate [23]. The high potassium or shoshonitic volcanic series have been associated with the evolution of mature volcanic arcs and generally have been conceptualized to occur in continental areas far from the trench and/or back-arc, where the Benioff zone tends to be deeper and is commonly related to processes of continental extension [23–25]. However, it has recently been observed that these rocks can also occur in other areas along the volcanic arc, where compressional stress regimes can coexist temporal and spatially with extensional-transtensional tectonics, which results in thermal equilibrium in the lithospheric mantle associated with the complex evolution of the subduction processes [24, 26–29].

A lot of coeval calc-alkaline and shoshonitic magmatic series in subduction zones have been studied in several arcs worldwide, including the Eolian arc [30, 31], Central Andes arc in the Puna region [32, 33], Mariana arc [34], Fiji-Tonga arc [35], Sunda arc [36, 37] and Trans-Mexican Volcanic Belt [7, 8, 36–42]. Although the petrological and geochemical variations of the shoshonitic series have been relatively well characterized, the structural and the distinctive morphological relationships have been scarcely studied.

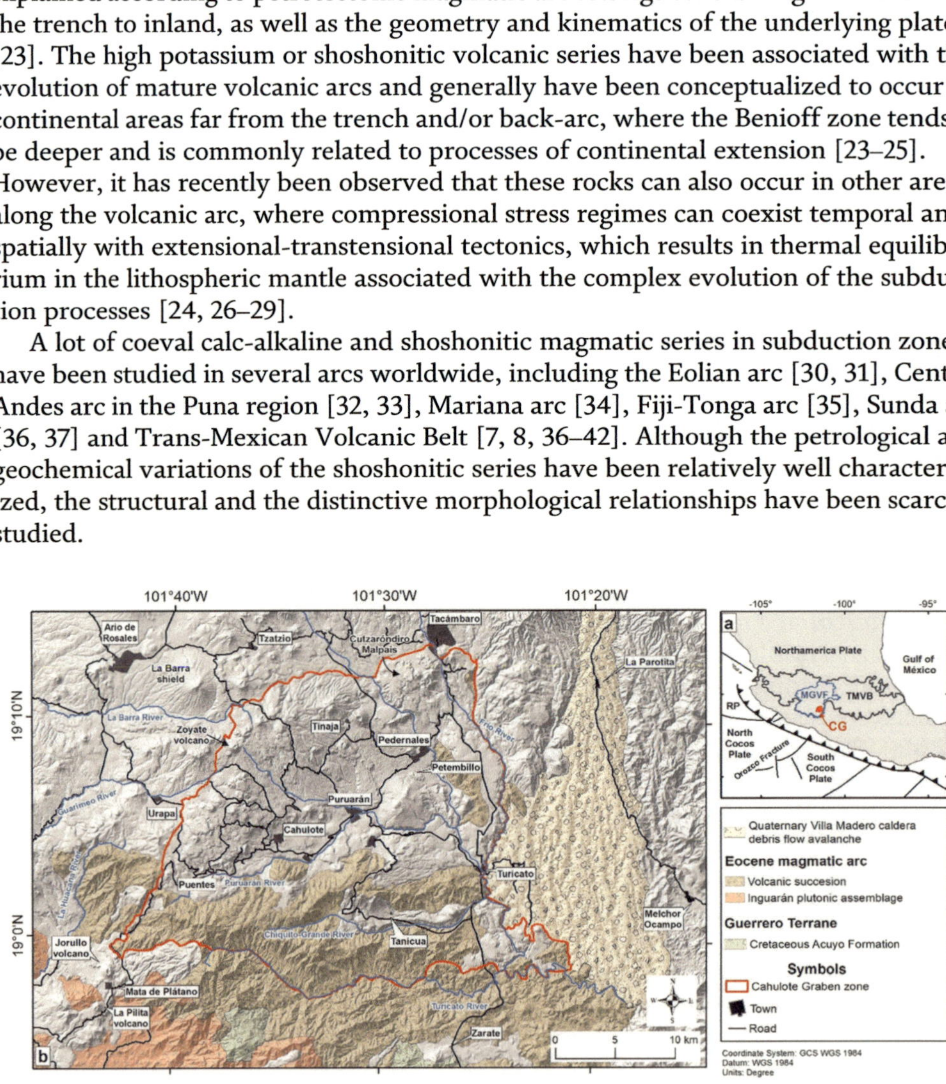

Figure 1.
Location of the Cahulote Graben zone within the Michoacán Guanajuato volcanic field (MGVF) and Transmexican Volcanic Belt (TMVB) (inset a) also shows the local basement of the Cretaceous Acuyo Formation and the magmatic products of the Sierra Madre del Sur province (b).

In this work, we present a detailed study of Cahulote Graben (CG) as a part of the Jorullo-Tacámbaro region (**Figure 1**). We have applied a holistic approach, which includes morphostructural, volcano morphometry, petrography, geochemistry and geochronology data. The main objective is to analyze regional tectonic relationships with volcanotectonic structures and correlate the spatial and temporal distribution of the different magmatic suites of the CG. A volcanotectonic model is proposed to explain the emplacement mechanism of shoshonitic magmatism.

2. Tectonic framework

The geodynamic framework of the TMVB and MGVF is located in the southern portion of the North American (Mexican) continental plate, where the oceanic plates of Cocos and Rivera are subducting under the continental plate (**Figure 2**) [43, 44]. The kinematics of the Cocos plate is very complex, as it presents variations in timing, speed and subduction angle [44]. This plate is bounded by oceanic fractures that were formed by the prolongation of the transform faults of the Eastern Pacific ridge, which allow the Cocos plate to be divided into four segments [43–45]. The MGVF is built on the North Cocos segment between the Rivera-Orozco fractures [45].

Based on the analysis of satellite images and digital elevation models, as well as the layout of structures from previous works [5, 46–48], the MGVF is located at the intersection of two regional fault systems (**Figure 2**): (i) in the southern sector, the

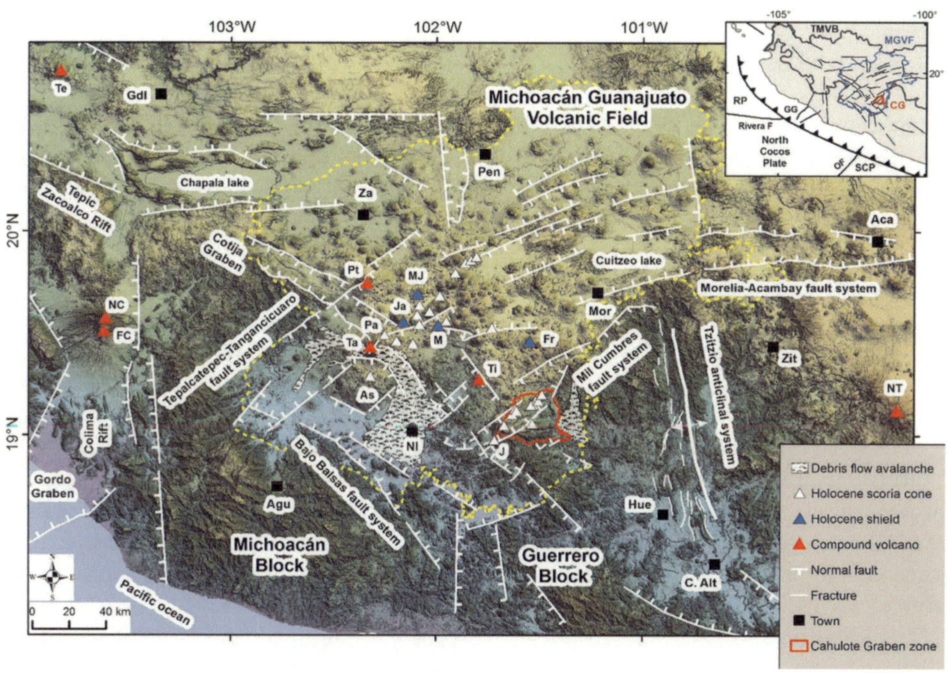

Figure 2.
Tectonic setting of the southwestern México and MGVF. Symbols and abbreviations: Red triangles: Relevant volcanoes from west to east; Te: Tequila; NC: Nevado de Colima; FC: Fuego de Colima; Ta: Tancítaro; Pa: Parícutin; Pt: Patamban; M: Metate; Ja: Jánamo; MJ: Mari Juata; Fr: Frijol; Ti: Tipitarillo; J: Jorullo; NT: Nevado de Toluca. Dark squares: City; Gdl: Guadalajara; Co: Coalcoman; Za: Zamora; Pen: Pénjamo; Mor: Morelia; Qro: Querétaro; Zit: Zitácuaro; Aca: Acambay. MGVF: Michoacán Guanajuato Volcanic Field.

NW-SE Bajo Balsas Fault System (BBFS) and (ii) throughout the MGVF, the NE–SW Mil Cumbres Fault System (MCFS).

The Bajo Balsas Fault System (BBFS) is formed by normal faulting and volcanotectonic alignments N120–150 direction spreading along ~180 km and covering ~7000 km^2 (**Figure 2**). The BBFS is an asymmetrical graben structure whose extension varies between 20 km to the NW close to Tepalcatepec and ~14 km to the SE close to Infiernillo dam. Furthermore, the BBFS exhibits a complex extensional geometry with a series of antithetical and synthetic structures. The relatively minor structure runs near the La Huacana-Inguarán area, defining the boundary of the Jorullo-Tacámbaro volcanic region [14].

The NE-SW Mil Cumbres Fault System (MCFS) is constituted by regional structures exposed along deep valleys that are intercepting the BBFS in the La Huacana-Inguarán region. These structures extend up to ~58 km from La Huacana to the Villa Madero and Morelia region (**Figure 2**) [49], where these structures are cut by the E-W Morelia-Acambay Fault System [50, 51]. The geometry and structure of the MCFS are closely resembling with the Tepalcatepec-Tangancícuaro Fault System located at the southwestern of the MGVF (**Figure 2**) [5, 22, 52]. The MCFS is related to the Tancítaro collapse events (~ 600 ka and ~ 240 ka) and debris flow avalanches collapses sliding southward [3]. Additionally, there is a strong correlation between the MCFS and multiple volcanic alignments in the MGVF. The geometric and structural arrangement of the CG developed closely associated with the evolution of the MCFS system and we argue that it is also linked to the highest density of volcanic vents within the MGVF and Jorullo-Tacámbaro region [1, 2, 22, 28].

3. Geological setting

The Cahulote Graben (CG) is located at the front of the Transmexican Volcanic Belt (TMVB), and it is a part of the Jorullo-Tacámbaro volcanic region, in the southeastern portion of the MGVF (**Figures 1** and **2**). The MGVF is considered one of the largest monogenetic volcanic fields in the world [53], with roughly 2000 volcanic vents [1, 2, 54].

Lithological basement of MGVF and CG is compounded by Jurassic-Cretaceous volcanic-sedimentary sequences [55, 56], which are cut or covered by Eocene plutonic or volcanic rocks, respectively [4]. Eocene volcanic rocks crop out an area of $\sim$40% km^2 to the southern of the CG, underlying the isolated younger lavas flows of the MGVF front. The eastern portion of the CG zone is bounded by debris avalanche deposits derived from the collapse of the Pliocene Villa Madero caldera [57].

On the Jorullo-Tacámbaro region, 114 Quaternary (<2 Ma) volcanoes were mapped, where basaltic and basaltic andesites are predominant, but alkaline rocks had already been previously reported [9, 10, 14] and shoshonitic rocks were firstly described [10, 14]. Notably, the area exhibits anomalously high rates of monogenetic activity, with 16 cones/100 km^2 [14].

Calc-alkaline and alkaline exhibit an arc-type geochemical that is similar isotopically, suggesting that they are scarcely influenced by crustal contamination [10]. According to Guilbaud et al. [10], alkaline and transitional magmas seem to be derived from a lower degree of partial melting. However, what stands out about the Jorullo-Tacámbaro region is the high-rate volcanic activity on the front of the TMVB, the coexistence between calc-alkaline and alkaline volcanism and undoubtedly the record of the historical Jorullo volcano (1759–1774) [4, 9, 15–18].

Structural control and higher rates of lithospheric extension have been invoked to explain the high density of monogenetic volcanism, such as the coexistence of calc-alkaline, transitional and alkaline magmatic series [10, 14]. However, related morphostructural and tectonic structures have not been mapped or considered in the discussion. In the CG, shoshonitic rocks are relatively abundant [9, 10, 16, 17], and highlighted the remarkable spatial matching with NE-SW extensional structures, suggesting close temporal correlation, which is the main goal of this study.

4. Methods

The spatial analysis and morphostructural mapping of the CG zone, along with the identification and typological classification of its eruptive centers, were conducted using a combination of digital elevation models. These included LiDAR (Light Detection and Ranging) with a resolution of 5 m, ALOS PALSAR at 12.5 m, and SPOT 6 satellite images with a resolution of 1.5 m for panchromatic and 6 m for multispectral data. Additionally, INEGI's digital topographic and hydrographic cartography, on a scale of 1:50,000, was utilized in this analysis.

In this work, a morphostructural domain is considered as a relief or morphogenetic unit that can be delimited and defined by having relatively distinctive geomorphological features, within a continuous space, and is related to contemporary geological processes (isotopic or relative time).

The typological classification of the different eruptive centers is based on the classification of volcanic morphology used by [58], supplemented by other authors [59–61]. It is worth noting that volcanic geoforms previously recognized as fissural effusion or coneless lava [13, 62] have been classified as spatter in terms of Sumner et al. [60] and Head and Wilson [63].

The mapping of volcanic buildings and tectonic structures was carried out in order to find the temporal relationships and later stratigraphy was correlated with a wide geochronological database [10, 14, 43] (see Supplementary **Table 1**). At the same time, based on the vast geochemical rock database (Supplementary **Table 2**), new representative samples of all volcanic events and previously identified volcanic typology were collected. Major elements of whole-rock compositions were acquired by XRF analyses with a Rigaku ZSX Primus II spectrometer at the Laboratorio Nacional de Geoquímica y Mineralogía, Instituto de Geología (LANGEM), UNAM, according to Ref. [64]. Whole-rock trace elements concentrations were determined by ICP-MS at the Laboratorio de Estudios Isotópicos (LEI), Centro de Geociencias, UNAM, following the procedures described by Ref. [65]

5. Results

5.1 Cahulote Graben and morphostructural domains

The Cahulote Graben is a volcanic area with small-volume volcanoes, mainly spatters and scarce scoria cones, shields and domes frequently running in volcanic alignments associated with tectonic structures (**Figure 3**). The highest altitude is located in the western sector (c. El Zoyate 2441 masl), while the lowest altitude is

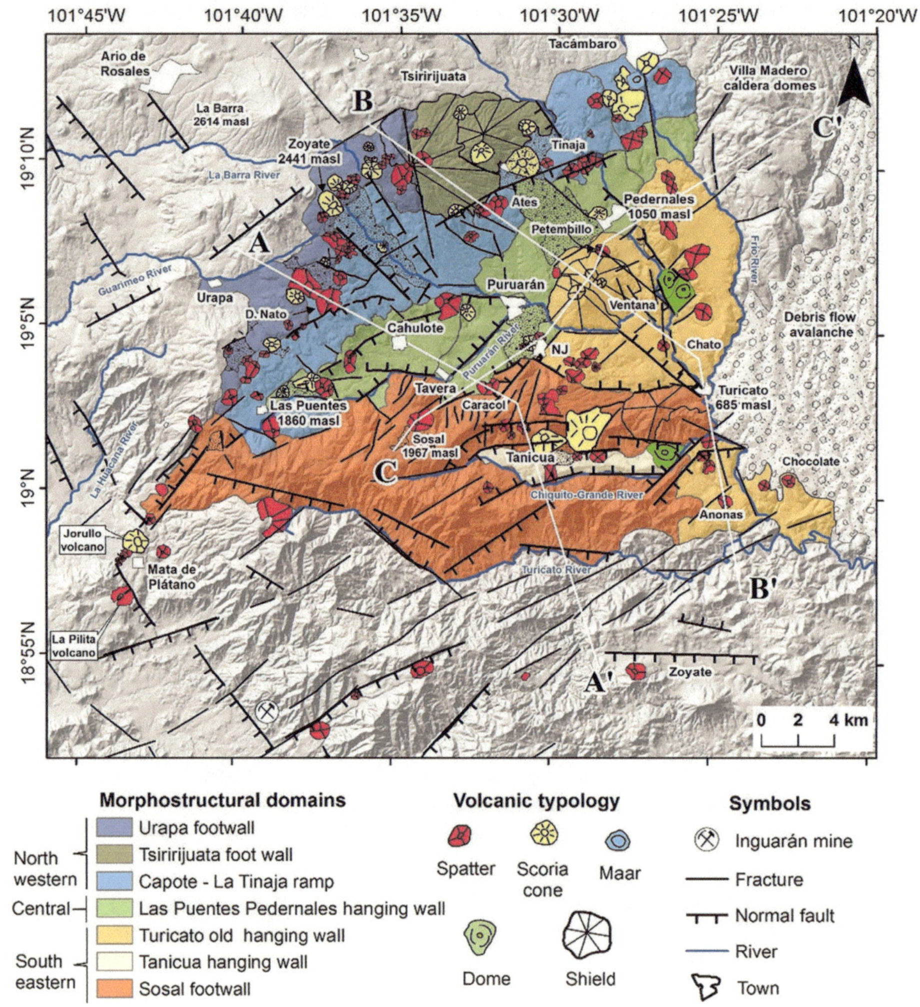

Figure 3.
Morphostructural domains and volcanic typology of the Cahulote Graben zone. Note location of the A-A´, B-B´, C-C' profiles shown in ***Figure 4****. Abbreviations. NJ: Nueva Jerusalen.*

localized at the Turicato town with 685 masl (**Figure 4**). Despite Turicato denoting the lowest point of the whole graben structure, it is worth noting that the CG exhibits a consistent topographical slope from north to southward, reaching a minimum elevation of 935 masl at Zárate town on the southern edge of the zone.

We divide the CG zone into three morphostructural domains: (i) north western; (ii) central and (iii) south eastern (**Figure 3**). All domains have relatively the same orientation NE-SW trend, since they correspond to the distribution of the structural geometry of the regional tectonic graben. Morphostructural domains have been divided into seven subdomains related to the geometry of the extensional fault blocks (e.g., footwall, hanging wall and ramp). Main parameters: height, length, area, drainage pattern, # vents, volcanic typology, tectonic and volcanic alignment structures are synthetized in **Table 1**.

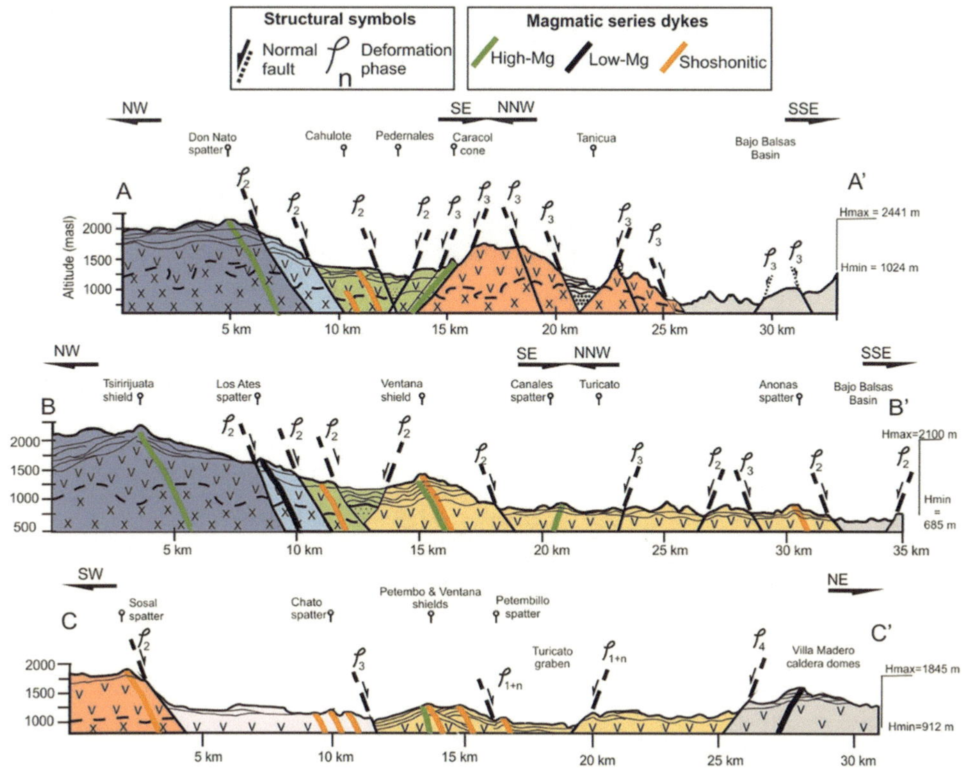

Figure 4.
Morphostructural cross profiles of the Cahulote Graben domains, subdomains and deformation systems. Map location is shown in ***Figure 3****.*

5.1.1 North western domain (NW-D)

This domain constitutes a regional tectonic block footwall type. Its main feature regards to a set of ramps with slopes >35° to the southeast, which are dislocating ancient stepped volcanic buildings. Its drainage pattern varies from radial to sub-dendritic and supports the emplacement of 60 volcanic buildings of diverse morphology.

5.1.1.1 Urapa/Tsiririjuata footwall (UTf-Sd)

An echelon NE-SW alignment of volcanic blocks is the main feature of this subdomain. This domain covers an area of 93 km^2 with heights range of 2441–1880 masl. The drainage is subdendritic to radial in young volcanic products and parallel in Tsiririjuata footwall. The tectonic blocks dip to the SW and are dislocated by NNW-SSE structures (**Figure 3**). In the UTf-Sd, there are 42 volcanic vents: at Tsiririjuata footwall outcrops older volcanism (1–0.3 Ma), whereas the Urapa footwall toward west outcrops younger volcanism (< 100 ky).

5.1.1.2 Puentes-Tinaja ramp (PTr-Sd)

This subdomain is composed mainly of lava flows coming from volcanic vents of the Urapa footwall in the SW sector, while the NE sector has a renewed landscape due to the emplacement of Holocene volcanoes like La Palma (4830–5170 B.P.), La Tinaja

Morphostructural domains		Trend	Altitude (masl)	L x W (km)	Area (km^2)	Dip	Drainage pattern	Volcanic type						VT structures
								SC	Sp	Sh	Do	Ma	Sum	
North western	Urapa footwall	N34	2441–1880	18 x 4	56	SW	Subdendritic	11	24	0	0	0	35	NE
	Tsiririjuata footwall	N50	2260–1864	8 x 6	37	SW	Radial/parallel	2	3	2	0	0	7	NW/NNW
	Puentes-Tinaja ramp	N44	2100–1415	32 x 1.5 to 7	121	SW	Dendritic/Parallel	7	10	0	0	1	18	NE/NW
Central	Cahulote-Pedernales hanging wall	N45	1750–1007	25 x 7	62	NE	Parallel/radial/ subdendritic	6	12	0	0	0	18	NE/E/NW
South eastern	Turicato old hanging wall	N322	1490–685	17 x 12	134	SE	Parallel/Radial	0	19	2	2	0	23	NW/E
	Tanicua hanging wall	N88	1242–826	11x2	14	E	Parallel/subdendritic	0	3	0	1	0	4	E
	Sosal footwall	N62	1960–1240	28 x 11	205	NE	Parallel/subdendritic	2	8	1	0	0	11	E/NNE/NE
							TOTAL	28	79	5	3	1	116	

Abbreviations: L x W = Length x Width; SC: Scoria Cone, Sp: Spatter, Sh: Shield, Do: Dome, Ma: Maar.

Table 1.
Morphostructural domains features from the Cahulote Graben zone.

(5115–5325 B.P.) and Malpais de Cutzaróndiro (2270–2360 B.P.) [14, 66]. This domain covers an area of 121 km^2 with heights ranging from 2100 to 1415 masl and has 18 vents. Dendritic drainage was developed on older lavas while subdendritic to radial drainage did so on overlying younger lavas.

5.1.2 Central domain (CD)

This domain constitutes the main Cahulote-Pedernales hanging-wall NE-SW block of the CG zone and denotes the depocenter of the tectonic basin with slopes <10°. The CD covers an area of 62 km^2 with a height range of 1750–1007 masl and is composed mainly of lavas coming from the northwestern domain. Parallel drainage predominates in this domain along central structures, but radial-subdendritic drainage develops on volcanoes and lavas. Eighteen volcanoes (12 spatter and 6 scoria cones) were emplaced along NE-SW and E-W tectonic structures.

5.1.3 Southeastern domain (SE-D)

This domain contains the eastern footwall of the CG zone, also holds an old NW-SE hanging wall and an E-W hanging wall. Its drainage pattern varies from dendritic, parallel and radial, containing at least 38 volcanic vents.

5.1.3.1 Turicato old hanging wall (TOhw-Sd)

Turicato old hanging wall extended from NE to SE sectors of the CG zone. Its morphology feature is an elongated NW-SE old hanging-wall associated with the Bajo Balsas Fault System. The TOhw-Sd covers an area of 62 km^2 with heights ranging from 1490 in la Ventana shield summit to 685 masl in the Turicato town (B-B′ in **Figure 4**), and is composed of 23 volcanic vents. This subdomain is characterized by the presence of shield volcanoes and domes NW-SE aligned while spatters are aligned in NE and NNW trends. The drainage is parallel to the regional structures, but centrifugal radial drainage develops in the shield volcanoes.

5.1.3.2 Tanicua hanging wall (Tahw-Sd)

It is located on the SE edge of the study zone and represents a symmetric hanging wall between SE-D and Eocene volcanic footwall. It has an orientation E-W, covers an area of 14 km^2 with a height range of 1242–826 masl and dips to the E, suggested by the Chiquito-Grande river, which joins the Turicato river to the east. The subdomain is relatively small and only contains three spatters and one dome. This domain shows parallel drainage along the Chiquito-Grande river and subdendritic at the intersection of the E-W and NE-SW faults at the western region of the domain.

5.1.3.3 Sosal footwall (Sofw-Sd)

This is the southern limit of the CG, covering an area of 205 km^2. It is composed of a trachytic-rhyolite Eocene volcanic sequence, which forms the basement blocks overlapped by 11 volcanic vents (A-A’ section in **Figure 4**; **Table 1**). Gentle slopes (0–10°) develop over lavas and pyroclasts. In the piedmont, the slopes are slightly steeper (15–25°) and very steep up to 60° on scarps. It shows an altitudinal range of 1240–

1960 masl and exhibits dendritic and parallel drainage which is conditioned by E-W faults and fractures.

5.2 Volcanic typology and morphometry

A total of 116 volcanic vents were identified on the Cahulote Graben: 79 spatter (68%), 28 scoria cones (24%), 5 small shield volcanoes (4.3%), 3 domes (2.5%) and 1 maar (0.8%) (**Figure 3**). A statistical synthesis of the typological variation is presented in **Table 2**, while the detailed morphometric parameters are available in Supplementary **Table 1**.

The spatter cones are mainly overlying old volcanic structures, and rarely, they are isolated volcanic buildings (e.g., C. Verde; **Figure 5c**). The average geometric parameters of the spatter cones are height of 124 m, basal diameter of 666 m and volume of 0.14 km^3. Throughout CG, lava flows associated with spatters reach lengths of up to 6 km (e.g., Verde, Don Nato and Rancho Viejo). However, the estimated volume could be larger, as it is common to observe volcanic structures constituted by agglutinated lava-pyroclasts or relatively covered by pyroclastic deposits (e.g., Los Pozos and La Palma).

The scoria cones are usually isolated (e.g., Caracol; **Figure 5a**). These commonly show a semi-circular crater that is well-defined, and only the Caracol volcano shows a horseshoe shape open to the NW (**Figure 5a**). The cinder cones have a mean height of 212 m, a basal diameter of 918 m and a volume of 0.11 km^3.

The shield volcanoes evidently represent ancient volcanic structures. They are partially buried by recent small-volume volcanism and they are considered as continental small shields [61]. They have a mean height of 608 m, basal diameter of 3437 m, and a volume of 0.43 km^3. Although, within the CG zone only can be observed five small shields (**Figure 5c**), to the northwest and out of the CG zone, many other aligned Pleistocene shield volcanoes are observed (e.g., La Barra; **Figure 3**), which show very resembling morphometric features correlating with the shield volcanoes of the CG area.

Only three domes were observed in the CG zone. They are commonly isolated structures outcropped in the southeastern region between Turicato and Pedernales towns (**Figure 5b**). Domes show a mean height of 324 m, a basal diameter of 1003 m, and a volume of 0.09 km^3. In the CG zone only one maar volcano is recognized in the NE edge with a crater diameter of 680 m and crater height of 48 m.

Volcano type	This work	H&C,1985	G, 2012	DCo (m)	HCo (m)	Volume (km^3)
Spatter	79	11	24	250–1877	44–389	0.002–0.83
Cinder cone	28	37	47	454–1732	95–438	0.002–0.38
Dome	3	2	5	816–1189	209–440	0.03–0.13
Shield	5	1	2	3399–3476	588–628	0.28–0.61
Maar	1	0	1	928	113	—
Total volcanic vent	116	51	79			
Total Volume (km^3)	16.5	1.9	12.35			

Table 2.
Morphometric parameters of the Cahulote Graben volcanoes and the comparison with previous works of Hasenaka & Carmichael [13] (H&C, 1985) and Guilbaud et al. [14] (G, 2012).

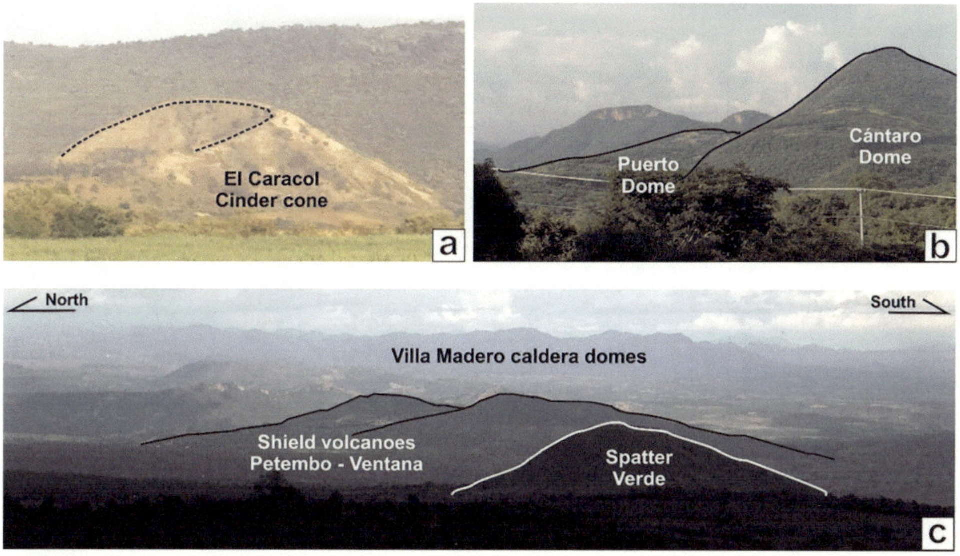

Figure 5.
Morphology Landscape pictures of the Cahulote Graben. a. El Caracol horseshoe-shaped cinder cone; b. Puerto and Cántaro domes; c. Petembo and Ventana shield volcanoes and Verde spatter. In the background is observed the Villa Madero domes alignment.

5.3 Magmatic series in the Cahulote Graben

Geochemistry and petrographic published data of CG, Jorullo and la Pilita area were compiled (see Supplementary **Table 2**). Geochemistry data base was rearranged with 30 new selected geochemistry samples obtained in this study (**Table 3**). Detailed petrographic description is presented in Supplementary Table 3.

Four magmatic series were recognized and mapped: (i) calc-alkaline high-Mg, (ii) calc-alkaline low-Mg, (iii) intraplate-like and iv) shoshonitic series. The shoshonitic series nomenclature is assumed in terms of Mueller et al. [29] and Peccerillo et al. [67], where they are potassic volcanic-arc rocks with high Na_20 + K_2O (> 5 wt%), normative hypersthene + olivine, low iron enrichment, high LILE (e.g., K, Rb, Sr and Ba), high but variable Al_2O_3 (9–20 wt%) and generally low TiO_2 < 1.5 wt%. Intraplate-like is stated like Gomez-Tuena et al. [68].

Spatial distribution of magmatic series in the CG are shown in the geological map as well as the location of the geochemical and geochronological analysis samples (**Figure 6**). The calc-alkaline high-Mg (42.7 vol.%), low-Mg (30.6 vol.%) and shoshonitic magmas(25.7 vol.%) are the most abundant magmatic series in the CG zone, while the intraplate-like (1 vol.%) are relatively scarce. Shoshonites are relatively more abundant in the Central domain than in the southeastern domain appearance. A single intraplate-like vent is present on the Puentes-Tinaja ramp of the North western domain (**Figures 6** and 7).

All magmatic series data were plotted on binary diagrams SiO_2 versus K_2O, MgO, TiO_2, P_2O_5 major elements (**Figure 8a–d**) and versus Cr and V trace elements (**Figure 8e,f**). In order to compare the CG data, the data of the volcanoes of Parícutin and Jorullo (calc-alkaline medium-K) and La Pilita volcano (shoshonitic) have been projected [4, 19]. Trace element patterns normalized to N-MORB [69] is presented in **Figure 9**.

Sample	G7	G14	G25	G30	G36	G4	G5	G6	G8	G10	G11	G13	G15A	G17	G18	G22	G35	G44	G45	G16	G20	G26	G27	G31	G33	G37	G38	G39	G41	G43	T11 [10]
Serie			Shoshonitic										High Mg												Low Mg						Intraplate
SiO_2	51.94	51.20	56.93	51.62	53.79	52.34	54.85	53.40	55.42	56.94	55.12	51.41	51.74	53.80	54.92	57.14	53.63	54.67	53.84	61.20	58.57	55.55	56.82	56.60	56.77	55.84	61.54	56.78	55.77	56.82	52.39
TiO_2	0.94	1.07	1.08	1.34	0.94	0.84	0.95	0.81	0.76	0.69	0.72	0.95	1.39	0.80	0.80	0.71	0.75	0.77	0.80	0.62	0.76	1.17	1.10	1.07	1.11	1.32	0.81	1.06	1.04	1.05	1.86
Al_2O_3	15.25	16.68	15.87	14.78	17.52	17.35	17.84	16.22	17.02	17.19	17.53	17.02	16.72	18.01	17.76	17.34	15.84	16.51	16.31	17.07	17.41	17.00	17.32	17.19	17.19	18.05	16.25	17.11	17.40	17.46	16.48
Fe_2O_3t	7.75	8.47	5.89	8.46	7.35	8.21	7.41	7.95	7.24	6.30	6.91	8.68	9.01	7.48	7.22	6.47	7.42	7.32	7.35	5.23	5.85	7.81	6.98	7.26	7.22	7.27	5.02	6.88	6.69	7.02	11.55
MnO	0.10	0.12	0.08	0.10	0.09	0.12	0.10	0.11	0.10	0.09	0.10	0.13	0.12	0.10	0.10	0.09	0.11	0.10	0.10	0.07	0.08	0.12	0.10	0.10	0.09	0.10	0.07	0.10	0.10	0.10	0.18
MgO	9.13	7.90	4.89	6.83	6.49	8.34	6.45	8.88	7.09	6.91	7.55	8.13	7.41	7.31	6.76	6.43	9.36	7.74	8.96	4.23	4.37	5.01	4.50	4.82	4.02	4.37	3.41	4.72	4.70	4.54	2.98
CaO	8.10	8.73	6.38	8.03	7.24	8.64	7.30	8.24	7.25	6.60	7.43	9.74	8.10	7.70	7.50	6.66	8.60	8.32	8.28	5.67	6.23	7.77	7.28	7.31	7.02	7.07	6.24	7.13	6.90	7.36	8.31
Na_2O	3.78	3.77	3.28	4.69	4.32	3.36	3.89	3.36	3.71	3.85	3.76	3.17	3.35	3.69	3.61	3.89	2.75	3.05	3.09	4.09	4.18	3.45	3.71	3.62	3.62	3.79	3.84	3.19	5.05	3.67	3.08
K_2O	2.09	1.68	4.32	3.07	1.40	0.83	1.00	0.85	0.82	1.15	0.85	0.67	1.32	0.91	0.93	1.22	1.27	1.15	1.14	1.63	1.69	1.29	1.58	1.46	1.48	1.57	1.78	1.82	1.58	1.56	1.74
P_2O_5	0.66	0.48	0.52	0.95	0.53	0.13	0.24	0.14	0.16	0.17	0.16	0.15	0.42	0.14	0.16	0.19	0.17	0.17	0.17	0.18	0.24	0.27	0.29	0.29	0.29	0.32	0.26	0.29	0.33	0.29	0.59
PXC	0.26	-0.09	0.77	0.14	0.31	-0.14	-0.02	0.06	0.44	0.10	-0.12	-0.04	0.42	0.05	0.23	-0.14	0.12	0.21	-0.03	0.03	0.62	0.57	0.32	0.29	1.19	0.32	0.78	0.93	0.46	0.15	1.01
Total	99.74	100.09	99.23	99.86	99.69	100.14	100.02	99.94	99.56	99.90	100.12	100.04	99.58	99.95	99.77	100.14	99.88	99.79	100.03	99.97	99.38	99.43	99.68	99.71	98.81	99.68	99.22	99.07	99.54	99.85	99.16
Sc	20.71	24.02	15.51	17.97	16.54	18.16	17.03	23.97	19.04	15.07	17.30	32.18	21.84	19.32	18.80	14.54	25.23	22.78	23.98	12.42	14.04	21.54	18.29	17.37	18.47	18.60	16.11	17.81	18.12	18.73	33.50
V	156.83	184.02	164.11	178.91	143.65	130.96	141.92	175.90	118.11	119.73	138.30	211.68	176.59	153.19	145.12	119.70	175.24	151.77	169.26	100.52	114.32	150.80	144.50	131.21	136.87	145.57	115.78	136.56	136.56	143.46	259.00
Cr	375.28	237.12	160.09	242.13	194.86	285.27	271.30	450.76	287.65	216.77	327.34	407.51	315.31	265.93	262.23	195.94	374.37	236.70	344.82	154.96	151.26	147.75	111.77	119.34	114.25	99.24	132.69	124.36	118.22	101.46	35.50
Co	36.29	33.75	21.71	32.35	28.03	24.73	32.94	37.43	29.62	26.59	28.90	36.35	35.31	30.79	28.12	24.83	35.60	32.78	34.50	17.51	20.63	32.77	24.39	24.82	26.35	24.02	17.83	24.00	26.74	23.72	28.40
Ni	264.07	109.32	106.10	125.22	136.90	213.79	156.86	210.98	187.10	222.30	208.04	116.99	182.85	160.36	160.71	162.94	182.89	196.82	170.88	99.19	127.96	88.85	62.20	88.72	91.73	70.38	92.82	77.86	79.48	65.90	17.00
Cu	56.34	67.19	46.14	75.27	43.81	34.75	26.17	56.09	29.98	30.77	47.61	45.33	46.72	44.26	32.72	29.52	45.90	33.25	31.75	22.67	27.64	47.42	36.27	39.71	43.80	38.95	27.53	39.66	43.04	41.93	115.00
Zn	102.29	90.52	82.69	131.26	104.81	46.75	73.79	64.95	73.27	66.22	67.08	66.08	77.77	70.09	68.25	66.11	63.49	59.34	63.65	66.41	73.13	69.92	71.22	67.55	65.92	74.80	60.62	68.83	71.36	71.69	111.00
Ga	19.73	18.93	21.72	23.30	20.58	12.17	18.68	17.14	19.28	18.00	18.15	16.83	17.93	18.21	17.96	17.71	17.11	17.49	17.34	19.44	19.72	19.06	19.29	18.07	19.27	19.67	20.42	18.85	19.47	19.13	17.00
Li	14.24	12.56	10.58	18.54	14.40	6.99	10.62	9.10	11.62	13.99	11.63	8.61	8.94	11.17	10.53	12.57	7.81	10.50	9.48	19.91	15.05	11.74	10.32	11.89	12.48	12.97	11.27	9.46	13.09	14.64	n.d.
Be	2.46	1.70	2.62	2.90	1.73	0.59	1.29	0.83	1.04	1.03	0.92	0.79	1.55	0.92	0.92	1.03	0.89	0.94	0.90	1.24	1.42	1.52	1.46	1.33	1.44	1.51	1.42	1.44	1.52	1.44	n.d.
Rb	15.85	17.41	103.27	32.29	13.57	7.61	11.05	9.47	9.52	17.59	10.92	8.21	17.34	12.27	9.90	18.01	18.53	19.46	18.08	23.24	31.76	24.56	27.31	27.74	28.16	27.48	23.45	31.11	28.61	27.04	33.00
Sr	1476.34	1093.56	1136.25	1934.81	931.57	279.76	572.10	497.85	543.33	506.39	525.47	492.26	542.38	509.49	494.76	549.74	719.18	835.52	733.82	679.83	709.42	787.04	770.37	695.95	724.87	755.15	1228.78	723.12	739.68	768.56	285.00
Y	15.44	20.40	16.59	16.13	13.36	12.74	18.41	15.76	14.51	12.37	12.26	19.97	22.72	15.65	15.66	12.72	15.19	14.89	15.28	11.40	13.78	32.95	21.42	21.04	24.73	22.08	16.61	21.35	26.47	21.57	65.00
Zr	164.71	137.91	391.53	186.98	118.46	66.69	113.33	88.22	110.42	109.33	91.48	75.60	172.60	91.74	89.22	99.79	101.86	96.71	98.40	125.18	136.75	156.73	164.59	154.53	164.45	176.90	140.63	163.74	169.21	162.09	231.00

Sample	G7	G14	G25	G30	G36	G4	G5	G6	G8	G10	G11	G13	G15A	G17	G18	G22	G35	G44	G45	G16	G20	G26	G27	G31	G33	G37	G38	G39	G41	G43	T11 [10]
Serie			**Shoshonitic**									**High Mg**													**Low Mg**						**Intraplate**
Nb	12.00	7.71	7.35	16.01	11.46	1.65	6.78	2.80	3.80	4.01	3.03	2.48	16.50	2.84	2.89	3.55	3.05	3.70	3.30	4.83	7.27	8.46	9.49	8.44	9.08	10.62	4.74	9.40	9.63	9.32	9.60
Cs	0.14	0.50	2.59	0.72	0.39	0.26	0.20	0.23	0.32	0.66	0.37	0.14	0.40	0.52	0.28	0.41	0.58	0.69	0.34	0.84	1.26	0.38	0.79	1.10	0.48	0.63	0.39	1.19	0.60	0.79	1.70
Ba	859.50	619.27	1172.28	1095.65	446.43	163.32	327.63	276.01	305.37	369.73	266.16	238.64	349.15	268.09	282.07	401.84	298.67	319.69	312.71	535.75	502.86	497.31	514.37	422.30	486.63	463.18	441.37	443.24	470.16	463.51	481.00
La	44.50	35.89	38.37	67.34	32.12	5.26	15.57	9.43	11.38	11.90	9.31	8.49	20.15	8.69	9.26	12.73	12.81	13.60	12.79	15.23	20.32	23.26	21.27	18.83	27.94	20.05	25.26	20.23	25.05	21.10	29.80
Ce	84.81	70.39	82.10	127.36	57.75	11.82	34.78	19.92	24.02	24.07	19.75	17.86	42.76	18.67	19.52	25.30	27.15	28.72	26.92	30.80	38.76	45.07	42.50	38.88	52.40	42.34	47.09	40.03	45.95	41.58	63.40
Pr	10.36	8.75	11.16	14.80	6.58	1.66	4.60	2.77	3.38	3.19	2.70	2.59	5.57	2.65	2.66	3.29	3.66	3.81	3.60	3.93	4.93	5.99	5.58	5.17	7.45	5.54	6.76	5.41	6.33	5.54	8.62
Nd	37.61	32.78	44.24	54.50	23.08	7.80	19.24	12.21	14.63	13.35	11.87	11.73	22.57	11.99	11.79	13.63	15.45	15.73	15.16	15.75	19.16	24.61	22.21	21.00	29.11	22.35	26.57	21.77	25.38	22.11	37.20
Sm	6.01	5.99	8.49	8.48	3.88	1.98	4.26	2.84	3.29	2.88	2.69	2.87	4.93	2.93	2.81	2.89	3.30	3.26	3.21	3.13	3.72	5.26	4.64	4.46	6.07	4.76	4.85	4.60	5.27	4.58	9.92
Eu	1.58	1.63	2.17	2.11	1.10	0.65	1.30	0.89	1.01	0.91	0.87	0.95	1.45	0.94	0.92	0.91	0.99	0.98	0.97	0.96	1.13	1.59	1.35	1.30	1.64	1.39	1.36	1.33	1.52	1.34	1.98
Tb	0.60	0.70	0.78	0.76	0.47	0.35	0.59	0.44	0.46	0.39	0.38	0.50	0.70	0.45	0.44	0.40	0.46	0.45	0.45	0.39	0.46	0.82	0.64	0.62	0.81	0.67	0.55	0.64	0.74	0.64	1.70
Gd	4.59	5.01	6.37	6.11	3.29	2.20	4.30	2.85	3.14	2.70	2.60	3.11	4.69	2.99	2.87	2.70	3.07	3.05	3.03	2.77	3.31	5.42	4.33	4.19	5.49	4.47	4.07	4.34	5.09	4.33	10.60
Dy	2.91	3.72	3.49	3.31	2.45	2.21	3.45	2.72	2.64	2.25	2.23	3.27	4.07	2.78	2.74	2.27	2.70	2.64	2.66	2.12	2.50	4.96	3.75	3.59	4.57	3.86	2.89	3.73	4.32	3.75	10.10
Ho	0.56	0.72	0.59	0.58	0.48	0.46	0.57	0.56	0.52	0.45	0.44	0.69	0.80	0.56	0.56	0.46	0.54	0.54	0.55	0.42	0.49	1.03	0.74	0.71	0.87	0.76	0.56	0.74	0.86	0.74	2.06
Er	1.56	1.99	1.54	1.60	1.31	1.26	1.31	1.55	1.39	1.20	1.18	1.96	2.19	1.50	1.54	1.23	1.49	1.45	1.51	1.11	1.31	2.83	2.03	1.95	2.35	2.10	1.53	2.02	2.38	2.03	6.19
Yb	1.36	1.72	1.21	1.27	1.19	1.24	1.67	1.49	1.29	1.12	1.11	1.94	2.03	1.42	1.49	1.17	1.43	1.33	1.45	1.01	1.17	2.56	1.90	1.80	2.10	1.95	1.35	1.88	2.15	1.91	5.58
Hf	4.20	3.45	12.30	4.82	2.98	1.62	2.66	2.18	2.76	2.64	2.27	1.85	3.73	2.30	2.23	2.44	2.56	2.38	2.47	3.08	3.32	3.62	3.77	3.51	3.76	3.98	3.62	3.73	3.85	3.70	6.50
Ta	0.57	0.40	0.40	0.78	0.54	0.11	0.41	0.17	0.24	0.26	0.19	0.16	0.97	0.18	0.18	0.23	0.21	0.26	0.21	0.31	0.45	0.55	0.58	0.53	0.63	0.65	0.29	0.59	0.60	0.57	0.70
W	0.37	0.30	0.41	0.41	0.29	0.85	0.81	0.38	0.19	0.41	0.31	0.35	0.43	0.36	0.36	0.36	0.30	0.38	0.35	0.50	0.28	0.37	1.19	0.87	0.31	0.30	0.23	0.41	0.43	0.41	n.d.
Tl	0.01	0.11	0.56	0.29	0.06	0.03	0.03	0.03	0.04	0.08	0.03	0.03	0.07	0.06	0.03	0.05	0.10	0.01	0.05	0.08	0.10	0.08	0.13	0.14	0.12	0.11	0.09	0.20	0.12	0.12	n.d.
Pb	12.15	8.03	10.49	15.02	7.74	2.63	4.55	3.82	3.59	5.42	4.13	4.07	4.72	4.24	4.64	6.02	4.19	3.30	4.22	7.99	7.58	5.27	6.91	6.00	6.35	6.76	6.55	6.65	6.54	6.76	10.00
Th	4.24	2.63	4.43	5.81	2.70	0.39	0.98	0.72	0.81	0.97	0.65	0.57	1.78	0.63	0.69	1.09	1.63	1.75	1.57	1.41	2.95	1.87	2.34	2.02	2.21	2.31	2.64	2.26	2.29	2.30	4.78
U	1.11	0.70	2.22	1.60	0.57	0.15	0.33	0.24	0.24	0.34	0.24	0.19	0.61	0.25	0.23	0.35	0.55	0.51	0.50	0.46	0.92	0.63	0.77	0.67	0.72	0.76	0.66	0.77	0.78	0.77	1.55

Table 3.
Major (wt%) and trace elements (ppm) of the Cahulote Graben lavas.

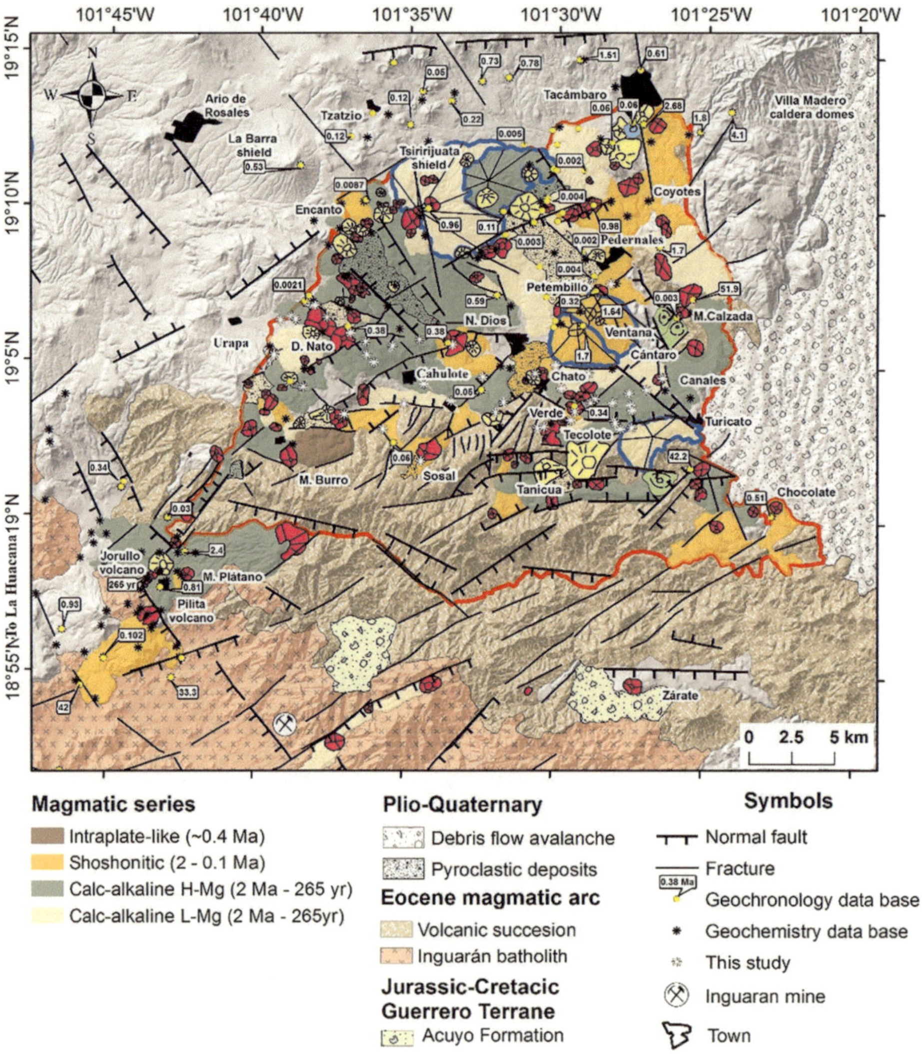

Figure 6.
Magmatic series map of the Cahulote Graben and volcanic vent morphology, ages and major tectonic structures related.

5.3.1 Calc-alkaline H-Mg and L-Mg series

H-Mg lavas prevail in the Urapa footwall and mainly erupt from spatter cones (**Figure 6**). These are characterized by their pāhoehoe shelly flow type, which flowed mainly toward the southeast and up to 6 km in length. The L-Mg lavas are ʻAʻā and blocky type; they are concentrated in the Urapa/Tsiririjuata footwall, and they were erupted by scoria cones, domes and a few spatter cones (**Figure 6**).

The calc-alkaline H-Mg rocks are porphyritic (10–20 vol.% phenocrysts; **Figure 7c**), while calc-alkaline L-Mg lavas are commonly aphanitic (<5 vol.% phenocrysts). Calc-alkaline H-Mg rocks display an assemblage of Plg + Gl + Ox + Ol + Cpx with scarce phenocrysts of Opx. In contrast, calc-alkaline L-Mg lavas show an assemblage of Opx + Cpx + Plg ± Amp.

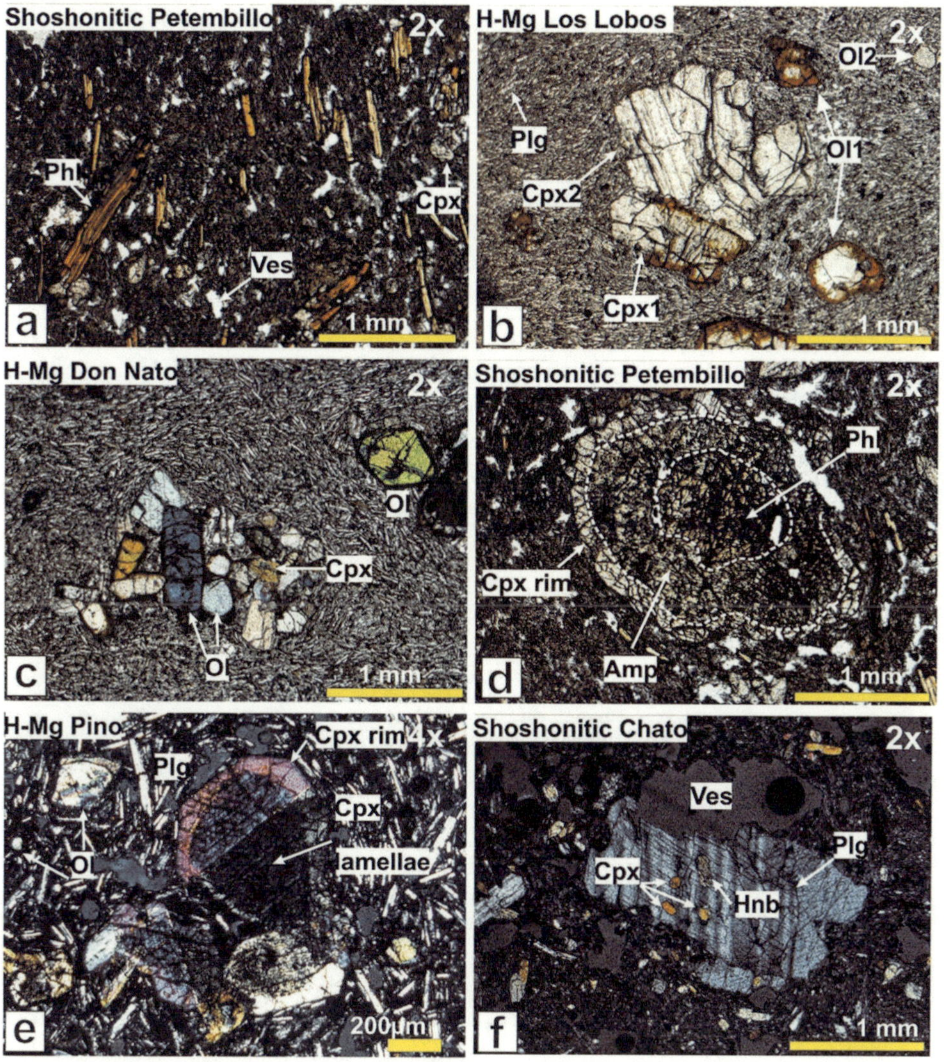

Figure 7.
Photomicrograph of thin sections of the CG. a. Traquitic-vitreous texture with phenocrysts of phlogopite (CG-25, Petembillo); b. Two families of olivines and pyroxenes (CG-04, Los Lobos); c. Glomeroporphyritic cluster of Ol + Cpx (CG-06, Don Nato); d. Phlogopite antecryst with reaction rims of amphibole and clinopyroxene (CG-25); e. Cpx with exsolution lamellae and Cpx reaction rim (CG-36, Pino); f. Poikilitic texture of plagioclase xenocryst with pyroxenes chadacrysts (CG-30, Chato).

Calc-alkaline rocks exhibit a variable SiO_2 composition but all rocks fit at medium potassium series (**Figure 8a**). The calc-alkaline H-Mg (>6 wt.%) and L-Mg (<6 wt.%) rocks display SiO_2 ranges of 48.5–58 wt.% and 56–63.5 wt.% respectively. All calc-alkaline series show a typical magmatic differentiation trend with a decrease of MgO, TiO_2 and V, and increasing of K_2O, Na_2O, La, Ba as the content of SiO_2 rises. TiO_2 content ranges from 0.6 to 1.4 wt.%. However, the L-Mg lavas from the Turicato hanging-wall and the H-Mg lavas from shield volcanoes are slightly enriched in TiO_2 and immobile elements.

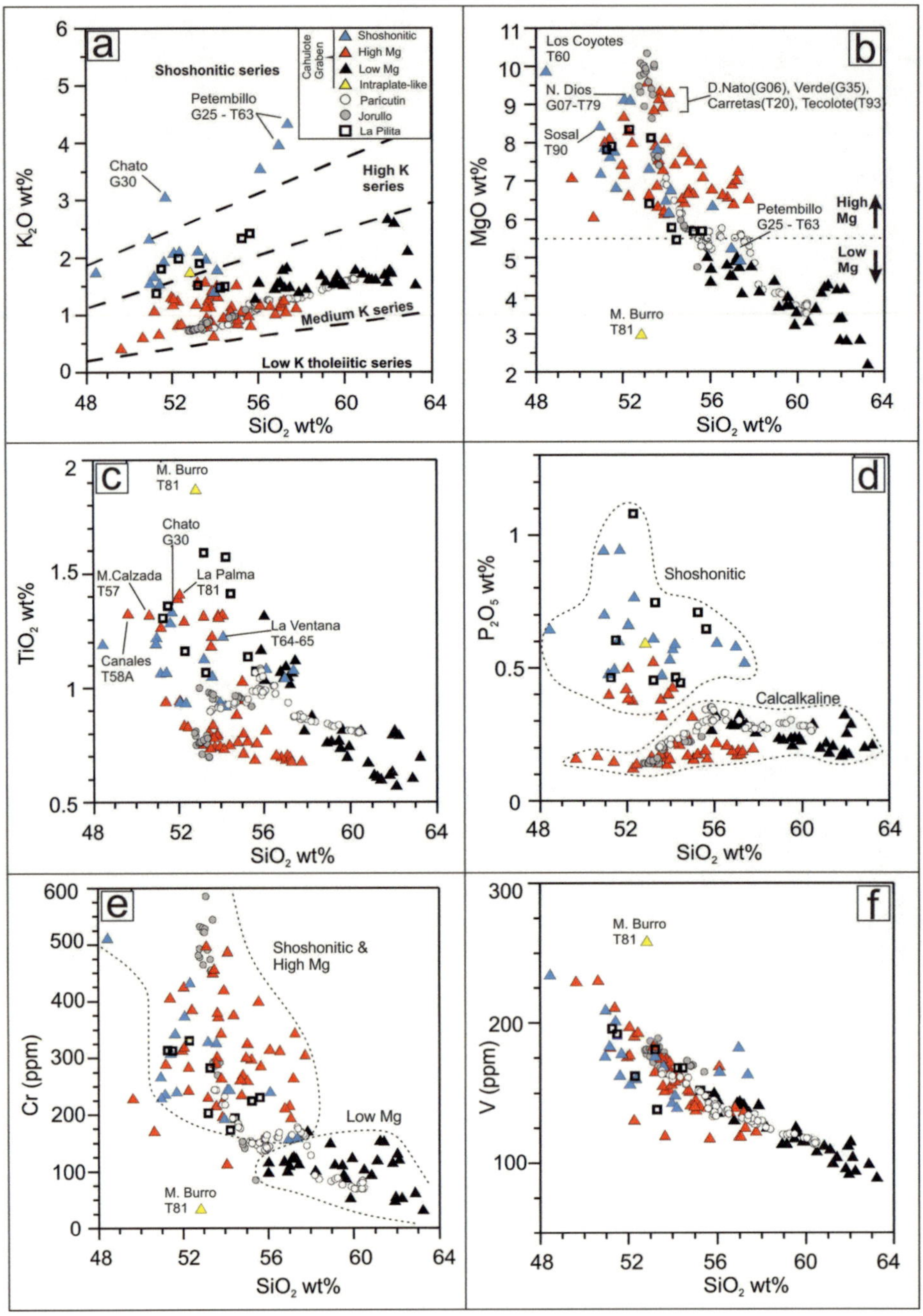

Figure 8.
Bivariate diagram for the shoshonitic, H-Mg, L-Mg calc-alkaline and intraplate-like magmatic series. All diagrams use the SiO2 as a differentiation index. The K2O diagram (a) shows the Peccerillo and Taylor (1976) fields.

5.3.2 Shoshonitic series

The shoshonitic lavas come from spatter cones and rarely from shield volcanoes, usually with blocky lavas as the main products. Shoshonites were emplaced along NE-

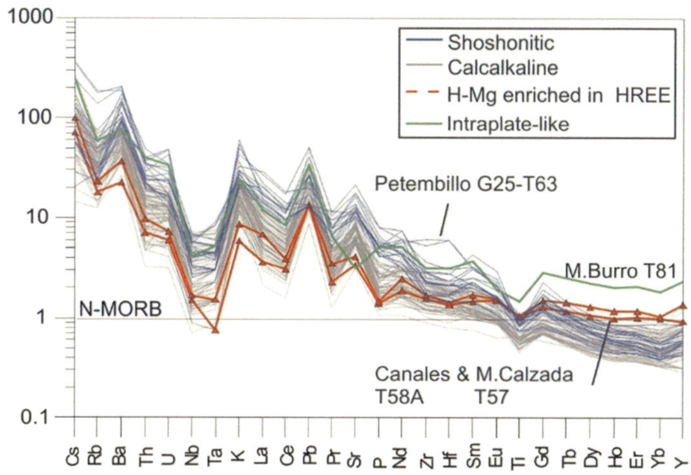

Figure 9.
N-MORB-normalized (Sun and McDonough 1989) multi-element diagrams for Cahulote Graben magmatic series.

SW volcanic alignments in the central hanging wall (**Figure 6**). These lavas are commonly aphyric with less than 10 vol.% phenocrysts. Mineral assemblage is Cpx + Ol + Plg ± Phl ± Ox ± Amp ± Lct. Only the Petembillo spatter is porphyritic with 20 vol.% phenocrysts and mineral association of Phl + Cpx + Ol + Plg + Amp (**Figure 7a, d**). The groundmass is composed of glass (>50 vol.%) + feldspar (35 vol.%) + oxides (<10 vol.%) ± Ol/Cpx (<5 vol.%) and apatite like mineral accessory. The feldspar and feldspathoids are present as microphenocrysts or microliths.

Shoshonitic rocks display SiO_2 ranging from 48.5 to 58 wt.% and high concentrations of K_2O varying from 1.1 to 4.8 wt.%, and ergo, they belong to the high-K series (**Figure 8a**). Compared with medium potassic lavas, the shoshonitic magmas show higher values of MgO (6.5–9 w.t%), TiO_2 (1–1.5 wt.%), P_2O_5 (0.5–1 wt.%), and lower Al_2O_3 values with 14–16 wt.%.

Also, shoshonitic lavas are characterized by high LILE as Sr, Rb, Ba, Pb, LREE and slightly enrichment on immobile elements (**Figure 9**). Especially, phlogopite bearing trachyandesite of Petembillo spatter show higher Zr values (>400 ppm) and Hf (>10 ppm), while each other shoshonitic samples show Zr values <200 ppm and Hf < 5 ppm (**Figure 9**). However, all shoshonitic and H-Mg calc-alkaline lavas contain high values of Ni (100–265 ppm) and Cr (200–510 ppm; **Table 3**).

5.3.3 Intraplate-like series

Only one intraplate-like volcano was noticed in the Puentes-Tinaja ramp, which corresponds to a spatter cone known as Mesa El Burro (**Figure 6**). Mainly, Plg + Ol-like phenocrysts and groundmass mineral assemblage were described by Guilbaud et al. [10].

Intraplate-like rock displays SiO_2 ~ 53 wt.% and shows high concentrations of K_2O (1.5 wt.%), the highest TiO_2 values with ~1.9 wt.% and the lowest MgO values with 3 wt.% (**Figure 8**). Also shows the lowest Ni (17 ppm), Cr (35 ppm), and Sr (285 ppm) and enriched in V (259 ppm), Cu (115 ppm), Zr (231 ppm; **Table 3**). Intraplate-like lava shows Sr and moderate negative Eu anomaly, associated with the high plagioclase phenocryst content. The multi-element diagram shows a flat pattern of REE with high values of HREE compared with the other samples (**Figure 9**; Supplementary Table 2).

5.4 Deformation systems in the Cahulote Graben

For this work, we assume a deformation system as a volcanotectonic process that depends on the coetaneous interaction between magmatic reservoir systems and regional and local tectonics. The analysis of deformation approaches through the relationship between (i) morphostructural domains, (ii) tectonic structures and (iii) volcanic alignments as indicators of tectonic stress orientations [70–73]. Structural and volcanic alignments mapping of the CG zone are presented in **Figure 10**. Although NE–SW system is the main structural and volcanotectonic regime in the CG area, at least three more systems are observed interrelating in time and space, and we will define their main features.

First deformation system NW-SE (S1 + n) is the oldest brittle deformational event and is related to the regional graben of the Bajo Balsas structures [5, 52]. In the CG zone, we observe structures striking N125–150, <5 km length, and they are cut by other younger deformational systems or they are relatively covered by younger volcanic pyroclasts or lavas. The NW-SE (S1 + n) is related to (i) a narrow Pedernales-Turicato Graben ~ 5 km length, which presents geothermal activity and (ii) a graben ~ 2 km length located NE Cahulote town. The Pleistocene NW-SE shield volcanic alignment dated with >0.5 Ma [14] and widespread between CG zone and Ario de Rosales could be associated with NW-SE (S1 + n) system.

Structures of *the second deformation NE-SW system (S2)* are ubiquitous in the CG and are related to the main landscape associated to the NE-SW Mil Cumbres regional system. Perhaps all of the morphostructural domains of the CG zone are associated with the kinematic development of this system (**Figure 3**) and probably have assisted the emplacement of more than 80% of the CG volcanic vents (**Figure 6**). The main structure of the NE-SW system (S2) correlates with the extensional fault length of 66 km drawing from La Huacana to Tacámbaro towns, which is parallel to NE Mil Cumbres fault extending up to Morelia city (**Figure 2**). The NE-SW system (S2) is closely related to interconnected faults and volcanic alignments, crater and vent elongation, this system is associated with the emplacement Jorullo-La Pilita volcanoes and successively crossing the whole CG zone (**Figure 10a,b**). Along NE-SW system (S2) were emplaced 25 Holocene volcanic vents. Therefore, NE-SW system (S2) is the main volcanotectonic system of the CG.

In order to understand the geometry of the CG zone, in **Figure 4**, we present three profiles that orthogonally cross the CG zone and the NE-SW system (S2) structures. The displacement of tectonic blocks is closely related to morphostructural domains. Based on morphostructural mapping, CG could be interpreted as an asymmetric extensional-transtensional structure mainly striking with the NE-SW system (S2). Recurrently, CG-tilted structural blocks are arranged in the form of antithetic or synthetic fault structures dipping toward southwest. On the other hand, the intersection between NE-SW system (S2) structures with the old NW-SE (S1 + n) structures led to small hanging-wall or dilatational zones length-3–5 km (**Figure 10a**), which apparently are closely correlated with alignments volcanoes like c. Verde, c. Petembo and c. El Sosal. In the northeastern of the Cahulote-Pedernales, the hanging-wall domain was emplaced along this intersection and matched with the main shoshonitic volcanoes of the CG (**Figures 3** and **10b**). The NE-SW structures hold the majority of

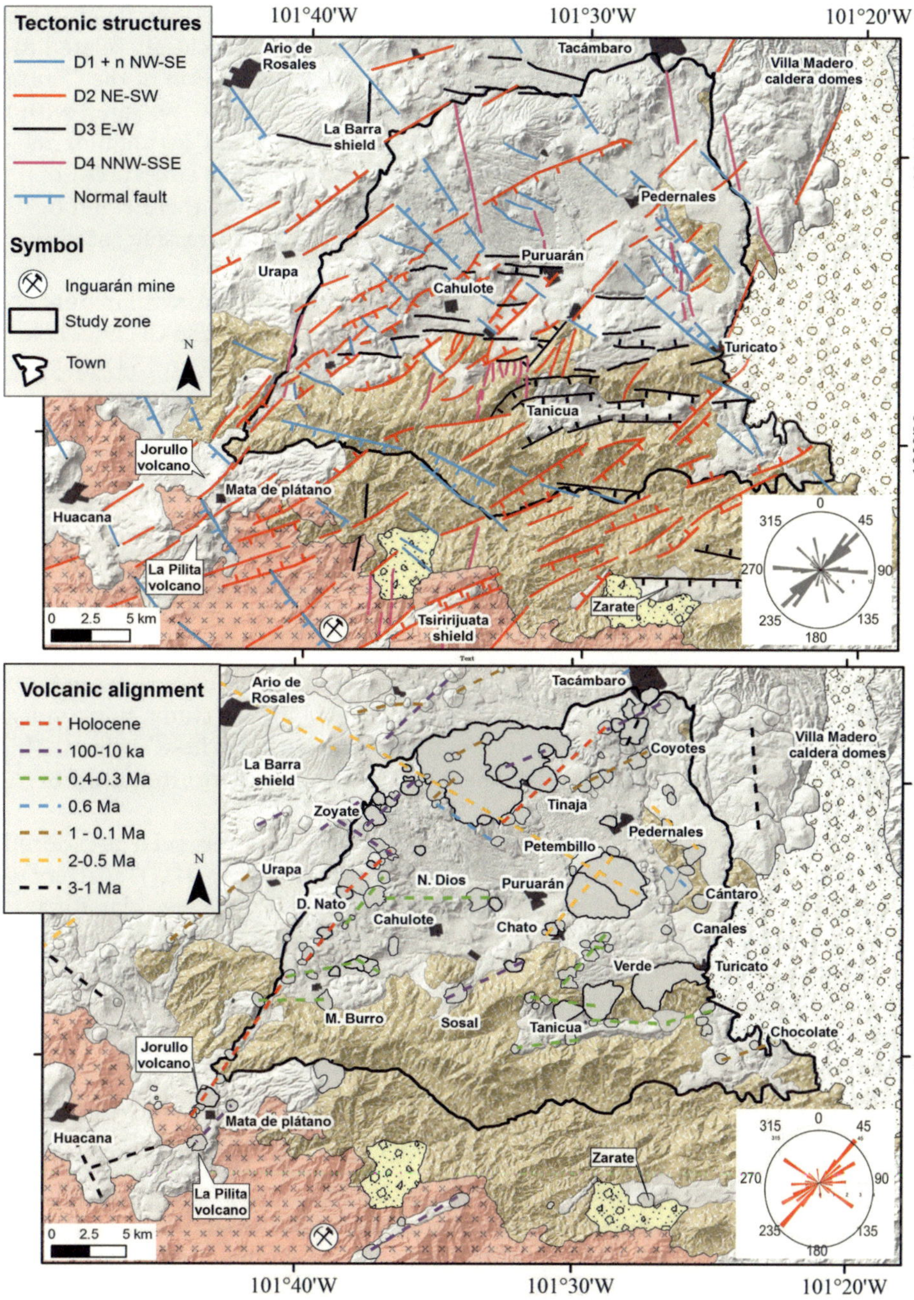

Figure 10.
a. Deformation systems of the Cahulote Graben; B. Temporal volcanic alignments and rose diagrams of tectonic and volcanic structures.

the volcanic alignments (60%). These are abundant in the northwestern domain sector and have an echelon geometry along 40 km, from Jorullo volcano to Tacámbaro town (**Figure 10b**).

The E-W deformation system (S3) is less widespread, with structures ranging from ~2 to 5 km in length on the S-SE region of the CG. This system is closely related to the Tanicua hanging wall in the southeastern domain and the dislocation of the Central domain at Cahulote town (**Figures 3** and **4**). Volcanic alignments associated with the E-W system (S3) include La Palma, Caramicuas and Nombre de Dios alignments. Although the E-W system (S3) cut the NE-SW system (S2), their structures are relatively less abundant than NE-SW system (S2) structures, and there could be conjugate structures or could indicate a relative changing or progressive rotation of the main NE-SW tectonic system.

The NNW-SSE system (S4) structures are relatively uncommon, with mainly length < 2 km. They only were observed in the eastern sector of the CG zone. No vertical displacement was observed. However, the NNW-SSE system (S4) plays a central role in the emplacement of the domes alignment of Villa Madero caldera in the northeastern bounding with the Cahulote zone (**Figure 10b**).

6. Discussion

6.1 Volcanic typology, morphometry, volume and eruption rate

Table 2 shows the count and classification of the volcanic typology of the CG zone, where they are compared with those identified in the emblematic work made by Hasenaka and Carmichael [1, 13] and Guilbaud et al. [14]. In our study, more volcanic buildings (~40%) have been identified than those identified by these authors. This difference is due in part to the fact that these authors do not report the count of spatter volcanic vents that we have found, 79 versus 12.

Previous authors have used different classifications for spatters such as fissured lavas or lavas without cones (or vents). Indeed, in some cases (e.g., [1]), spatters were not considered in the identification of volcanic vents. In this work, we classify as spatters volcanic vents or cones with basal diameters (Dco) < 200 meters and height (Hco) < 100 m [58]. In the Cahulote Graben, these measurements reach up to 1880 m in diameter and 390 m in height (**Table 2**). The average data for small-volume volcanoes is particularly relevant for this classification.

The total eruptive volume calculated for the Cahulote Graben, 16 km^3, is relatively similar to 12.35 km^3 estimated by Guilbaud et al. [14] but seven times more than 1.8 km^3 estimation of Hasenaka and Carmichael [13]. The difference can be explained because the authors only took into account the volcanic vent volume and not the extent of the lava flows. However, it is worth noting that any volume estimation is relatively underestimated due to the continuous erosion of the landscape or buried by younger volcanoes and because the pyroclastic products volume is not calculated, which could represent up to 60% of the total volume eruption [74].

When compared with other world databases, the volume calculated for the CG is similar to that of the other subduction monogenetic volcanic fields, such as the Anatolian Central (24 km^3) [75]; Antofagasta de la Sierra (21 km^3) [76] and is higher than other Mexican volcanic fields, such as Colima (1.4 km^3) [77], Los Tuxtlas (2.1 km^3) [78] and, this is smaller than the Tancítaro - Nueva Italia field with 326 km^3 [6] or Chichinautzin field with 470 km^3 [79].

In relation to volcanic density and rate eruption, according to [18, 28], the Jorullo-Tacámbaro region in the MGVF shows the highest density with 16/100 km^2 volcanoes and the CG with the Holocene eruption rate (1 vent / 400 yr), so it would be the most volcanic active area of the MGVF [1, 2, 14]. In **Table 4**, we present new density and eruption rate data for the CG. In addition to #volcanic vents, area and volume, it is included other parameters like volcano typology, magmatic series, deformational phase and age. The total eruption rate calculated is 6.49 km^3/Ma. However, eruption volume data show that in the Cahulote Graben, the eruption rate was intermittent and variable.

The volcanic typology in relation to age reveals notable trends in eruptive volume. The largest volumes were emplaced in the CG during three distinct periods: (i) 2–1 Ma, with 1.29 km^3 produced by shield volcanism; (ii) 0.4–0.3 Ma, yielding 4.9 km^3 from spatter eruptions; and (iii) < 0.011 Ma (Holocene), generating 3.43 km^3 from scoria and spatter cones. In summary, two significant peaks of volcanic activity were recorded, with spatter-type structures dominating magma emplacement during these periods.

6.2 Petrological and geochemical variations of the Cahulote Graben

From a petrogenetic perspective, it is useful to classify as spatter cones because this small-volume morphology can indicate mafic, alkaline or shoshonitic compositions [80]. Our petrographic and geochemical data of the shoshonitic magmatic series reflects a rapid ascent of melt batches and scarce interaction with the crust according to previous authors [10, 18, 80, 81], who also indicate that shoshonitic products are associated with the low degree of partial melting of the magma parent source. We argue that aphyric textures, composition and rheological features of the shoshonitic rocks promote the fast rise of magmas, which are assisted by lithospheric stress, inducing crystallization at upper crustal depths as the mafic and lamprophyre dikes [82].

The relatively fast and two-stage ascent of magmas from deep reservoirs can be interpreted based on the mineral assemblage and textures observed in the shoshonites of the CG. The remark of two generations of olivines, pyroxenes with lamellae exsolution and common skeletal and zoned textures with reaction rims of Fe-Ti oxides suggest rapid cooling (**Figure 7**). Likewise, the plagioclases show unstable textures such as skeletal, sieve and zoned.

Furthermore, some textures suggest equilibria variations and probably material exchange (mixing) between more than one magma system during its ascent is plausible. All samples show glomeroporphyritic mafic clusters (**Figure 7c**). These clusters constitute <10% of the mineral assemblage and are composed of Ol + Cpx + Plg or Cpx + Ol + Phl and sometimes these clusters are monomineral composed exclusively of clinopyroxene or olivine. Canales spatter shows a diabase xenolith with Plg + Opx association aside from olivine, clinopyroxene, plagioclase and phlogopite antecrysts (**Figure 7d-e**).

To examine geochemical variations within the CG and simplify comparison with other regional magmatic series, in **Figure 11**, we present the CG data plotted alongside regional magmatic calc-alkaline, shoshonitic and intraplate-like series from Zitácuaro, Mascota, Tepic-Zacoalco rift and Valle de Santiago volcanic regions. Parícutin, Jorullo and La Pilita volcanoes geochemical data are also included (Supplementary Table 2).

Major and trace elements ratios support the stated petrographic above observations. The Ti/K versus silica plot (**Figure 11a**) highlights the high potassium character of the shoshonites from CG, Mascota and Zitácuaro volcanic fields. Although CG shoshonites are lower in silica compared to Zitácuaro, they are higher in silica than Mascota data. The nearly constant Ti/K ratio suggests an origin from a consistent and well-defined deep magma source. The H-Mg lavas of CG exhibit a broad Ti/K ratio, and although its potassium content is relatively comparable with other regional

Geological time	Time (Ma)	Volcanic Phase	Volcano typology	Magmatic series	Volume (km^3)	Eruptive rate (km^3/ka)	Alignments system	Tensional Stress (⊠3)
Holocene	< 0.011	6	Spatter and cinder cones	Low Mg, High Mg	3.43	0.34	NE/NNW	N130-N140
Upper Pleistocene	0.1–0.03	5	Spatter and cinder cones	Low Mg, High Mg and shoshonitic	0.92	0.01		N130-N140
Medium Pleistocene	0.4–0.3	4	Spatter	Shoshonitic, High Mg and high Ti	4.9	0.49	NE	N140-N150
	0.6–0.4	3	Spatter and cinder cones	Low Mg, High M	0.37	0.001	NE/EW	N175-N190
	1–0.6	2	Domes, cinder and spatter cones	Low Mg, High Mg and shoshonitic	0.38	0.001	NE	N140-N150
Lower Pleistocene	2–1 Ma	1	Shields and couleés	Low Mg - Shoshonitic	1.29	0.001	NW	N40-N60
				Total volume	11.29			
				Total eruptive rate CG (Vol/ 1.74 Ma)	6.49 km^3/Ma			

Table 4.
Density and eruption rate data for the CG. In addition to #volcanic vents, area and volume, it is included other parameters like volcano typology, magmatic series, deformational phase and time.

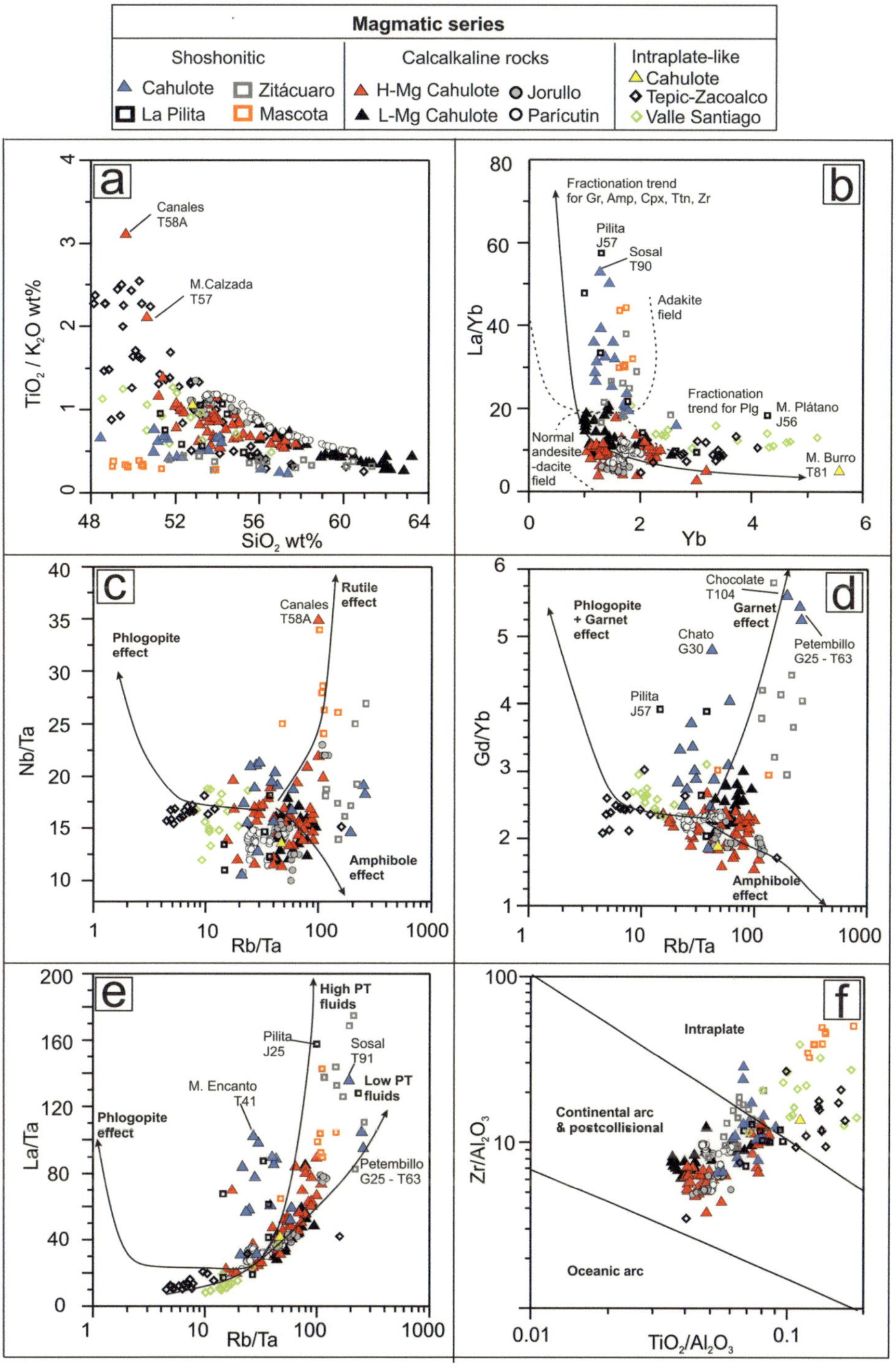

Figure 11.

Variations diagrams of major and trace elements. (a) TiO2/K2O versus SiO2 and (b) La/Yb versus Yb highlight an adakite-like trend for shoshonitic rocks. The strongest garnet signatures are observed in the Sosal and La Pilita volcanoes. (c) Rb/Ta ratios correlate positively with Nb/Ta, (d) Gd/Yb and (e) La/Ta ratios in potassic volcanic rocks, whereas High Mg and Low Mg usually follow the opposite trend with some exceptions [11]. f. Petrotectonic diagram from Muller [2].

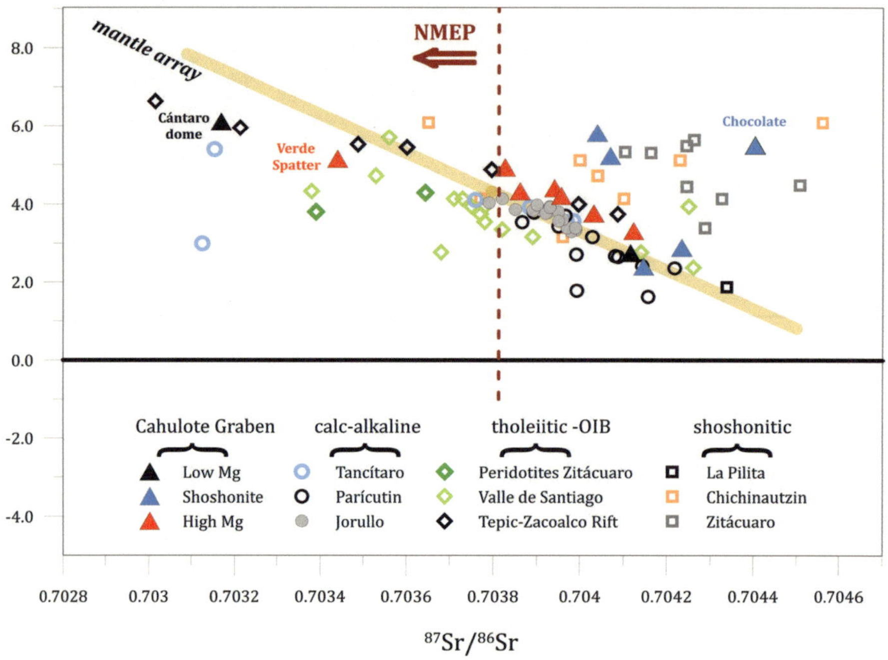

Figure 12.
$^{87}Sr/^{86}Sr$ isotope ratio versus εNd values for the lavas of the Cahulote Graben (compiled data of [10]). Also are plotted data from other volcanic regions or individual volcanoes such as Jorullo [18], Parícutin [19], Pilita [9], Zitácuaro [7, 83], Tepic-Zacoalco [68] and Valle de Santiago [68]. NMEP. North Mexican Extensional Province mantle type proposed by Luhr et al. [84].

magmas, the significant variation in Ti content suggests differences in parental partial fusion source. In contrast, the L-Mg lavas display silica richer contents with close Ti/K ratios, suggesting a relatively shallower magma source deep.

According to La/Yb and Gd/Yb ratios, which show higher enrichment in HREE from CG shoshonites (**Figure 11b-d**), it could suggest garnet fractionation from a lower crustal melting source and/or similar to adakites-like magmas. It is worth noting that although CG shoshonites contain phlogopite, their effect is not noticeable in Nb/Ta and Gd/Yb ratios (**Figure 11c-d**). In addition, considering that all magmatic series of the CG do not show Eu anomaly, it seems that the volume of plagioclase fractionation did not play an important role. In contrast, H-Mg series agree with an amphibole effect suggested by Gb/Yb plot (**Figure 11d**) and L-Mg series show lower Yb content, plotting indistinctly between the effect of amphibole and the effect of garnet.

Finally, the Zr/Al versus Ti/Al diagram show a positive correlation between both ratios in the whole CG volcanic series. The shoshonitic series exhibits high Ti and Zr contents while the low-Mg series shows lower contents. This difference is in agreement with the high ionic charge, immobile nature and high incompatibility of both elements, Zr and Ti. These positive correlations support the contrasting depth of the source magmas estimated.

6.3 Nd - Sr isotopic signatures of the Cahulote Graben

The isotopic data compiled from Guilbaud et al. [10] were grouped and rearranged with the magmatic series of the CG and plotted (Supplementary Table 2). **Figure 12**

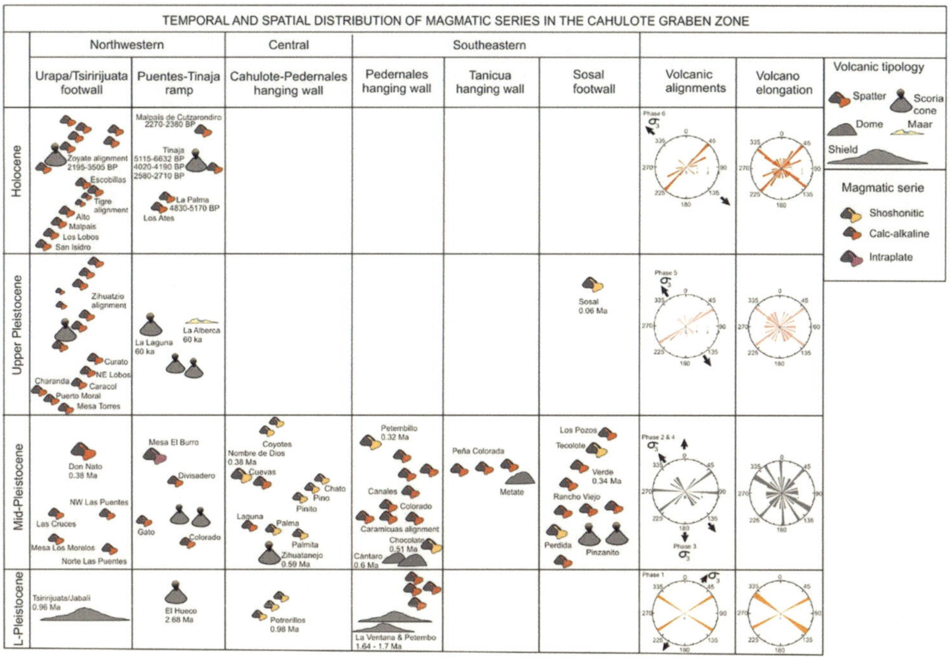

Figure 13.
Geological sketch that shows the spatio-temporal volcanic evolution of the CG. Volcano typology and magmatic series are illustrated by different symbols and colors, respectively. The rose diagrams of volcano elongation and volcanotectonic alignments for each volcanotectonic phase are also shown.

shows the $^{87}Sr/^{86}Sr$ versus εNd isotopic ratios of the CG along the data of the geochemical database from the Transmexican Volcanic Belt (Supplementary Table 2). Three trends are predictable in the distribution and comparison of the data.

a. Most of the values of $^{87}Sr/^{86}Sr$ from the H-Mg and L-Mg series lie between 0.7038 and 0.7041, while their corresponding εNd lie between +2.7 and + 4.9. This trend is comparatively similar to the calc-alkaline samples from Jorullo, Tancítaro and Parícutin volcanoes; particularly, Jorullo shows isotopic signatures like the H-Mg and L-Mg series of the CG zone.

b. Two samples from the H-Mg and L-Mg magmatic series of the CG: Cántaro dome and Verde spatter show higher εNd and lower Sr radiogenic contents: εNd (+5.1 – +6.1) and $^{87}Sr/^{86}Sr$ (0.7031–0.7034). This signature fit well with Tancítaro and intraplate-like from Valle de Santiago, Tepic-Zacoalco rift and Zitácuaro peridotite xenoliths and is resembling to the trend of the North Mexican Extensional Province (NMEP) mantle type proposed by Luhr et al. [84].

c. The shoshonitic lavas from CG show higher radiogenic Sr contents, from 0.704039 to 0.704234, and εNd from +2.1 to +5.8. The Chocolate volcano exhibits the highest radiogenic Sr ratio, 0.704404, and a + 5.5 εNd value. The radiogenic Sr and Nd trends in of the shoshonitic series of the GC have a good fit with the shoshonitic rocks reported in the Zitácuaro, Chichinautzin and La Pilita volcanic regions [21, 22]. The radiogenic Sr and Nd values of the shoshonites deviate from the values of the mantle array, suggesting a distinctive contribution of a mantle source. This observation is consistent with the Zr/Ti discriminant ratio (**Figure 11f**) and fits well with Tepic-Zacoalco rift, strongly suggesting an extensive tectonic regime for the shoshonites of the CG.

6.4 Deformation and morphostructural partitioning sites as indicators of magmatic series

The CG is a volcanic area where ascent magmatism has been assisted to reach the surface by an asymmetric extensional-transtensional structures mainly developed during NE-SW system (S2). We divide the CG zone in three morphostructural domains, which all domains have relatively the same orientation NE-SW trend and they correspond to the distribution of the structural geometry footwall or hanging-wall tectonic graben.

Morphostructural domains are described and successively correlated with overlapped volcanic typology, tectonic structures and volcanic alignments as indicators of tectonic stress orientations [70, 71, 73]. In addition, geochronological and compositional data of magmatic series of volcanic rocks provide comprehensive information for interpreting the relationships between morphostructural sites and magmatic series, especially shoshonitic magmatism.

Figure 13 and **Table 4** concentrate on all the information obtained in the holistic approach carried out in CG. Calc-alkaline volcanism is focused at the footwall and ramp sites of the Northwest domain sites, where NW-SE S1 of the Bajo Balsas system is better exposed and shield-type volcanism developed during ~2–1 Ma. Shoshonitic volcanism is mainly distributed in the NE-SW hanging-wall site of the Central domain

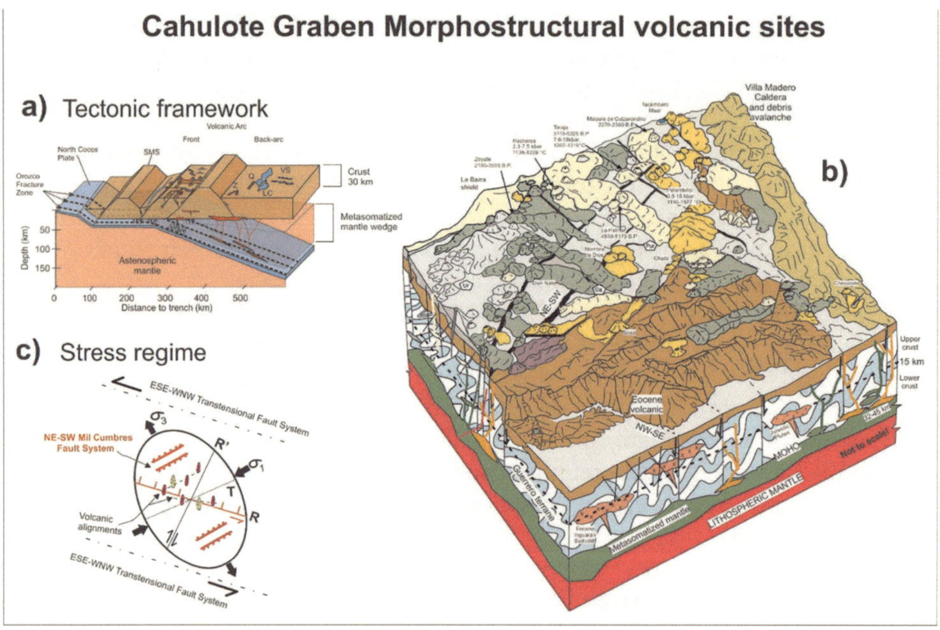

Figure 14.
Cahulote Graben morphostructural partitioning sites. (a) Tectonic framework model of the Cocos plate subduction zone in the central portion of the TMVB. (b) Volcanotectonic block model of the CG and morphostructural sites of the volcanic vents; (c) Riedel system sketch for the extensional-transtensional regime and partitioning emplacement volcanic structures.

and to a lesser extent in the hanging-wall of the southeastern domain and mainly associated with spatter typology (**Figure 3** and **Figure 14**). However, shoshonitic magmatism were emplaced at least during four magmatic or volcanic pulses: (i) Lower Pleistocene ~ 2–1 Ma; (ii) Medium Pleistocene ~ 0.61–0.3 Ma; (iii) Upper Pleistocene ~ 0.1–0.06 Ma; and (iv) Holocene <11 ka (**Figure 13** and **Table 4**). Between 0.4–0.3 Ma the magma flux increased significantly, emplacing at least 20 spatters (i.e., 1 vent each 5 ka) making 4.9 km^3 of lavas which were emplaced with the highest eruptive rate with 0.49 km^3/ka. All volcanoes of this phase are placed on the NE-SW hanging-wall sites and they are relatively aligned along tectonic structures, suggesting that the NE-SW clustering on the volcanic alignments has a close correlation between zones of high magma flux (magmatic footprint [74]) and active extensional system during different volcanotectonic stage [12].

The average elongation of vents (**Figure 1**; Supplementary Table 1) is consistent with the deformation systems of the CG. However, the spatter cones show three distinct elongation directions, suggesting an interaction between the last three systems, S2-S4, where the intersections of these structures significantly assist the conditions for the spatter emplacement. These intersections can be interpreted using a geometric Riedel analysis (**Figure 14**), allowing the identification site by site of possible geometric relationships and good conditions to feeder dikes' magma emplacement. A good example regards a pull-apart zone closely the Cahulote town (**Figures 3** and **6**), where a stress dilatational regime prevails, leading to the emplacement of at least ten volcanoes, of which 7/10 are shoshonitic volcanic structures (**Figure 6**).

Additionally, the intersection of these structures can be related to the reactivation of relatively older deformational systems, either from the NW-SE (S1) or NE-SW (S2) structures. A good example of this intersection is located at the Turicato hanging wall (**Figures 3** and **6**). A small clustering of shoshonitic volcanism is observed at the intersection between the reactivation of NW (S1) structures, which are tilted and cut by the NE system (S2). This zone matches the emplacement of the Petembillo shoshonitic volcano (0.32 ka; [14]).

In summary, in the CG there are morphostructural partitional sites that conditionate the emplacement of different magmatic series and volcano types (**Figure 14b**). The shoshonitic, H-Mg calc-alkaline and intraplate-like are emplaced along the hanging-wall structures and at the intersection of three regional structures (NE-SW, NW-SE and E-W).

6.5 Tectonic implications of shoshonites at Cahulote Graben

From a geodynamic perspective, the shoshonitic rocks have been associated with extensional in orogenic regions at the end of subduction phases or where there is a change between two subduction regimes with different geodynamic processes [24, 26, 29]. The central TMVB (i.e., MGVF) is associated with the oblique convergence of Cocos plate's subduction (**Figure 14a**), characterized by an active WNW-ESE left-lateral transtensional deformation, with a regional 6_3 trending NW-SE and 6_2 or 6_1 trending NE-SW (**Figure 14c**) [85].

In MGVF, the volcanic alignments were associated with different subduction timing of the Cocos plate and extensional structures. The old stratovolcanoes and shields volcanoes (3–0.5 Ma) are trending parallel aligned to the trench (NW-SE), and they are associated with the evolution of old volcanic activity [2, 14, 85]. On the other hand, NE-SW Late Pleistocene-Holocene volcanic alignments are mainly constituted by small-volume volcanoes where the spatter cone typology turns out to be the distinctive morphology of the shoshonitic magmatic series. They have been associated with two deformational processes: i) related to horizontal σ_1 and directly associated with complex compressional-extensional stress regime of the Coco's Plate; ii) the second deformational process is related to the NE-SW alignments, extensional-transtensional structures with oblique regional σ_3 and σ_2 trending NE-SW and E-W structures (**Figure 14c**) [84]. According some authors [2, 86, 87], this stress regime could be hypothetically associated with fragmentation zones of the Cocos plate. We argue that the CG volcanism is related to both deformational stages, where NW-SE Balsas system has produced the first thinning and extensional conditions, and subsequently, the transtensional NE-SW system assists the magma ascent, allowing the last upper crust rising through R' or tensional (T) structures of Riedel system (**Figure 14c**).

Diverse geophysics models suggest that the Orozco Fracture Zone (OFZ) extends inland and separates into two microplates, the northern and southern Cocos plates (**Figure 14**) [47, 88–90]. It is also argued that the OFZ has separated inside continent two Michoacán and Guerrero crustal blocks, and it is associated with transtensional tectonics for the coexistence of the same magmatic series in Zitácuaro -Valle de Bravo volcanic region [7, 8, 83, 91–94]. This scenario could lead to the ascent of second-stage growth of the hydrous shoshonitic and mafic magmas through deep structures of the Cahulote Graben (**Figure 14**).

7. Conclusions

The spatter cones morphology is the most abundant in the CG (60%), and this typology seems to be the distinctive morphology of the shoshonitic magmatic series. Its ubiquitous presence could be morphotectonic indicators associated with extensional faults that caused the adiabatic ascent of the deep shoshonitic and mafic magmatic reservoirs.

The volcanic and tectonic alignments as well as the volcanic vent elongation, are power tools to identify the different deformation systems. Furthermore, the volcanic alignment versus ages reflects different stress directions and volcanotectonic phases through time.

The NE-SW Mil Cumbres system is the dominant structural system and the main morphostructural partitioning sites in the CG, where volcanic structures have been emplaced by mimetic or reactivated structures during Late Pleistocene and Holocene, NE-SW extensional-transtensional system is considered as an active volcanotectonic system.

The coexistence of calc-alkaline, shoshonitic and intraplate-like magmatic series in the Cahulote Graben is associated with different morphostructural partitioning sites, magma ascent pathways and several emplacement stages and magma mixing processes, which are related to the concomitance of extensional-transtensional tectonic.

The tectonic setting of the CG is a fertile scenario for the high magma eruption rate, which is consistent with the highest density volcanoes reported for this sector of the MGVF and TMVB.

Acknowledgements

This work was performed as part of the Masterwork of the first author on a grant from the Consejo Nacional de Humanidades Ciencia y Tecnología (CONAHCYT).

We are grateful Erandi Mejía, José Barajas, Gilberto López, and Almonte's family for hospitality in field campaigns, to Rufino Lozano (Laboratorio Nacional de Geoquímica y Mineralogía, Instituto de Geología, UNAM) for XRF analyses, and to Ofelia Pérez (Laboratorio de Estudios Isotópicos, Centro de Geociencias, UNAM) for LA-ICP-MS analyses. We would like to thank ENES Morelia, the Instituto de Investigaciones en Ciencias de la Tierra, and Felipe García Tenorio from the Instituto de Geofísica Unidad Morelia for their support in sample preparation. Special thanks to Diana Cinthia Soria Caballero for the beautiful drawings of **Figure 14**.

Supplementary Information

Using the following link, you can download Supplementary Table 1 (Volcanic Morphometry and Age), Supplementary Table 2 (Petrography), and Supplementary Table 3 (Geochemical Rock Database): https://intech-files.s3.amazonaws.com/a043Y000011KFdTQAW/a09Tc000001BAptIAG/Videos%20%28Full%20Chapter%29-Morphostructural%20partitional%20sites%20as%20indicator%20of%20%282024-11-12%2014%3A33%3A07%29.zip

Author details

Daniel Alberto Gómez-Calderón[1], Pedro Corona-Chávez[2]*, María del Sol Hernández-Bernal[1] and Francisco Abraham Paz-Moreno[3]

1 Escuela Nacional de Estudios Superiores Unidad Morelia, Universidad Nacional Autónoma de México, Morelia, Michoacán, México

2 Instituto de Investigaciones en Ciencias de la Tierra, Universidad Michoacana de San Nicolás de Hidalgo, México

3 Departamento de Geología, Universidad de Sonora, Sonora, México

*Address all correspondence to: pedro.corona@umich.mx

References

[1] Hasenaka T, Carmichael ISE. The cinder cones of Michoacán—Guanajuato, Central Mexico: Their age, volume and distribution, and magma discharge rate. Journal of Volcanology and Geothermal Research. 1985;**25**(1–2): 105-124. DOI: 10.1016/0377-0273(85) 90007-1

[2] Connor CB. Structure of the Michoacán-Guanajuato volcanic field, Mexico. Journal of Volcanology and Geothermal Research. 1987;**33**(1–3):191-200. DOI: 10.1016/0377-0273(87)90061-8

[3] Ownby S, Granados HD, Lange RA, Hall CM. Volcán Tancítaro, Michoacán, México, 40Ar/39Ar constraints on its history of sector collapse. Journal of Volcanology and Geothermal Research. 2007;**161**(1–2):1-14. DOI: 10.1016/j.jvolgeores.2006.10.009

[4] Guilbaud MN, Siebe C, Layer P, Salinas S, Castro-Govea R, Garduño-Monroy VH, et al. Geology, geochronology, and tectonic setting of the Jorullo volcano region, Michoacán, México. Journal of Volcanology and Geothermal. 2011;**201**:97-112. DOI: 10.1016/j.jvolgeores.2010.09.005

[5] Corona-Chávez P. Revisitando los volcanes de la Región del Parícutin: morfología, petrología y edad. Ciencia Nicolaita. 2018;**74**:31-58. DOI: 10.35830/cn.v0i74.414

[6] Ownby SE, Lange RA, Hall CM, Delgado-Granados H. Origin of andesite in the deep crust and eruption rates in the Tancítaro–Nueva Italia region of the central Mexican arc. Geological Society of America Bulletin. 2011;**123**(1–2):274-294. DOI: 10.1130/B30124.1

[7] Blatter DL, Carmichael ISE, Deino AL, Renne PR. Neogene volcanism at the front of the central Mexican Volcanic Belt: Basaltic andesites to dacites, with contemporaneous shoshonites and high-TiO_2 lava. Geological Society of America Bulletin. 2001;**113**(10):1324-1342. DOI: 10.1130/0016-7606(2001)113<1324: NVATFO>2.0.CO;2

[8] Blatter DL, Lang Farmer G, Carmichael IS. A north–south transect across the central Mexican volcanic belt at⊠ 100° W: Spatial distribution, petrological, geochemical, and isotopic characteristics of quaternary volcanism. Journal of Petrology. 2007;**48**(5):901-950. DOI: 10.1093/petrology/egm006

[9] Luhr JF, Carmichael., I.S.E. Contemporaneous eruptions of calc-alkaline and alkaline magmas along the volcanic front of the Mexican Volcanic Belt. Geofisica Internacional. 1985;**24**(1): 203-216. DOI: 10.22201/igeof.00167169p.1985.24.1.1071

[10] Guilbaud MN, Siebe C, Rasoazanamparany C, Widom E, Salinas S, Castro Govea R. Petrographic, geochemical and isotopic (Sr–Nd–Pb–Os) study of Plio-quaternary volcanics and the tertiary basement in the Jorullo-Tacámbaro area, Michoacán-Guanajuato volcanic field, Mexico. Journal of Petrology. 2019;**60**(12):2317-2338. DOI: 10.1093/petrology/egaa006

[11] Mazzarini F, Ferrari L, Isola I. Self-similar clustering of cinder cones and crust thickness in the Michoacan–Guanajuato and sierra de Chichinautzin volcanic fields, trans-Mexican Volcanic Belt. Tectonophysics. 2010;**486**(1–4):55-64. DOI: 10.1016/j.tecto.2010.02.009

[12] Pérez-López R, Legrand D, Garduño-Monroy VH, Rodríguez-Pascua MA, Giner- Robles JL. Scaling laws of the

size-distribution of monogenetic volcanoes within the Michoacán-Guanajuato volcanic field (Mexico). Journal of Volcanology and Geothermal Research. 2011;**201**(1–4):65-72. DOI: 10.1016/j.jvolgeores.2010.09.006

[13] Hasenaka T, Carmichael E, I. S. A compilation of location, size, and geomorphological parameters of volcanoes of the Michoacan-Guanajuato volcanic field, Central Mexico. Geofisica Internacional. 1985;**24**(4):577-607. DOI: 10.22201/igeof.00167169p.1985.24.4.2179

[14] Guilbaud MN, Siebe C, Layer P, Salinas S. Reconstruction of the volcanic history of the Tacámbaro-Puruarán area (Michoacán, México) reveals high frequency of Holocene monogenetic eruptions. Bulletin of Volcanology. 2012; **74**(5):1187-1211. DOI: 10.1007/s00445-012-0594-0

[15] Johnson ER, Wallace PJ, Cashman KV, Granados HD, Kent AJ. Magmatic volatile contents and degassing-induced crystallization at Volcán Jorullo, Mexico: Implications for melt evolution and the plumbing systems of monogenetic volcanoes. Earth and Planetary Science Letters. 2008;**269**(3–4):478-487. DOI: 10.1016/j.epsl.2008.03.004

[16] Weaver S, Wallace PJ, Johnston AD. A comparative study of continental vs. intraoceanic arc mantle melting: Experimentally determined phase relations of hydrous primitive melts. Earth and Planetary Science Letters. 2011;**308**(1–2):97-106. DOI: 10.1016/j.epsl.2011.05.040

[17] Weaver S, Wallace PJ, Johnston AD. Experimental constraints on the origins of primitive potassic lavas from the trans-Mexican volcanic belt. Contribution to Mineralogy and Petrology. 2013;**166**:825-843. DOI: 10.1007/s00410-013-0921-2

[18] Rasoazanamparany C, Widom E, Siebe C, Guilbaud MN, Spicuzza MJ, Valley JW, et al. Temporal and compositional evolution of Jorullo volcano, Mexico: Implications for magmatic processes associated with a monogenetic eruption. Chemical Geology. 2016;**434**:62-80. DOI: 10.1016/j.chemgeo.2016.04.004

[19] Larrea P, Widom E, Siebe C, Salinas S, Kuentz D. A re-interpretation of the petrogenesis of Paricutin volcano: Distinguishing crustal contamination from mantle heterogeneity. Chemical Geology. 2019;**504**:66-82. DOI: 10.1016/j.chemgeo.2018.10.026

[20] Albert H, Larrea P, Costa F, Widom E, Siebe C. Crystals reveal magma convection and melt transport in dyke-fed eruptions. Scientific Reports. 2020; **10**(1):11632. DOI: 10.1038/s41598-020-68421-4

[21] Gómez-Vasconcelos MG, Luis Macías J, Avellán DR, Sosa-Ceballos G, Garduño-Monroy VH, Cisneros-Máximo G, et al. The control of preexisting faults on the distribution, morphology, and volume of monogenetic volcanism in the Michoacán-Guanajuato volcanic field. Geological Society of America Bulletin. 2020;**132**(11–12):2455-2474. DOI: 10.1130/B35397.1

[22] Gómez-Calderón DA, Soria-Caballero DC, Corona-Chávez P. Tectónica y sismicidad en la Región Volcánica Parícutin-Tancítaro, Michoacán, México. Ciencia Nicolaita. 2021;**82**:23-38. DOI: 10.35830/cn.vi82.556

[23] Condie KC. Plate Tectonics and Crustal Evolution. 4th ed. Oxford: Butterworth Heinemann; 2003. 278 p.

DOI: 10.1016/B978-0-7506-3386-4.X5000-9

[24] Morrison GW. Characteristics and tectonic setting of the shoshonite rock association. Lithos. 1980;**13**(1):97-108. DOI: 10.1016/0024-4937(80)90067-5

[25] Müller D, Rock NMS, Groves DI. Geochemical discrimination between shoshonitic and potassic volcanic rocks in different tectonic settings: A pilot study. Mineralogy and Petrology. 1992; **46**(4):259-289. DOI: 10.1007/BF01173568

[26] Joplin GA. The shoshonite association: A review. Journal of the Geological Society of Australia. 1968;**15**(2):275-294. DOI: 10.1080/00167616808728699

[27] Gomez-Tuena A, Straub SM, Zellmer GF, editors. Orogenic Andesites and Crustal Growth. London: Geological Society of London Special Publication; 2013

[28] Gan C, Zhang Y, Barry TL, He J, Wang Y. Jurassic metasomatised lithospheric mantle beneath South China and its implications: Geochemical and Sr-Nd isotope evidence from the late Jurassic shoshonitic rocks. Lithos. 2019; **320–321**:236-249. DOI: 10.1016/j.lithos.2018.09.007

[29] Muller D, Groves D. Potassic igneous rocks and associated gold-copper mineralization. In: Mineral Resource Reviews. 5th ed. Cham: Springer; 2019. DOI: 10.1007/978-3-319-92979-8

[30] De Astis G, Peccerillo A, Kempton PD, La Volpe L, Wu TW. Transition from calc-alkaline to potassium-rich magmatism in subduction environments: Geochemical and Sr, Nd, Pb isotopic constraints from the island of Vulcano (Aeolian arc). Contributions to Mineralogy and Petrology. 2000;**139**: 684-703

[31] Peccerillo A, Association internationale de volcanologie et de chimie de l'intérieur de la Terre. Cenozoic Volcanism in the Tyrrhenian Sea Region. New York, NY, USA: Springer; 2017. p. 399

[32] Déruelle B. Petrology of quaternary shoshonitic lavas of northwestern Argentina. In: GSA Special Papers 265. Andean Magmatism and its Tectonic Setting. Boulder: Geological Society of America; 1991. pp. 201-216. DOI: 10.1130/SPE265-p201

[33] Fernandez-Turiel JL, Saavedra J, Perez-Torrado FJ, Rodriguez-Gonzalez A, Rejas M, Guillou H, et al. New ages, morphometric and geochemical data on recent shoshonitic volcanism of the Puna, central volcanic zone of Andes: San Jerónimo and negro de Chorrillos volcanoes. Journal of South American Earth Sciences. 2021;**109**:103270. DOI: 10.1016/j.jsames.2021.103270

[34] Bloomer SH, Stern RJ, Fisk E, Geschwind CH. Shoshonitic volcanism in the northern Mariana arc: 1. Mineralogic and major and trace element characteristics. Journal of Geophysical Research: Solid Earth. 1989;**94**(B4): 4469-4496

[35] Leslie RA, Danyushevsky LV, Crawford AJ, Verbeeten AC. Primitive shoshonites from Fiji: Geochemistry and source components. Geochemistry, Geophysics, Geosystems. 2009;**10**(7): Q03020

[36] Nicholls IA, Whitford DJ. Potassium-rich volcanic rocks of the Muriah complex, Java, Indonesia: Products of multiple magma sources? Journal of Volcanology and Geothermal Research. 1983;**18**(1–4):337-359

[37] Lange RA, Carmichael IS. Hydrous basaltic andesites associated with minette and related lavas in western Mexico. Journal of Petrology. 1990;**31**(6):1225-1259

[38] Lange RA, Carmichael IS. A potassic volcanic front in western Mexico: The lamprophyric and related lavas of san Sebastian. Geological Society of America Bulletin. 1991;**103**(7):928-940

[39] Allan JF, Carmichael ISE. Lamprophyric lavas in the Colima graben, SW Mexico. Contributions to Mineralogy and Petrology. 1984;**88**:203-216

[40] Righter K, Carmichael IS. Hawaiites and related lavas in the Atenguillo graben, western Mexican volcanic belt. Geological Society of America Bulletin. 1992;**104**(12):1592-1607

[41] Righter K, Rosas-Elguera J. Alkaline lavas in the volcanic front of the western Mexican volcanic belt: Geology and petrology of the Ayutla and Tapalpa volcanic fields. Journal of Petrology. 2001;**42**(12):2333-2361

[42] Vigouroux N, Wallace PJ, Kent AJ. Volatiles in high-K magmas from the western trans-Mexican volcanic belt: Evidence for fluid fluxing and extreme enrichment of the mantle wedge by subduction processes. Journal of Petrology. 2008;**49**(9):1589-1618

[43] Ferrari L, Orozco-Esquivel T, Manea V, Manea M. The dynamic history of the trans-Mexican volcanic belt and the Mexico subduction zone. Tectonophysics. 2012;**522–523**:122-149. DOI: 10.1016/j.tecto.2011.09.018

[44] Manea VC, Manea M, Ferrari L. A geodynamical perspective on the subduction of Cocos and Rivera plates beneath Mexico and Central America. Tectonophysics, Moho: 100 Years After Andrija Mohorovicic. 2013;**609**:56-81. DOI: 10.1016/j.tecto.2012.12.039

[45] Guzmán-Speziale M, Zúñiga FR. Differences and similarities in the Cocos–North America and Cocos–Caribbean convergence, as revealed by seismic moment tensors. Journal of South American Earth Sciences. 2016;**71**: 296-308. DOI: 10.1016/j.jsames.2015.10.002

[46] Frausto P. Análisis morfotectónico del sector oriental del curso bajo del río Balsas. Michoacán, México: Universidad Nacional Autónoma de México, tesis de licenciatura; 1995, 159 p

[47] Bandy WL, Hilde TWC, Yan CY. The Rivera-Cocos plate boundary: Implications for Rivera-Cocos relative motion and plate fragmentation. Geological Society of America Special Papers. 2000;**334**:1-28. DOI: 10.1130/0-8137-2334-5.1

[48] Rosas-Elguera J, Lopez-Martinez M, Alva-Valdivia L, Gogichaishvili A, Urrutia-Fucugauchi J. The Cotija half-graben: A reconaissance geologic and paleomagnetic study. In: AGU Fall Meeting Abstracts; 10-14 December 2001; San Francisco: American Geophysical Union; 2001. p. S31B-S0616B

[49] Gómez-Calderón DA. Coexistencia de magmas shoshoníticos y calcialcalinos en la Región Volcánica Graben del Cahulote de Santa Ana Pedernales. In: Petrografía, geoquímica e implicaciones tectónicas [thesis]. Morelia, Michoacán: Universidad Nacional Autónoma de México; 2021

[50] Suter M, López-Martínez M, Quintero-Legorreta O, Carrillo-Martínez M. Quaternary intra-arc extension in the central trans-Mexican volcanic belt.

Geological Society of America Bulletin. 2001;**113**(6):693-703. DOI: 10.1130/0016-7606(2001)113%3C0693:QIAEIT%3E2.0.CO;2

[51] Garduño-Monroy VH, Pérez-López R, Israde-Alcántara I, Rodríguez-Pascua MA, Szynkaruk E, Hernández-Madrigal VM, et al. Paleoseismology of the southwestern Morelia-Acambay fault system, Central Mexico. Geofisica Internacional. 2009;**48**(3):319-335. ISSN 0016-7169

[52] Romero-Mosqueda SJ. Análisis morfoestructural de las relaciones petrológicas del graben de Peribán, Michoacán. Guanajuato [thesis]. Guanajuato: Universidad de Guanajuato; 2017

[53] Valentine GA, Connor CB. Basaltic volcanic fields. In: The Encyclopedia of Volcanoes. 2nd ed. San Diego: Academic Press; 2015. pp. 423-439. DOI: 10.1016/B978-0-12-385938-9.00023-7

[54] Demant A. Interpretación geodinámica del volcanismo del Eje Neovolcánico Transmexicano. Revista Mexicana de Ciencias Geológicas. 1981;**5**(2):217-222

[55] Campa MF, Coney PJ. Tectono-stratigraphic terranes and mineral resource distributions in Mexico. Canadian Journal of Earth Sciences. 1983;**20**(6):1040-1051. DOI: 10.1139/e83-094

[56] Centeno-García E. Mesozoic tectono-magmatic evolution of Mexico: An overview. Ore Geology Reviews. 2017;**81**:1035-1052. DOI: 10.1016/j.oregeorev.2016.10.010

[57] Gómez-Vasconcelos MG, Pantoja UO, Garduño￾Monroy VH, Hernández-Madrigal VM, Layer PW. La caldera de Villa Madero: vulcanismo explosivo del cuaternario en el borde sureste del campo volcánico Michoacán Guanajuato. Ciencia Nicolaita. 2015;**65**:12-35. DOI: 10.35830/cn.vi65.256

[58] Kereszturi G, Németh K. Monogenetic basaltic volcanoes: Genetic classification, growth, geomorphology and degradation. In: Updates in Volcanology - New Advances in Understanding Volcanic Systems. London, UK: IntechOpen; 2012. DOI: 10.5772/51387

[59] Francis P, Oppenheimer C, Gaffney ES, Damjanac B, Valentine GA. Localization of volcanic activity: 2. Effects of preexisting structure. Earth and Planetary Science Letters. 2007;**263**(3–4):323-338. DOI: 10.1016/j.epsl.2007.09.002

[60] Sumner JM, Blake S, Matela RJ, Wolff JA. Spatter. Journal of Volcanology and Geothermal Research. 2005;**142**(1–2):49-65. DOI: 10.1016/j.jvolgeores.2004.10.013

[61] Grosse P, Kervyn M. Morphometry of terrestrial shield volcanoes. Geomorphology. 2018;**304**:1-14. DOI: 10.1016/j.geomorph.2017.12.017

[62] Hasenaka T, Carmichael IS. The cinder cones of Michoacan-Guanajuato, Central Mexico: Petrology and chemistry. Journal of Petrology. 1987;**28**(2):241-269. DOI: 10.1093/petrology/28.2.241

[63] Head JW III, Wilson L. Basaltic pyroclastic eruptions: Influence of gas-release patterns and volume fluxes on fountain structure, and the formation of cinder cones, spatter cones, rootless flows, lava ponds and lava flows. Journal of Volcanology and Geothermal Research. 1989;**37**(3–4):261-271. DOI: 10.1016/0377-0273(89)90083-8

[64] Lozano-Santa Cruz R, Bernal JP. Characterization of a new set of eight geochemical reference materials for XRF major and trace element analysis. Revista Mexicana de Ciencias Geológicas. 2005; **22**:329-344. ISSN 1026-8774

[65] Mori L, Gómez-Tuena A, Cai Y, Goldstein SL. Effects of prolonged flat subduction on the Miocene magmatic record of the central trans-Mexican volcanic belt. Chemical Geology. 2007; **244**:452-473. DOI: 10.1016/j.chemgeo.2007.07.002

[66] Mahgoub AN, Böhnel H, Siebe C, Salinas S, Guilbaud MN. Paleomagnetically inferred ages of a cluster of Holocene monogenetic eruptions in the Tacámbaro-Puruarán area (Michoacán, México): Implications for volcanic hazards. Journal of Volcanology and Geothermal Research. 2017;**347**:360-370. DOI: 10.1016/j.jvolgeores.2017.10.004

[67] Peccerillo A, Frezzotti ML. Magmatism, mantle evolution and geodynamics at the converging plate margins of Italy. Journal of the Geological Society. 2015;**172**(4):407-427. DOI: 10.1144/jgs2014-085

[68] Gómez-Tuena A, Mori L, Straub SM. Geochemical and petrological insights into the tectonic origin of the transmexican volcanic belt. Earth-Science Reviews. 2018;**183**:153-181. DOI: 10.1016/j.earscirev.2016.12.006

[69] Sun SS, McDonough WF. Chemical and isotopic systematics of oceanic basalts: Implications for mantle composition and processes. Geological Society, London, Special Publications. 1989;**42**(1):313-345. DOI: 10.1144/GSL.SP.1989.042.01.19

[70] Le Corvec N, Spörli KB, Rowland J, Lindsay J. Spatial distribution and alignments of volcanic centers: Clues to the formation of monogenetic volcanic fields. Earth-Science Reviews. 2013;**124**: 96-114. DOI: 10.1016/j.earscirev.2013.05.005

[71] Tibaldi A, Bonali FL, Corazzato C. Structural control on volcanoes and magma paths from local-to orogen-scale: The Central Andes case. Tectonophysics. 2017;**699**:16-41. DOI: 10.1016/j.tecto.2017.01.005

[72] Uslular G, Le Corvec N, Mazzarini F, Legrand D, Gençalioğlu-Kuşcu G. Morphological and multivariate statistical analysis of quaternary monogenetic vents in the central Anatolian volcanic province (Turkey): Implications for the volcano-tectonic evolution. Journal of Volcanology and Geothermal Research. 2021;**416**:107280. DOI: 10.1016/j.jvolgeores.2021.107280

[73] Nakamura K. Volcanoes as possible indicators of tectonic stress orientation —Principle and proposal. Journal of Volcanology and Geothermal Research. 1977;**2**(1):1-16. DOI: 10.1016/0377-0273(77)90012-9

[74] Valentine GA, Perry FV. Tectonically controlled, time-predictable basaltic volcanism from a lithospheric mantle source (central basin and Range Province, USA). Earth and Planetary Science Letters. 2007;**261**:201-216. DOI: 10.1016/j.epsl.2007.06.029

[75] Gürsoy H, Piper JDA, Tatar O, Mesci L. Palaeomagnetic study of the Karaman and Karapinar volcanic complexes, Central Turkey: Neotectonic rotation in the south-central sector of the Anatolian block. Tectonophysics. 1998;**299**(1–3): 191-211. DOI: 10.1016/S0040-1951(98)00205-4

[76] Báez W, Carrasco Nuñez G, Giordano G, Viramonte JG, Chiodi A.

Polycyclic scoria cones of the Antofagasta de la sierra basin, southern Puna plateau, Argentina. Argentina Geological Society of London, Special Publication. 2017;**446**:311-336. DOI: 10.1144/SP446.3

[77] Carmichael ISE, Frey HM, Lange RA, Hall CM. The Pleistocene cinder cones surrounding Volcan Colima, Mexico re-visited: Eruption ages and volumes, oxidation states, and sulfur content. Bulletin of Volcanology. 2006;**68**(5): 407-419. DOI: 10.1007/s00445-005-0015-8

[78] Jácome-Paz MP, Torres-Orozco R, Espinasa-Pereña R, de la Fuente Rivera JR, Sánchez JOR, Delgado-Granados H. Review of geology and geomorphology of the Xalapa monogenetic volcanic field, eastern trans-Mexican volcanic belt. Journal of Volcanology and Geothermal Research. 2022;**432**: 107689. DOI: 10.1016/j.jvolgeores.2022.107689

[79] Arce JL, Munoz-Salinas E, Castillo M, Salinas I. The~ 2000 yr BP Jumento volcano, one of the youngest edifices of the Chichinautzin volcanic field, Central Mexico. Journal of Volcanology and Geothermal Research. 2015;**308**:30-38. DOI: 10.1016/j.jvolgeores.2015.10.008

[80] McGee LE, Smith IEM. Interpreting chemical compositions of small scale basaltic systems: A review. Journal of Volcanology and Geothermal Research. 2016;**325**:45-60. DOI: 10.1130/G37100Y.1

[81] Larrea P, Albert H, Ubide T, Costa F, Colás V, Widom E, et al. From explosive vent opening to effusive outpouring: Mineral constraints on magma dynamics and timescales at paricutin monogenetic volcano. Journal of Petrology. 2021;**62**(4):egaa112. DOI: 10.1093/petrology/egaa112

[82] Rock NMS. Lamprophyres. New York: Springer; 1991. DOI: 10.1007/978-1-4757-0929-2

[83] Blatter DL, Carmichael ISE. Hornblende peridotite xenoliths from Central Mexico reveal the highly oxidized nature of subarc upper mantle. Geology. 1998a;**26**(11):1035-1038. DOI: 10.1130/0091-7613(1998)026%3C1035:HPXFCM%3E2.3.CO;2

[84] Luhr JF, Kimberly P, Siebert L, Aranda-Gómez JJ, Housh TB, Mattietti GK. México's quaternary volcanic rocks: Insights from the MEXPET petrological and geochemical database. In: Neogene-Quaternary Continental Margin Volcanism: A Perspective from México. Boulder: Geological Society of America Special Papers; 2006. p. 402, 1–44. DOI: 10.1130/2006.2402(01)

[85] Ego F, Ansan V. Why is the central trans-Mexican volcanic belt (102–99 W) in transtensive deformation? Tectonophysics. 2002;**359**(1–2):189-208. DOI: 10.1016/S0040-1951(02)00511-5

[86] Ban M, Hasenaka T, Delgado-Granados H, Takaoka N. K-Ar ages of lavas from shield volcanoes in the Michoacán-Guanajuato volcanic field, Mexico. Geofisica Internacional. 1992;**31**(4):467-473. DOI: 10.22201/igeof.00167169p.1992.31.4.1367

[87] Mooser F. The Mexican volcanic belt structure and tectonics. Geofisica Internacional. 1972;**12**(2):55-70. DOI: 10.22201/igeof.00167169p.1972.12.2.1024

[88] Carr MJ, Stoiber RE, Drake CL. The segmented nature of some continental margins. In: The Geology of Continental Margins. Berlin, Heidelberg: Springer; 1974. pp. 105-114. DOI: 10.1007/978-3-662-01141-6_9

[89] Lonsdale P. Structural patterns of the Pacific floor offshore of peninsular California. In: Dauphin JP, Simoneit BRT, editors. The Gulf and Peninsular Province of the Californias. Tulsa: American Association of Petroleum Geologists; 1991. pp. 87-125

[90] Dougherty SL, Clayton RW, Helmberger DV. Seismic structure in Central Mexico: Implications for fragmentation of the subducted Cocos plate. Journal of Geophysical Research: Solid Earth. 2012;**117**:B09316. DOI: 10.1029/2012JB009528

[91] Stubailo I, Beghein C, Davis PM. Structure and anisotropy of the Mexico subduction zone based on Rayleigh-wave analysis and implications for the geometry of the trans-Mexican volcanic belt. Journal of Geophysical Research: Solid Earth. 2012;**117**::B05303. DOI: 10.31223/osf.io/qvr94

[92] Blatter DL, Carmichael ISE. Plagioclase-free andesites from Zitacuaro (Michoacán), Mexico: Petrology and experimental constraints. Contributions to Mineralogy and Petrology. 1998;**132**:121-138. DOI: 10.1007/s004100050411

[93] Mukasa SB, Blatter DL, Andronikov AV. Mantle peridotite xenoliths in andesite lava at El Peñon, central Mexican volcanic belt: Isotopic and trace element evidence for melting and metasomatism in the mantle wedge beneath an active arc. Earth and Planetary Science Letters. 2007;**260**(1–2):37-55. DOI: 10.1016/j.epsl.2007.05.013

[94] Blatter DL, Hammersley L. Impact of the Orozco fracture zone on the central Mexican volcanic belt. Journal of Volcanology and Geothermal Research. 2010;**197**(1–4):67-84. DOI: 10.1016/j.jvolgeores.2009.08.002

Chapter 4

Complex Magma Chamber Dynamics and Volcanic Activity in the Karapınar Volcanic Field: Insights from Mineral and Whole-Rock Chemistry – A Review

Gülin Gencoglu Korkmaz and Hüseyin Kurt

Abstract

The thermal and material conditions of magma chambers influence magma mobilization processes and temporal dynamics, which are crucial for assessing volcanic hazards. Crystals preserve historical magma reservoir information, facilitating alignment with computational simulations. The Karapınar Volcanic Field (KPVF) in Central Anatolia, part of the Cappadocia Volcanic Province, serves as an ideal natural laboratory for studying monogenetic volcanism, characterized by scoria cones along a fault line, maars, and lava flows. Basaltic, basaltic andesitic, and andesitic lavas, along with their pyroclastics, are widespread in this region. The basaltic and basaltic andesitic lava flows incorporate various enclaves, such as magma mixing enclaves, cumulate enclaves, and xenocrysts. They also contain several ocelli, embayed-quartz xenocrysts, and densely sieved plagioclase megacrysts. Meke and Yılan maars exemplify intriguing maar volcanism, with basaltic andesitic lavas containing numerous enclaves. Interestingly, these mafic rocks contain zircon grains (generally found in felsic magmas). The presence of different types of enclaves, mineral textural properties, and zircon-bearing mafic rocks indicate complex magma chamber dynamics in the area. Moreover, whole-rock and mineral chemistry from KPVF reveal complex magmatic histories, emphasizing the presence of multiple feeding and plumbing systems that potentially trigger volcanic activity. Temporally and spatially, mineral chemistry analyses can provide high-resolution insights into the volcano-tectonic history of the region.

Keywords: basalt, complex magmatic history, enclave, plagioclase, zircon

1. Introduction

Magmatic rocks are essential for understanding geological processes due to their compositions, which reflect a range of magmatic events, including crystallization, magmatic recharge, and the assimilation of surrounding rocks. While whole-rock geochemistry provides significant insights into these processes, it often integrates multiple magmatic histories, which can obscure specific magmatic events [1, 2]. To overcome this limitation, detailed mineral-scale chemical analyses—particularly trace element zoning within crystals—offer a more refined understanding of magmatic processes and magma evolution [3–7]. These analyses allow for the differentiation of various magmatic events and provide deeper insights into the processes shaping magma evolution. Volcanic textures and mineral chemistry are critical for revealing information about magmatic processes and volcanic activities. Textural features such as vesicles, crystal sizes, and mineral compositions reflect the conditions during magma ascent and cooling, while variations in xenoliths and mineral compositions can indicate processes such as contamination and magma mixing [1, 8]. By integrating textural and geochemical analyses, we can enhance our understanding of how both internal factors (e.g., magma composition) and external factors (e.g., crustal interactions) influence volcanic systems and eruption dynamics [4].

In Central Anatolia, Cenozoic volcanism is widespread, with the southeastern part of the Cappadocia Volcanic Province divided into two main subgroups: the Neogene Karacadağ Volcanic Complex (KCVC) and the Quaternary Karapınar Volcanic Field (KPVF) (**Figure 1**) [11]. Keller [13] proposed that the products of the Karapınar

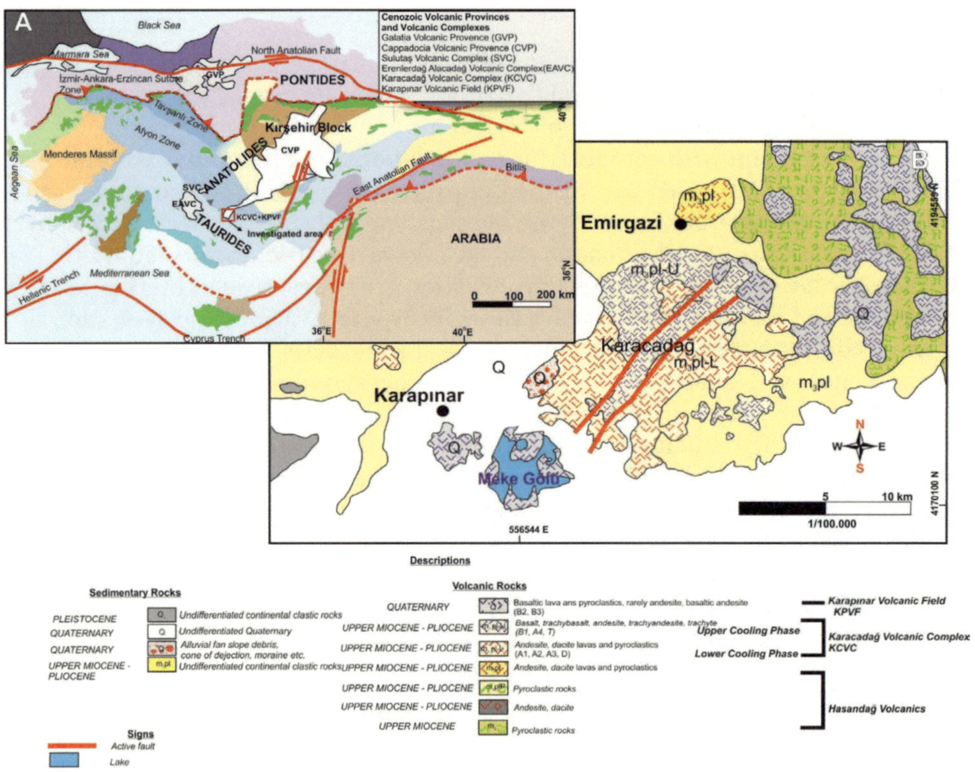

Figure 1.
(A) Simplified tectonic map of Turkey taken from Van Hinsbergen, Maffione [9] and (B) Geological map of the investigated area [10] is modified from Gencoğlu Korkmaz, Kurt [11] The terms "Upper and Lower Cooling Phase" refer to the volcanic rocks within the Karacadağ Volcanic Complex (KCVC). These phases represent different rock groups, such as A1, A2...; B1, B2..., which correspond to various rock types, including andesite and basalt. The mineralogical and geochemical characteristics of these rock groups are discussed in detail (refer to Refs. [11, 12]).

Volcanic Field overlie the older Karacadağ volcanic rocks and lack a direct genetic relationship with them. Previous studies [11, 12, 14–16] have explored complex magma chamber processes using mineral chemistry, whole-rock geochemistry, and isotopic data. This study aims to elucidate the intricate magmatic processes within the Karapınar Volcanic Field through comprehensive petrological and geochemical analyses of basaltic scoria cones. By synthesizing data from previous studies [11, 12, 14–16], we have integrated the first findings of zircon from basaltic rocks—typically associated with felsic and intermediate magmatic rocks—to investigate the complex processes within the magma chamber. This study thus provides a novel perspective on how zircon can occur in basaltic rocks, offering new insights into the intricate dynamics of magma chamber processes. This compilation offers significant insights into the complex magmatic processes involved in rock evolution. By providing a detailed review of open-system and complex magma chamber processes, our study makes substantial contributions to the understanding of regional volcanic history and magmatic processes in Central Anatolia.

2. Volcanic morphology: Interaction of internal and external factors

Volcanic systems are influenced by a complex interplay of internal and external factors that dictate the type of volcanism, the nature of volcanic products, and the resulting volcanic landforms [17, 18]. Internal factors, such as magma chemistry, viscosity, and volume, govern the dynamics within magma chambers, affecting processes like fractional crystallization, magma mixing, and contamination. These processes, in turn, influence the eruptive style and the compositional diversity of volcanic products. Conversely, external factors, including the availability of external water, topography, and climate, can significantly alter eruption dynamics, particularly in systems where phreatomagmatic interactions dominate [19]. The interplay between these factors is well-documented in various volcanic fields around the world, such as the Auckland Volcanic Field in New Zealand and the Eifel Volcanic Field in Germany, where both internal magmatic processes and external environmental conditions contribute to the diverse volcanic landscapes observed today: the Karapınar Volcanic Field (KPVF) in Central Anatolia, part of the larger Cappadocia Volcanic Province, provides a natural laboratory for investigating how these internal and external factors interact to produce a diverse array of volcanic features. In particular, the Meke Maar and other monogenetic centers within the KPVF exemplify the intricate balance between magmatic processes and environmental controls, offering valuable insights into the conditions that favor the formation of specific volcanic landforms and eruption styles [17].

Monogenetic volcanism, characterized by discrete, short-lived eruptive events from small-volume volcanic systems, provides a distinct framework for examining volcanic processes [19]. This type of volcanism allows for detailed analysis of isolated volcanic events, offering insights into how individual eruptions contribute to our understanding of volcanic processes. The small scale and short-lived nature of these systems enable researchers to study each eruption independently, providing a unique perspective on volcanic dynamics. The growth of intra-crater scoria cones within maar or tuff ring craters offers a compelling case study of how internal and external factors interact to shape volcanic features [17, 18]. The formation of intra-crater scoria cones within maar or tuff ring craters provides a valuable framework for understanding the interaction between internal and external volcanic factors [17]. These cones develop due to the specific interplay between magmatic conditions, such as magma volume, composition, and viscosity, and external environmental factors, including water availability and landscape elevation. Internal magmatic parameters influence the fragmentation of magma and eruption style, while viscosity affects magma mobility and escape during eruptions. External factors, particularly the presence of water, can enhance phreatomagmatic activity, leading to more explosive eruptions. In the case of Meke Maar, the high availability of external water has driven significant phreatomagmatic processes, shaping the volcanic features such as the scoria cone and tuff ring. This interaction underscores the complex relationship between internal and external factors in volcanic systems, offering insights into the dynamic processes that influence volcanic eruptions and formations. In this context, the Meke Maar in the Karapınar region exemplifies the dynamics of such systems (**Figure 1**). The interaction between internal magmatic parameters—such as magma volume, composition, and viscosity—and external environmental factors—particularly the availability of external water and landscape elevation—plays a crucial role in determining eruption style and volcanic architecture [17, 18]. In the case of Meke Maar, the high availability of external water has led to dominant phreatomagmatic processes. The formation

Figure 2.
In the Karapınar volcanic field, variations in volcanic landforms and their products are influenced by ratios of environmental factors (such as water) and the composition of the magma.

of the scoria cone within Meke Maar illustrates the intricate interaction of volcanic processes in this unique geological setting. Initially, Meke Maar was created through phreatomagmatic eruptions, where the violent interaction between groundwater and rising magma led to explosive activity, resulting in a broad and shallow crater. This stage reflects a dominant influence of external water, which controlled the eruptive style and shaped the maar's characteristics. The presence of substantial external water was crucial in driving these phreatomagmatic processes, leading to the formation of the maar crater. Subsequently, a shift in magmatic conditions occurred, possibly due to changes in magma chemistry or the depletion of external water, which allowed for the formation of a scoria cone within the existing maar. This basaltic scoria cone, formed by less explosive, gas-rich magma eruptions, highlights the transition from a water-dominated eruptive phase to a more magmatic one. The occurrence of a scoria cone within a maar does not indicate a contradiction but rather underscores the complexity of volcanic systems. It shows how different eruptive phases can coexist and evolve within the same volcanic environment, demonstrating the dynamic interplay between internal magmatic conditions and external environmental factors [17]. **Figures 2** and **3** provide different volcanic edifices in the area. The Meke Maar volcanic features highlight the base surge deposits and enclave-bearing lavas within the maar walls. The base surge deposits provide critical evidence of the intense magma fragmentation and pyroclastic

Figure 3.
In the Karapınar volcanic field, variations in volcanic landforms and their products are influenced by ratios of environmental factors (such as water) and the composition of the magma. (A) Enclave-bearing lava flows in Meke Maar, (B) general view of the base surge deposits (small dunes, thin layering, antidunes, low-angle stratification) in Meke Maar, (C-D) tuff deposits around Acigöl Maar.

transportation dynamics driven by phreatomagmatic activity. Additionally, the enclave-bearing lavas within the maar walls illustrate the significant role of magmatic processes such as magma mixing and contamination, in shaping the volcanic architecture, offering insights into the complex interaction between internal magmatic forces and external water-driven phreatomagmatic processes.

3. Geological setting and magmatism in the area

The Late Pliocene-Quaternary volcanism in the Konya region is characterized by a diverse range of volcanic features, including volcanic cones, explosion craters, scoria cones, and maars. In the Karapınar area, Quaternary-aged volcanic formations are predominantly basaltic scoria cones, lava flows, and maars, most of which are part of the calc-alkaline series [20]. These volcanic structures are aligned along a significant fault line, indicating structural control rather than random distribution [21, 22]. The volcanic units consist of basalt, basaltic andesite, trachy-basalt, trachy-andesite, and andesite, with associated pyroclastic flows, basic tuffs, and fall deposits [13, 23–30]. Güllü and Kadıoğlu [31] inferred that the volcanic rocks around Karapınar-Meke Lake and Acıgöl craters share a common origin with deeper magma sources. Several studies have examined the mantle source of the volcanics in this area. Recent studies [32, 33] have further advanced the understanding of regional magmatism. Reid, Schleiffarth [32] proposed that the mafic volcanics in the southwestern Hasan Mountain are derived from a peridotitic source with minimal lithospheric thinning, based on geodynamic modeling. McNab, Ball [33] argued that high asthenospheric temperatures significantly influence regional dynamic topography and basaltic magmatism, based on drainage analysis and geochemical data from basaltic rocks. A recent comprehensive study [34] has synthesized geochemical and age data from the Late Oligocene to the Quaternary and proposed that the slice rupture process commenced 25 My ago, following the collision of the Arabian-Anatolian block 30 My ago. They associated the 5 My delay between the collision and subsequent magmatism with the gradual steepening and rifting of the lithosphere. Refs. [35, 36] suggested that peridotite source lithologies and contributions from a metasomatized lithospheric mantle are significant for basalt formation at the Eğrikuyu site, arguing that both peridotite and metasomatized mantle played crucial roles in the volcanic processes. Gençoğlu Korkmaz and Kurt [14] identified two distinct geochemical types of Karapınar basalts: KB1, which is mildly alkaline with high MgO and primitive olivine, and KB2, which is calc-alkaline and more evolved. Olivine mineral chemistry is widely recognized as a reliable indicator of mantle source characteristics [6, 12, 37–40]. Their analysis of olivine chemistry revealed that basalts with MgO contents exceeding 10 wt% and olivine with forsterite (Fo) contents ranging from 88 to 90 suggest a pyroxenitic mantle source contribution [12]. This implies that the mantle beneath the region is not solely peridotitic but also includes components derived from pyroxenite, indicating a more complex mantle composition and geodynamic history. The data suggest that KB1 basalts originate from a heterogeneous mantle source, comprising approximately 70% pyroxenite and 30% peridotite, and are influenced by asthenospheric upwelling associated with the rollback and break-off of the Cyprus slab [12].

The morphological diversity of maar eruptions and variations in magma compositions are pivotal in developing a comprehensive model of monogenetic volcanism, as highlighted by Uslular, Gençalioğlu-Kuşcu [41]. Their investigation of the Eğrikuyu Monogenetic Field (EMF), one of the five monogenetic volcanic clusters within the

Quaternary Central Anatolian Volcanic Province (CAVP), primarily focuses on scoria cones and a few maars. The study utilizes morphometric data from 77 scoria cones in the EMF to analyze their size distribution, revealing a power-law distribution based on base diameters. This distribution indicates that the cones formed through self-organized critical processes, providing insights into eruption mechanisms, and aiding in volcanic hazard assessment. The variability in scoria cone formation is attributed to factors such as magma composition, vent location, and environmental conditions. Furthermore, the study underscores the significance of understanding the morphometric parameters and spatial distribution of scoria cones, which are essential for estimating crustal thickness, determining relative ages, analyzing clustering and alignment, and elucidating the mechanisms of their occurrence. The relationship between fracture systems and the spatial distribution of scoria cones, often characterized by power-law distributions, parallels many natural processes, emphasizing the broader implications of this research.

4. Method

In this study, we aimed to elucidate the complex magma chamber processes in the Karapınar region by compiling the data from previous studies [11, 12, 14–16]. Additionally, we integrated new findings to enhance our understanding. Specifically, we have observed zircon grains from the Karapınar basalts and examined them by using scanning electron microscopy (SEM). SEM analysis and cathodoluminescence (CL) imaging were conducted at the Earth Sciences Institute of Heidelberg University (Germany) to determine the preliminary chemical composition and morphology of the zircon grains.

5. Textural and mineralogical properties of the rocks: Evidence of multiple-complex processes

The Karapınar Volcanic Field (KPVF) contains basalt to basaltic andesite lavas that display a porphyritic texture. These lavas are composed predominantly of olivine, clinopyroxene, plagioclase, and minor amounts of Fe-Ti oxides, with occasional quartz xenocrysts. Various types of enclaves, including magma mixing enclaves (MMEs), magma segregation enclaves (MSEs), and xenocrysts, are present within these rocks, offering critical insights into the underlying magmatic processes (**Figures 4** and 5) [15]. The mineralogical and petrographic characteristics of these rocks and their enclaves have been thoroughly detailed in previous studies [11, 12, 14–16].

Magma mixing enclaves (MMEs), found in the basaltic compositions within Meke Scoria, Meke Maar, and Yılan Maar, exhibit a fine-grained texture with elongated, bladed, and acicular microphenocrysts of plagioclase, alongside rounded pyroxene and olivine crystals. Some MMEs also include embayed-ocelli quartz minerals, indicating rapid cooling due to temperature differences between the mafic intrusions and the evolved host magma. Magma segregation enclaves (MSEs), composed of basalt, dunite, and clinopyroxenite, display equigranular textures that suggest prolonged mafic magma intrusion into the host magma. These enclaves often have chilled margins and vesicles, indicative of rapid quenching. The whole-rock composition of MSEs is generally more primitive than that of their host rocks, suggesting formation from more primitive magma or as early crystallized phases within the same magmatic

Figure 4.
Microphotographs of the investigated Karapınar basalts showing enclave types. (A-B) Cognate enclave, (C-D) quartz and biotite xenocrysts, and (E-F) magma mixing enclave. Figures taken from [15]. The images are as follows: images A, D, and E are taken under plain polarized light; while images B, C, and F are taken under cross-polarized light.

system [15]. Xenocrysts such as quartz, plagioclase, amphibole, and biotite, show varied textures and degrees of alteration. For example, amphibole and biotite are often bladed, sieved, and altered, while quartz xenocrysts are typically embayed and ocelli. Plagioclase xenocrysts range from coarse to fine-grained, sometimes exhibiting dusty sieving from core-to-rim and varying in size from phenocrysts to megacrysts. These xenocrysts are likely incorporated into the magma from the surrounding wall rock during magma ascent [15].

For basaltic compositions, the Putirka [42] clinopyroxene-melt geothermobarometer was applied to estimate pressure-temperature (P-T) conditions [16, 43]. Clinopyroxene-melt pairs were evaluated for equilibrium with a $Kd_{(Fe\text{-}Mg)}$ value of 0.27 ± 0.03. The results suggest that clinopyroxene phenocrysts in KPVF basalts crystallized at temperatures ranging from 1173 to 1251°C and pressures of 9.0–12.0 kbar, corresponding to depths of 25–45 km within the lower crust. For the enclaves (maar wall-enclave-bearing basaltic andesite) within the KPVF, the core-to-rim temperature and pressure ranges are estimated at 1212–1227°C and 9.8–9.0 kbar, respectively [16, 43].

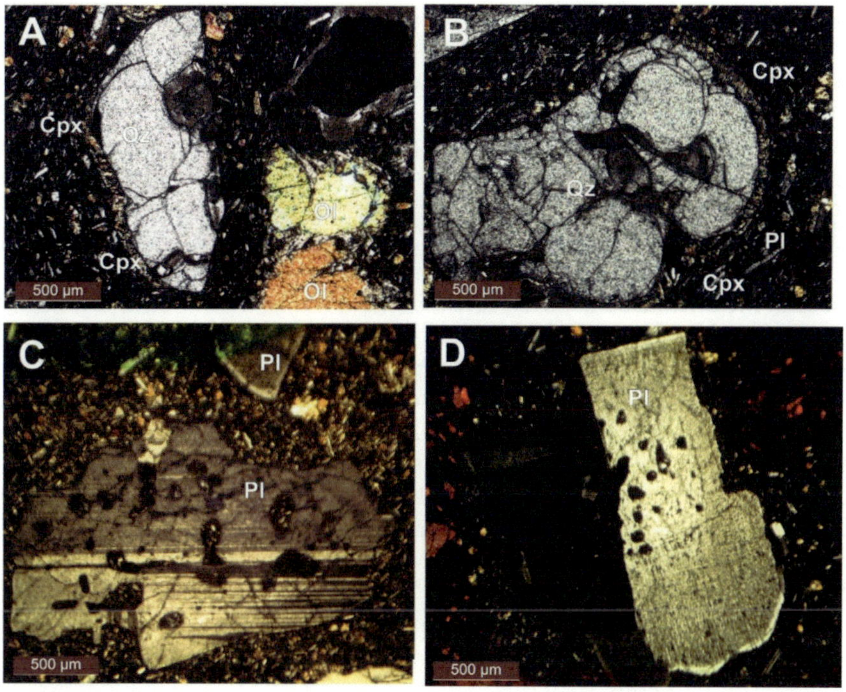

Figure 5.
(A-B) Embayed and ocelli quartz xenocrysts in the Karapınar Basalts (taken from Gencoğlu Korkmaz, Kurt [12] b) (C-D) coarse sieve-textured plagioclase megacrysts-xenocrysts from Karapınar basalts (taken from Gençoğlu Korkmaz and Kurt [14]). Microphotographs are taken under cross-polarized light.

The textural and chemical characteristics of the minerals provide vital clues about the magmatic history and environmental evolution of the region. Microscopic compositional variations in crystals are often linked to changes in the magmatic environment, reflecting complex processes, such as contamination, replenishment, and magma mixing [44–47]. For instance, the olivine crystals in basalts exhibit forsterite (Fo) contents ranging from Fo_{80} to Fo_{90}, with higher Fo values indicating early crystallization from more magnesian magmas [16]. Clinopyroxene crystals consist of augite and diopside and exhibit pronounced oscillatory zoning in elements such as Ca, Mg, Al, Ti, and Cr. Elevated TiO_2 levels are indicative of evolved magmas, whereas high Cr_2O_3 levels suggest early crystallization. These variations highlight multiple magmatic stages and processes, including magma mixing and recharge [16]. Plagioclases in the basalts from Meke exhibit anorthite (An) contents ranging from An_{60} to An_{40}, encompassing oligoclase, andesine, and bytownite. The relatively low An values observed in these plagioclases are more characteristic of more evolved, acidic magmas rather than the more common basaltic magmas, which typically show An values greater than An_{60}, such as those found in bytownite and anortite. This suggests that these plagioclase minerals likely originated from xenoliths or fragments of the country rock incorporated during magma ascent and convection processes, rather than being derived directly from the basaltic magma itself. The complex zoning patterns within plagioclase crystals reflect multiple episodes of crystallization and magma mixing [14]. The textural and chemical properties of the plagioclase phenocrysts, including reverse and oscillatory zoning in An, Mg, Fe, Mn, Sr., indicate significant recharging and assimilation processes within the magma chamber [14].

The presence of sieve textures and oscillatory zoning in plagioclase crystals suggests interactions with more primitive melts through magma mixing, self-mixing, and rapid decompression. Additionally, the presence of ocelli-quartz and embayed-quartz xenocrysts, along with biotite and amphibole xenocrysts, indicates that these components were likely plucked from the crust during magma ascent. The observation of sieve and cellular textures in phenocrysts like plagioclase further underscores the influence of rapid decompression and magma mixing processes. These textures, often associated with shallow magmatic environments, reflect the varying pressures and temperatures to which the magmas were subjected [48–50]. All of these processes contribute to the formation of texturally diverse crystals, highlighting the complex magmatic dynamics of the region.

In summary, the overall textural and compositional diversity observed in the Karapınar basalts is a direct result of complex magmatic processes, including magma mixing, self-mixing, and replenishment. These processes, driven by dynamic convection and new magma pulses within a shallow magma chamber, lead to the formation of heterogeneous crystal assemblages and diverse textures [48]. The evidence provided by the textural features, chemical zoning, mineral assemblages, and geothermobarometric data collectively supports the existence of a shallow, dynamic magma chamber in the Karapınar volcanic system. The dynamic activity within the shallow magma chamber at Karapınar, driven by convection and new magma pulses, results in repeated superheating events and the formation of heterogeneous crystal assemblages. These findings align with previous studies and enhance our understanding of the intricate magmatic processes occurring in this region.

6. New work on zircon chemical characteristics for complex magma chambers

Zircon ($ZrSiO_4$) is a robust accessory mineral commonly found in evolved igneous rock types, and has both chemical and physical durability. Given its ability to host geochemically significant indicator elements, such as rare earth elements (REE) and radioactive elements U-Th, zircon plays a crucial role in understanding the geological history of the Earth. Typically, zircon forms in acidic magmas and dissolves in

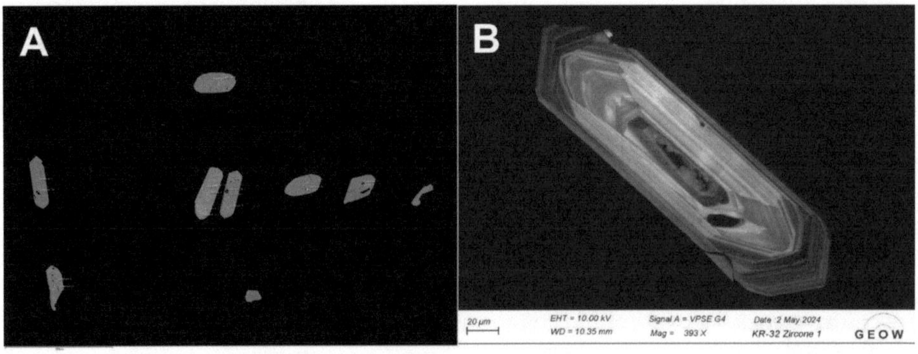

Figure 6.
Preliminary petrochemical work on zircon grains from Karapınar Basalts BSE images were generated by SEM at Heidelberg University Earth Sciences Institute (Germany). (A) BSE image shows some of the separated zircon grains, and (B) BSE image displays a selected zircon grain, highlighting concentric zoning patterns and their inclusions.

Sample/Zircon	**SiO_2**	**ZrO_2**	**HfO_2**	**ThO_2**	**UO_2**	**Total**					
KR-32-1	31.37	67.46	0.86	0.00	0.31	100.0					
KR-32-2	31.38	67.39	0.74	0.29	0.20	100.0					
KR-32-3	31.56	67.48	0.22	0.34	0.39	100.0					
KR-32-4	31.51	67.25	0.67	0.22	0.35	100.0					
KR-32-5 core	31.47	67.27	0.76	0.17	0.33	100.0					
KR-32-5 rim	31.61	65.76	1.02	0.45	1.16	100.0					
KR-32-6 core	31.46	66.69	0.77	0.47	0.61	100.0					
KR-32-6 rim	31.66	65.64	0.93	0.48	1.29	100.0					
KR-32-7 core	31.46	64.46	1.16	0.83	2.09	100.0					
KR-32-7 mid	31.38	66.24	0.93	0.38	1.07	100.0					
KR-32-7 mid	31.41	67.46	0.87	0.00	0.27	100.0					
KR-32-7 mid	31.50	63.97	1.22	0.88	2.43	100.0					
KR-32-7 rim	31.34	67.61	0.80	0.00	0.25	100.0					
Sample/Inclusion	Na_2O	Al_2O_3	SiO_2	Cl	K2O	CaO	TiO_2	MnO	FeO	BaO	Total
KR-32-1 inclusion glass	3.25	13.14	74.90	0.10	5.53	1.64	0.08	0.07	1.15	0.12	100.0
KR-32-3 inclusion glass	3.20	11.14	82.57		2.32	0.22	0.00		0.39	0.16	100.0
KR-32-4 inclusion glass	3.09	13.01	78.30	0.07	4.53	0.49	0.00		0.36	0.14	100.0

Table 1.
The first chemical results of the zircon grain from basaltic rocks in the KPVF.

mafic magmas due to its stability in high-temperature environments [51]. However, experimental studies suggest that zircons can survive short-term high-temperature conditions [51–53].

In the Karapınar basalts, the presence of zircon grains is particularly noteworthy, as zircon is rarely found in basaltic magmas (**Figure 6** and **Table 1**). The observation of zircon crystals in these high-temperature hybrid lavas is unexpected, given the typical conditions of magma mixing and recharge processes, which generally involve high temperatures. This finding suggests four possible scenarios: (1) The zircon crystals are xenocrysts originating from earlier stages of magmatic activity in the region; (2) they are crustal xenocrysts incorporated during magma ascent; (3) zircon crystals are preserved as inclusions within other phenocryst phases; or (4) the zircons formed as a result of mixing and hybridization events occurring shortly before the volcanic eruption. Further investigation into these scenarios will provide valuable insights into the dynamics of magma evolution and the preservation mechanisms of zircon crystals in hybrid volcanic rocks.

Melt inclusions are small droplets of melt trapped during crystallization at magmatic pressures and temperatures [54]. These inclusions represent juvenile melt enclosed by a growing crystal and can preserve the original temperature, pressure, and chemical composition of the magma during crystallization. Data from melt inclusions are used to interpret the history of magmatic systems [55]. The presence of melt inclusions in major rock-forming silicate minerals and their ease of identification and preservation in cleavage-free transparent minerals (such as quartz) make them widely useful in petrogenetic studies. Zircon, a common accessory mineral in silicic magmatic rocks, is found in a variety of geological settings and rock types (magmatic, metamorphic, and sedimentary), making it a suitable mineral for melt inclusion studies. However, accessory phases like zircon can significantly affect trace element concentrations in the melt phase during processes, such as fractionation, melt segregation, and/or retention in the residual melt [53, 56, 57]. Accurately determining trace element partition coefficients from melt inclusions in zircon is crucial for developing and testing models of melt formation and crystallization [56, 57]. The initial results reveal that the melt inclusions contain at least 74 wt% SiO_2, indicating that these zircons likely did not form within the basalts but were incorporated from a different environment during magma ascent. Further detailed trace element analysis of the zircons and their inclusions is required to ascertain their origin. This evidence supports the hypothesis of complex magma chamber processes in the region. Our future studies will aim to resolve this ambiguity by further elucidating the formation and preservation mechanisms of zircon crystals in hybrid volcanic systems.

7. Discussion

In the evolution of the investigated rocks, continental crustal contamination and magma mixing-encompassing repeated magma replenishment, self-mixing, and cryptic mixing-seem to play a more significant role than fractional crystallization. Additionally, dynamic processes within the magma chamber, such as the formation of enclaves with varying compositions and types, the recurrent development of compositional and textural zoning in crystals, and the emergence of instability textures, are believed to be crucial in triggering volcanic activity. These complex interactions highlight the importance of open-system magmatic processes in shaping the final characteristics of the rocks. In the previous study, we suggested that the Karapınar

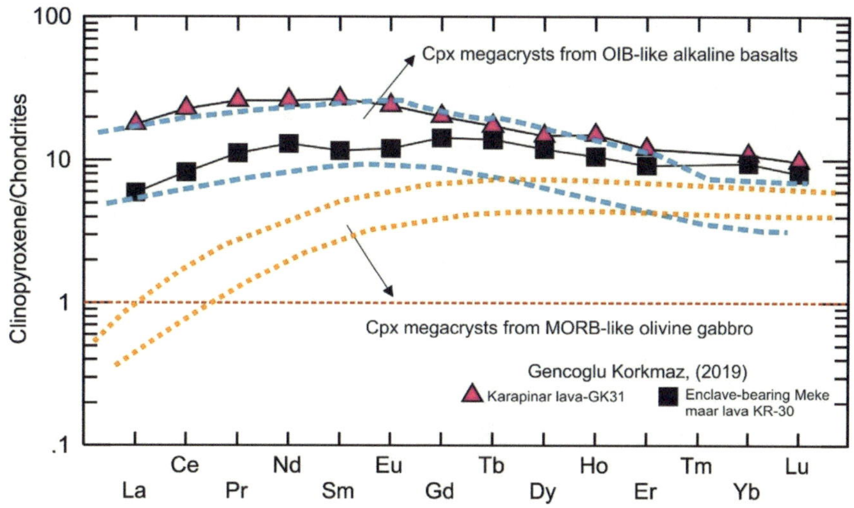

Figure 7.
Clinopyroxene trace element data of the basalts from the Karapınar area showing OIB basalts. Data were taken from Gençoğlu Korkmaz [16]. Typical compositions of MORB and OIB [58] are shown for comparison.

Basalts were contaminated by the Karacadağ Volcanics [11]. However, we now propose that this contamination extends beyond just the Karacadağ Volcanics. The size and abundance of quartz observed in the Karapınar Basalts are not characteristic of the Karacadağ Volcanics. Furthermore, the glass inclusions in zircon grains, with high silica content (>70 wt%), suggest contamination from quartz-rich crustal rocks in addition to the Karacadağ Volcanics, which are predominantly andesitic with only occasional trachytic-dacitic compositions (SiO_2 = 50–65 wt%) [11]. Therefore, we assert that crustal contamination has played a more significant role in the evolution of the Karapınar Basalts than previously anticipated.

The spatially mineral chemistry of olivine and pyroxene minerals, from core to rim, provides detailed and accurate insights into the source magmas and magma chamber processes. For instance, EPMA analyses of clinopyroxenes from the basalts of the Karapınar lavas confirm the OIB origin of the surface-exposed rocks in the area (**Figure 7**).

The chemical composition of olivine phenocrysts in the basalts from KPVF provides critical insights into the magmatic processes and source characteristics [12]. Olivines from the most primitive samples exhibit normal zoning, with core compositions of Fo_{89} transitioning to Fo_{86} at the rim, accompanied by a decrease in Ni and an increase in MnO and CaO contents [12]. In contrast, some of the olivines display both inverse and normal zoning patterns [12]. Enriched rims and growth zones in Li, Zn, Cr, Ti, Sc, and V contents in these olivines suggest the influence of recycled continental crust or magma recharging processes [40, 59]. The high Zn/Fe, Fe/Mn, and Co ratios, along with low Mn/Zn ratios in the olivines from the most primitive basalts in the area, indicate a pyroxenitic source contribution [6, 12, 40, 60].

8. Conclusion

The occurrence of plagioclase glomerocrysts, along with various enclaves and xenocrysts (e.g., quartz, biotite, and amphibole), suggests significant interactions between the ascending magma and the surrounding crustal materials. These xenocrysts were likely entrained during the ascent of magma from a shallow source. The presence of these components further supports the notion of a shallow magma chamber from which the basalts erupted. The geothermobarometric data, including clinopyroxene-melt geothermobarometry, provide temperature and pressure estimates consistent with conditions expected in shallow magma chambers. For instance, clinopyroxene phenocrysts in Karapınar basalts crystallized at temperatures between 1173 and 1251°C and pressures of 9.0–12.0 kbar. These conditions align with those found in lower crustal environments, typically associated with shallow magma chambers.

In conclusion, detailed spatial variations in elemental composition provide crucial insights into the intricate processes occurring within magma chambers, mantle sources, and the petrogenesis of the rocks. These variations, particularly when analyzed from the core to the rim of mineral phases, can reveal the history of magma evolution, including fractional crystallization, magma mixing, and contamination. The compositional zoning observed in minerals such as clinopyroxene and olivine can capture the dynamics of magma replenishment, crystallization conditions, the extent of crustal assimilation, source lithology, as well dynamics. Additionally, the study of trace elements and isotopic compositions within these minerals enables the distinction between lithospheric and asthenospheric mantle contributions, as well as the identification of cryptic processes like self-mixing and hybridization.

The interplay between internal magma dynamics and external environmental factors plays a crucial role in styling the eruptive behavior and geological outcomes of volcanic systems. Detailed spatial variations in elemental composition and mineral chemistry within the Karapınar Volcanic Field reveal the complexity of processes occurring within the magma chamber, including fractional crystallization, magma mixing, and crustal contamination. These processes, influenced by the chemical and physical properties of the magma, are further modulated by external factors such as the availability of external water, which can significantly impact eruption styles, as seen in the phreatomagmatic activity of the Meke Maar.

By integrating microscale and macroscale analyses, a comprehensive understanding of the region's geodynamic and magmatotectonic evolution emerges, highlighting the critical role of both internal magmatic conditions and external environmental influences in determining the volcanic architecture and hazards of the KPVF. The compositional diversity, particularly in the most primitive olivines, underscores the role of recharging events and decompression in the magmatic evolution of the Karapınar basalts. The presence of a heterogeneous mantle source, consisting of metasomatized peridotite and pyroxenite, beneath Anatolia is supported by the observed mineralogical and geochemical data, highlighting the complex interactions between different mantle lithologies in the petrogenesis of these basalts. This multi-scale approach not only enhances our understanding of volcanic activity but also provides broader insights into the Earth's interior dynamics and the mechanisms that govern the development and evolution of magmatic systems over geological timescales.

Author details

Gülin Gencoglu Korkmaz[1,2]* and Hüseyin Kurt[2]

1 Heidelberg University, Institute of Earth Sciences, Germany

2 Engineering and Natural Science Faculty, Geological Engineering Department, Konya Technical University, Türkiye

*Address all correspondence to: ggkorkmaz@ktun.edu.tr

References

[1] DePaolo DJ. Trace element and isotopic effects of combined wallrock assimilation and fractional crystallization. Earth and Planetary Science Letters. 1981;**53**(2):189-202

[2] Spera FJ. Open-system magma chamber evolution: An energy-constrained geochemical model incorporating the effects of concurrent eruption, recharge, variable assimilation and fractional crystallization (EC-E'RA FC). Journal of Petrology. 2004;**45**(12):2459-2480

[3] Zellmer GF, Annen C. An Introduction to Magma Dynamics. Vol. 304(1). London: Geological Society, Special Publications; 2008. pp. 1-13

[4] Ubide T et al. Antecrysts and their effect on rock compositions: The cretaceous lamprophyre suite in the Catalonian coastal ranges (NE Spain). Lithos. 2014;**206-207**:214-233

[5] Ubide T et al. High-resolution LA-ICP-MS trace element mapping of igneous minerals: In search of magma histories. Chemical Geology. 2015;**409**:157-168

[6] Jankovics MÉ et al. Olivine major and trace element compositions coupled with spinel chemistry to unravel the magmatic systems feeding monogenetic basaltic volcanoes. Journal of Volcanology and Geothermal Research. 2019;**369**:203-223

[7] Petrelli M, Laeger K, Perugini D. High spatial resolution trace element determination of geological samples by laser ablation quadrupole plasma mass spectrometry: Implications for glass analysis in volcanic products. Geosciences Journal. 2016;**20**(6):851-863

[8] Spera FJ et al. Dynamics of magma withdrawal from stratified magma chambers. Geology. 1986;**14**(9):723-726

[9] Van Hinsbergen DJ et al. Tectonic evolution and paleogeography of the Kırşehir block and the central Anatolian ophiolites, Turkey. Tectonics. 2016;**35**(4):983-1014

[10] Ulu Ü. Geological Maps of Turkey Karaman-M30 Sheet. Ankara, Turkey: General Directorate of Mineral Research and Exploration (in Turkish); 2009

[11] Gencoğlu Korkmaz G et al. Ar-Ar geochronology and Sr-Nd-Pb-O isotopic systematics of the post-collisional volcanic rocks from the Karapinar-Karacadag area (Central Anatolia, Turkey): An alternative model for orogenic geochemical signature in sodic alkali basalts. Journal of Geosciences. 2022;**67**(1):53-69

[12] Gencoğlu Korkmaz G et al. The role of peridotite and pyroxenite melts in the origin of the Karapinar basalts, Cappadocia Volcanic Province, Central Anatolia. Journal of Geosciences. 2022;**67**(4):311-329

[13] Keller J. Quaternary maar volcanism near Karapinar in Central Anatolia. Bulletin Volcanologique. 1974;**38**(1):378-396

[14] Gençoğlu Korkmaz G, Kurt H. Interpretation of the magma chamber processes with the help of textural stratigraphy of the plagioclases (Konya-Central Anatolia). European Journal of Science and Technology. 2021;**25**:222-237

[15] Gençoğlu Korkmaz G, Kurt H, Asan K. Classification and generation

of the enclaves in Karapınar-Karacadağ volcanic rocks (Central Anatolia). Turkish Journal of Geosciences. 2021;**2**:30-46

[16] Gençoğlu Korkmaz G. The Origin of the Enclaves in Emirgazi (Konya) and its Surrounding Volcanites and their Petrological Importance in the Genesis of the Regional Volcanism in Konya Technical University Graduate Education Institute. Turkey: Konya Technical University (in Turkish); 2019. p. 321

[17] Smith I, Németh K. Source to Surface Model of Monogenetic Volcanism: A Critical Review. London: Geological Society, Special Publications; 2017

[18] Németh K, Kósik S. Review of explosive hydrovolcanism. Geosciences. 2020;**10**(2):44

[19] Németh K. Phreatomagmatic volcanism at the Waipiata volcanic field, Otago, New Zealand [thesis]. New Zealand: University of Otago; 2001

[20] Türkecan A. Türkiye'nin Senozoyik volkanitleri. Ankara, Turkey: Maden Tetkik Arama Müdürlüğü (in Turkish); 2015

[21] Sür Ö. Türkiyenin, özellikle iç Anadolunun genç volkanik alanlarının jeomorfolojisi. Ankara: Ankara University Printing House; 1972

[22] Toprak V. Vent distribution and its relation to regional tectonics, Cappadocian Volcanics, Turkey. Journal of Volcanology and Geothermal Research. 1998;**85**(1):55-67

[23] Besang C et al. Radiometric age determinations on Neogene igneous rocks in Turkey. Geological Yearbook (in German). 1977;**25**:3-36

[24] Batum I. Geology and petrography of Acıgöl and Göllüdag volcanics at southwest of Nevsehir Central Anatolia, Turkey. Yerbilimleri. 1978;**4**:50-69

[25] Notsu K, Fujitani T, Ui T, Matsuda J, Ercan T. Geochemical features of collision-related volcanic rocks in central and eastern Anatolia, Turkey. Journal of Volcanology and Geothermal Research. 1995;**64**:171-192

[26] Ercan T. Orta Anadolu'daki Senozoyik Volkanizmasi. Maden Tetkik Ve Arama Dergisi. 1987;**107**:119-140

[27] Ercan T et al. Hasandagı-Karacadag (Orta Anadolu) dolaylarındaki Senozoyik yaslı volkanizmanın kökeni ve evrimi. Jeomorfoloji Dergisi. 1990;**18**:39-54

[28] Ercan T et al. Hasandagı–Karacadag (Orta Andolu) Kuvaterner volkanizmasına iliskin yeni jeokimyasal, izotopik ve radyometrik veriler. New geochemical, isotopic and radiometric data of quaternary volcanism of Hasandagı-Karacadag (Central Anatolia). Geological Bulletin of Turkey. 1992;**7**:8-21

[29] Temel A, Gundogdu MN, Gourgaud A. Petrological and geochemical characteristics of Cenozoic high-K calc-alkaline volcanism in Konya, Central Anatolia, Turkey. Journal of Volcanology and Geothermal Research. 1998;**85**:327-354

[30] Olanca K. Quaternary Volcanism of Karapınar-Konya Region: Geochemical Evolution. Turkey: Bulletin of Earth Sciences Application and Research Center, Hacettepe University; 1999. pp. 115-124

[31] Güllü B, Kadıoğlu YK. The petro-chemical properties of Meke and Acigol (Karapinar-Konya) volcanites. PAU Journal of Engineering Sciences. 2019;**25**(3):325-335

[32] Reid MR et al. Shallow melting of MORB-like mantle under hot continental lithosphere, Central Anatolia. Geochemistry, Geophysics, Geosystems. 2017;**18**(5):1866-1888

[33] McNab F et al. Neogene uplift and magmatism of Anatolia: Insights from drainage analysis and basaltic geochemistry. Geochemistry, Geophysics, Geosystems. 2018;**19**(1):175-213

[34] Rabayrol F, Hart CJR, Thorkelson DJ. Temporal, spatial and geochemical evolution of late Cenozoic post-subduction magmatism in central and eastern Anatolia, Turkey. Lithos. 2019;**336-337**:67-96

[35] Uslular G, Gençalioğlu-Kuşcu G. Mantle source heterogeneity in monogenetic basaltic systems: A case study of Eğrikuyu monogenetic field (Central Anatolia, Turkey). Geosphere. 2019;**15**(2):29

[36] Uslular G, Gençalioğlu-Kuşcu G. Geochemical characteristics of Anatolian basalts: Comment on "Neogene uplift and magmatism of Anatolia: Insights from drainage analysis and basaltic geochemistry" by McNab et al. Geochemistry, Geophysics, Geosystems. 2019;**20**(1):530-541

[37] Sobolev AV et al. The amount of recycled crust in sources of mantle-derived melts. Science. 2007;**316**(5823):412-417

[38] Gurenko AA et al. A composite, isotopically-depleted peridotite and enriched pyroxenite source for Madeira magmas: Insights from olivine. Lithos. 2013;**170-171**:224-238

[39] Zhang Y-H et al. Differential partial melting process for temporal variations of Shandong basalts revealed by melt inclusions and their host olivines. Gondwana Research. 2017;**49**:205-221

[40] De Hoog JCM, Gall L, Cornell DH. Trace-element geochemistry of mantle olivine and application to mantle petrogenesis and geothermobarometry. Chemical Geology. 2010;**270**(1):196-215

[41] Uslular G, Gençalioğlu-Kuşcu G, Arcasoy A. Size-distribution of scoria cones within the Eğrikuyu monogenetic field (Central Anatolia, Turkey). Journal of Volcanology and Geothermal Research. 2015;**301**:56-65

[42] Putirka K. Clinopyroxene Thermobarometers. 2008. Available from: http://www.fresnostate.edu/csm/ees/faculty-staff/putirka.html#downloads [Accessed: March 14, 2018]

[43] Kurt H, Korkmaz GG, Asan K. Textural and mineral chemistry data related to the Karapinar-Karacadag volcanic complex (Konya-Central Anatolia). In: Petrogenesis and Exploration of the Earth's Interior. Hammamet, Tunisia: Springer International Publishing; 2019. pp. 45-47

[44] Ubide T et al. The relevance of crystal transfer to magma mixing: A case study in composite dykes from the Central Pyrenees. Journal of Petrology. 2014;**55**(8):1535-1559

[45] Huppert HE et al. Replenished magma chambers: Effects of compositional zonation and input rates. Earth and Planetary Science Letters. 1982;**57**(2):345-357

[46] Streck MJ et al. The role of basalt replenishment in the generation of basaltic andesites of the ongoing activity at Arenal volcano, Costa Rica: Evidence from clinopyroxene and spinel. Bulletin of Volcanology. 2002;**64**(5):316-327

[47] Streck MJ. Mineral textures and zoning as evidence for open system

processes. Reviews in Mineralogy and Geochemistry. 2008;**69**(1):595-622

[48] Renjith ML. Micro-textures in plagioclase from 1994-1995 eruption, Barren Island volcano: Evidence of dynamic magma plumbing system in the Andaman subduction zone. Geoscience Frontiers. 2014;5(1):113-126

[49] Tepley FJ III, Davidson JP, Clynne MA. Magmatic interactions as recorded in plagioclase phenocrysts of chaos crags, Lassen volcanic center, California. Journal of Petrology. 1999;**40**(5):787-806

[50] Streck MJ et al. Plagioclase populations and zoning in dacite of the 2004-2005 Mount St. Helens eruption: Constraints for magma origin and dynamics. In: Sherrod DR, Scott WE, Stauffer PH, editors. Chapter 34 in a Volcano Rekindled: The Renewed Eruption of Mount St. Helens, 2004-2006. Reston, VA: U.S. Geological Survey Professional Paper; 2008. pp. 791-808

[51] Watson EB. Dissolution, growth and survival of zircons during crustal fusion: Kinetic principals, geological models and implications for isotopic inheritance. Earth and Environmental Science Transactions of the Royal Society of Edinburgh. 1996;**87**(1-2):43-56

[52] Boehnke P et al. Zircon saturation re-revisited. Chemical Geology. 2013;**351**:324-334

[53] Luo Y, Ayers JC. Experimental measurements of zircon/melt trace-element partition coefficients. Geochimica et Cosmochimica Acta. 2009;**73**(12):3656-3679

[54] Roedder E. Origin and significance of magmatic inclusions. Bulletin de Mineralogie. 1979;**102**(5):487-510

[55] Thomas R, Webster JD, Davidson P. Understanding Pegmatite Formation: The Melt and Fluid Inclusion Approach. Australia: University of Tasmania; 2006

[56] Thomas J et al. Determination of zircon/melt trace element partition coefficients from SIMS analysis of melt inclusions in zircon. Geochimica et Cosmochimica Acta. 2002;**66**(16):2887-2901

[57] Thomas J et al. Melt inclusions in zircon. Reviews in Mineralogy and Geochemistry. 2003;**53**(1):63-87

[58] Sun S, McDonough W. Chemical and Isotopic Systematics of Oceanic Basalts: Implications for Mantle Composition and Processes. Vol. 42(1). London: Geological Society, Special Publications; 1989. pp. 313-345

[59] Gómez-Ulla A, Sigmarsson O, Gudfinnsson GH. Trace element systematics of olivine from historical eruptions of Lanzarote, Canary Islands: Constraints on mantle source and melting mode. Chemical Geology. 2017;**449**:99-111

[60] Søager N et al. Olivine major and trace element compositions in southern Payenia basalts, Argentina: Evidence for pyroxenite-peridotite melt mixing in a back-arc setting. Journal of Petrology. 2015;**56**:1495-1518

Section 2

Complex Geophysical Approaches

Chapter 5

Imaging Subvolcanic Structures by Magnetotellurics

Özlem Hacıoğlu

Abstract

The generation of magma beneath volcanic edifices is a long-lived dynamic phenomenon that shapes the volcanic environment and can trigger eruptions. Analog and theoretical modeling are valuable tools used in the study of volcanic areas to better understand the processes driving volcanic activity, the behavior of volcanic systems, and the formation of volcanic structures. However, results obtained from analog and theoretical models and field studies need to be critically assessed against those obtained from other methods. Magnetotellurics (MT), a geophysical method that allows for the detailed mapping of electrical resistivity variations in the subsurface, provides reliable images of the interior structure of volcanoes (i.e., subvolcanic structures such as magma reservoirs, plumbing systems, hydrothermal systems, and collapsed calderas) based on the measurement of naturally occurring electric and magnetic field variations over time. This method is used not only to unveil the deep structure of volcanoes and prospect geothermal systems or ore deposits but also to assess volcanic hazards related to the size and depth of the magmatic source. This chapter presents a thorough examination of how MT helps reveal the hidden complexities beneath volcanoes, suggesting insights into magmatic processes, hydrothermal systems, structural controls, and the overall geoelectric architecture of volcanic edifices.

Keywords: volcano, magma reservoir, melt, hydrothermal system, geophysics, magnetotellurics, inversion

1. Introduction

Volcanoes are among Earth's most dynamic and fascinating geological features. Their ability to erupt and shape landscapes, coupled with the generation of magma beneath their edifices, makes them significant to both scientific communities and human societies, driving the need for comprehensive research and effective monitoring to mitigate potential hazards and better understand fundamental Earth processes.

A magma reservoir as the domains within the magmatic system encompasses regions where partially or wholly molten rock with varying proportions of melt, crystals, and exsolved volatiles extends from the source of magma generation to the surface. They may comprise zones with a very low fraction of intergranular melt, partially molten rock (mush), and melt lenses (or magma chambers) containing a high fraction of eruptible magma, along with pockets of exsolved magmatic fluids [1]. The parts of the magma reservoirs characterized by a significant abundance of crystals, sufficient to establish a semi-rigid structure, are termed as mush (e.g., [1, 2]). Melt pockets with a very low (or zero) crystal fraction may also exist in magma reservoirs. These bodies can form through various processes, namely melt extraction from crystal mushes due to compaction [3–5], convective or diffusive exchange between melts, or gas-driven filter pressing [6–8]. Alternatively, they can form by super-liquidus melts being injected into more crystal-rich material, such as a mush or solid country rock, which then cool and crystallize [2]. The magma reservoirs may extend vertically through the entire crust and perhaps beyond, referred to as transcrustal magmatic systems (i.e., transcrustal mush systems) (**Figure 1**), and they may evolve through the presence of multiple, stacked magma storage zones separated by sub-solidus materials [1, 9, 10]. However, the significant volume of melt is not prevalent within the crust; instead, it may be distributed within mushy regions in a magmatic system and extracted shortly before volcanic eruptions [1, 2, 9]. In some systems, magma may also cool and solidify within the Earth's crust, resulting in the formation of intrusive igneous rocks (i.e., plutons or batholiths), while in other magmatic systems, magma may erupt to form volcanoes.

The different domains that can encompass several multi-component phases (i.e., crystalline solids, melts, and fluids), namely magma (melt ± crystals and exsolved fluids), mush (melt + associated fluids within a crystalline framework), super-solidus (including unconnected melt pockets), and sub-solidus host rocks (no melt), interact

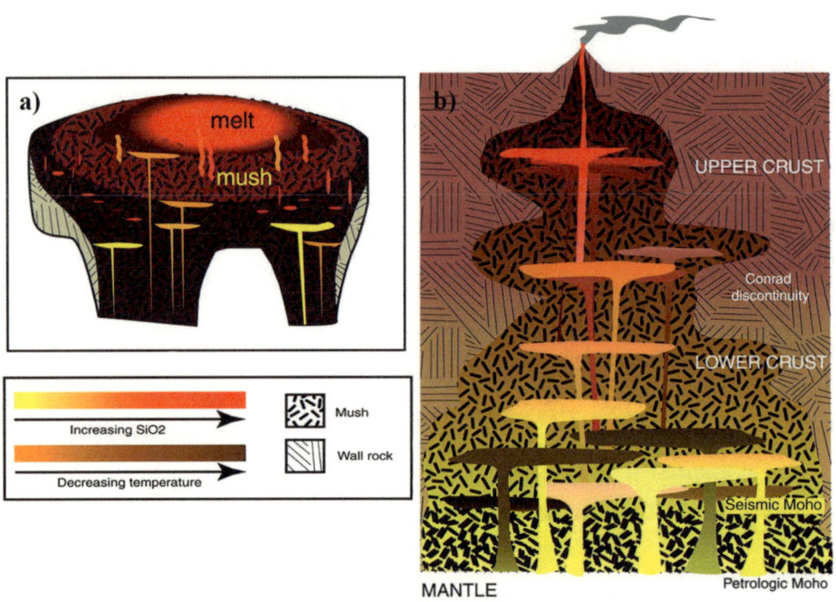

Figure 1.
An upper crustal magmatic system where silicic melts separate from underplated mafic magma, which is intermittently replenished from greater depths (a). A thermally mature transcrustal magmatic system where melt processing in the deep crust generates melts that are subsequently transferred to mid and upper crustal levels (b) (modified from Ref. [9]).

in complex ways and exhibit distinct rheological and physical properties in magmatic systems (e.g., [1, 11–13]). The host rocks, consisting of both pre-existing country rocks and fully solidified intrusions of the magmatic system, are affected by the transfer of magmatic heat, the influx of fluids originating from magma, and stress induced by magma. On the other hand, the external environment also plays a crucial role in shaping magmatic systems, as evidenced by the impact of tectonic processes and the cooling effects of convecting meteoric water [1]. The physical properties and dynamic behaviors of various domains (i.e., magma, mush, and host rocks) within magmatic systems are discussed by Spark et al. [1]. The transition from mush to magma and vice versa can occur depending on melt content or strain rate. Mush rheology is primarily dominated by the deformation of the crystalline framework, while in magma, it is predominantly controlled by the deformation of the melt. Particularly, interconnected melt friction significantly reduces shear viscosity and influences the mechanical behaviors in magmatic systems. The timescales on which reservoir processes occur also vary widely, ranging from the development of igneous systems (10^5–10^7 years) to the rapid separation of volatiles during explosive eruptions (tens of seconds) (e.g., [1]).

Hydrothermal systems are a key component of volcanic environments. These systems incorporate the circulation of hot water and magmatic fluids within the crust of the Earth, which may result in the formation of geothermal resources and mineral deposits. The interaction between geothermal fluids and the surrounding rocks can alter the physical and chemical properties of rocks, generating low-resistivity zones detectable by geophysical methods (e.g., [14, 15]). Understanding hydrothermal systems is therefore essential for comprehending the full dynamics of volcanic activity, including the interactions between magma and groundwater, the transport of heat and chemical elements, and the triggering of volcanic eruptions.

Geophysical methods are significant tools for imaging subvolcanic structures and gaining insights into the complex magmatic systems beneath volcanoes. These methods help scientists to infer the distribution and physical properties of subsurface materials, which are essential for understanding volcanic processes and assessing potential hazards. Among the various geophysical techniques, magnetotellurics (MT) have proven to be particularly effective in probing the deep structure of the Earth's crust and mantle, providing valuable insights into the presence of magma storage regions, geothermal reservoirs, and collapsed caldera structures (e.g., [14–20]). Accordingly, the method has been applied to several volcanoes, such as the Tongariro Volcanic Complex in the central North Island of New Zealand, the Jingpohu monogenetic volcanic field in northeast China, the Karadağ stratovolcano in Turkey, and Teide volcano in the Canary Islands, Spain, to identify low-resistivity zones, which typically indicate the presence of magma storage regions and hydrothermal systems and provide insights into the current state of the volcanoes, and their potential future behaviors (e.g., [21–24]).

This chapter aims to present how complex subvolcanic structures, including magma reservoirs, hydrothermal systems, volcanic plumbing, and collapsed calderas, are comprehensively revealed through MT imaging. The chapter details the use of MT in imaging subvolcanic features and assessing volcanic hazards, highlighting its effectiveness. Through case studies and a discussion of challenges and limitations, the chapter will emphasize the critical role of MT in advancing our understanding of volcanic processes and enhancing our ability to predict and mitigate volcanic hazards.

2. Basics of magnetotellurics

The MT method is a natural-source (passive), frequency-domain electromagnetic method used to image the electrical resistivity structure of the Earth from the near-surface down to the transition zone (~410 km) and beyond [25]. The method is based on measuring natural variations in the Earth's magnetic and electric fields to infer the electrical resistivity of subsurface materials. The frequency range of these natural electromagnetic (EM) fields extends from 10^{-4} to 10^{20} Hz [26], but the frequency range of interest in MT is from 10^{-4} to 10^{4} Hz. Frequencies below 1 Hz are generated by interactions between the Earth's magnetosphere and solar winds, while frequencies above 1 Hz are produced by meteorological activity, such as lightning discharges. The transition between these sources, around 1 Hz, is known as the MT dead-band, where the intensity of natural EM fields is very low [27].

The MT method measures time-varying EM fields at the Earth's surface, relying on the Earth's static magnetic field. The measured EM fields are a combination of primary fields from external sources and secondary fields from induced (telluric) currents. The method is governed by Maxwell's equations, characterizing the induction and propagation of EM waves in a magnetizable and polarizable medium without its own electric or magnetic source [28–30].

The Earth in the MT method is assumed to be a linear time-invariant system, where the transfer function (i.e., the Earth) characterizes a linear relationship between an input and a predictable output (i.e., a linear relationship between the measured EM field components) in the frequency domain. These transfer functions depend only on the electrical properties of the medium (i.e., independent from the EM source fields), identifying the resistivity distribution based on the measured frequency range. Therefore, the linear relationship between orthogonal electric (E_x, E_y) and magnetic (H_x, H_y, H_z) field components can be written as a tensor equation, complex impedance tensor **Z**, in the frequency domain:

$$\begin{bmatrix} E_x(\omega) \\ E_y(\omega) \end{bmatrix} = \begin{bmatrix} Z_{xx}(\omega) & Z_{xy}(\omega) \\ Z_{yx}(\omega) & Z_{yy}(\omega) \end{bmatrix} \begin{bmatrix} H_x(\omega) \\ H_y(\omega) \end{bmatrix} \tag{1}$$

where x and y denote the orthogonal components in horizontal directions, and ω is the angular frequency of the source signal: $\omega = 2\pi f$ [25]. $\mathbf{Z}$, a frequency-dependent second rank tensor, can relate the fields to Maxwell's equations and contains frequency-dependent information about the Earth (i.e., the electrical resistivity distribution of subsurface), where lower frequencies typically sample greater depths. For example, wide-band MT covers periods ranging from 10^{-3} to 10^{3} s and is capable of imaging subsurface structures at depths of a few kilometers to tens of kilometers. By using the above relation, the apparent resistivity (ρ_a) or the average resistivity of an equivalent half space and impedance phase (ϕ) at a particular frequency f can be estimated:

$$\rho_{aij} = \frac{1}{\omega\mu_0} \left| Z_{ij} \right|^2 \tag{2}$$

$$\phi_{ij} = \tan^{-1}\left(\frac{Im(Z_{ij})}{\mathrm{Re}(Z_{ij})} \right) \tag{3}$$

where μ_0 is the free-space value of magnetic permeability.

Additionally, the complex geomagnetic transfer function (or tipper, T), sensitive to the lateral resistivity contrasts, defines the relationship between the horizontal (H_x, H_y) and vertical (H_z) magnetic field components:

$$H_z = \begin{pmatrix} T_x & T_y \end{pmatrix} \begin{pmatrix} H_x \\ H_y \end{pmatrix} \tag{4}$$

More detailed information on the theory of MT can be found in Refs. [25, 27, 31, 32].

MT in volcanic settings provides an enormous amount of information from the electrical resistivity and has led to insights about the presence of fluids and melts as well as their pathways beneath the volcanic areas. The MT method is sensitive to variations in temperature, composition, and fluid content, which makes it significant for mapping magma reservoirs and associated structures, such as the presence of molten rocks, partial melt, and hydrothermal fluids, as well as for delineating the boundaries of different geological units (e.g., [14, 15, 17, 18, 33–37]).

3. Imaging magma storage by MT

Magma storage regions within the Earth's crust and mantle play a critical role in volcanic activity, the evolution of geothermal systems, and tectonic processes. Identifying the location and physico-chemical properties of magma reservoirs is also crucial for understanding the structure and dynamics of magmatic systems, providing insights into assessing hazards, exploring resources, and advancing scientific research. Geophysical imaging, particularly MT, helps to construct an improved image of the interior structure of volcanoes and their associated subvolcanic systems at a regional scale and validates the results obtained from analog and mathematical modeling, and field observations. Accordingly, several MT studies in volcanic environments have identified low-resistivity zones interpreted as magma storage regions containing melts (e.g., andesitic and dacitic melts) at depth (e.g., [17, 23, 34, 36, 38–45]).

The identification of magma reservoirs at depth using MT provides essential constraints on their geometry (i.e., shape and size) and, thus, the current thermodynamic state and possible future behaviors of volcanic edifices. The resolvable wavelength for a magmatic system is directly proportional to the period range of the collected MT data. Shallow-seated magma reservoirs, characterized by upper crustal low-resistivity zones from the surface to a few kilometers depth (<10 km), have been imaged by MT beneath various volcanic edifices around the world (e.g., [35, 43]). For example, at the Karadağ stratovolcanic complex, one of the young (< 45 ka) dormant stratovolcanoes in the Central Anatolian Volcanic Province (Türkiye), which comprises the north-northeast section of a paleomagnetic arc related to the Cyprus subduction zone, the electrical resistivity structure (**Figure 2**), derived from the three-dimensional (3D) inversion of broad-band MT data collected at 124 stations, reveals a deep and large zone of low resistivity (<10 Ωm). This zone (C3) has been associated with a shallow-seated andesitic magma reservoir composed of a near-solidus silicic crystal mush (i.e., a low melt fraction of silicic crystal mush), with the support of geochemical data [46], within the upper crust at a depth of 3 to 7 km beneath the volcano [23]. Petrological studies have suggested that andesitic melts may occur at upper crustal depths (<~6 km) (e.g., [47, 48]). The inversion of MT data from the intraplate monogenetic volcanic fields in northeast China and in central Mongolia has also shown that intraplate volcanoes are characterized by low-percent partial

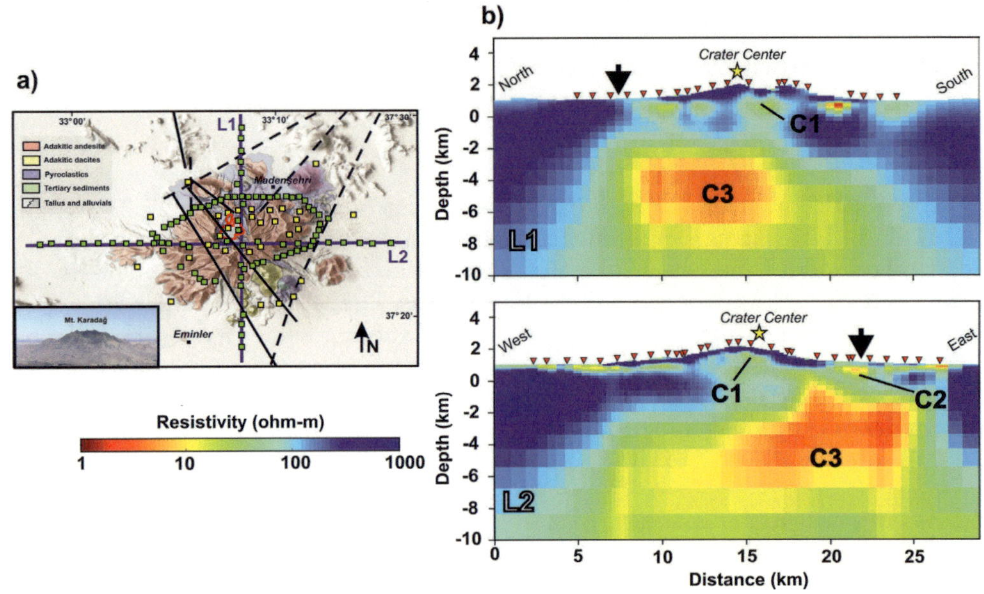

Figure 2.
Simplified geological map of the Karadağ stratovolcano (Central Anatolia, Türkiye), showing the locations of MT sites (green and yellow rectangles) (a). The solid black lines represent observed faults, while the broken lines represent interpreted faults. The red line marks the position of the crater center. The resistivity cross sections derived from the 3D inversion model at profile locations shown on the map (L1: North-south profile, L2: West-east profile) (b). The yellow star indicates the position of the crater center, and the red-inverted triangles mark the locations of the MT sites. The low-resistivity areas (C1, C2, and C3) are labeled in the sections (modified from Ref. [23]).

melt and low magma supply rates [22, 34]. At Socompa volcano, one of the Holocene volcanoes of the Central Volcanic Zone (CVZ) of the Andes, a possible shallow dacitic magma reservoir characterized by a low-resistivity area with resistivity values of 6–15 Ωm between 2 and 7 km depth below the surface of the volcano, has been imaged by 3D inversion of MT data [49]. This low-resistivity body is attributed to the magmatic heat source of the volcano and the associated hydrothermal system. Magma reservoirs at shallow crustal depths have also been identified in other stratovolcanoes of the Central Volcanic Zone of the Andes, such as Lastarria volcano [40], the San Pedro-Linzor Volcanic Chain [50], and Uturuncu volcano [51, 52].

The bulk resistivity of magma reservoirs is strongly sensitive to melt fraction and its geometry, the composition of the melt (mainly H_2O and Na_2O content), phase assemblages, as well as temperature and pressure conditions (e.g., [53–56]). The upper crustal magma reservoirs beneath the volcanic edifices may be linked to deeper crustal magma reservoirs that contain partially molten rocks (i.e., mush). A low-resistivity region at shallow depths can thus be explained by saline fluids of magmatic origin (and/or meteoric origin) (e.g., [41, 49, 51, 52]). Afanasyev et al. [57] have proposed that brines (i.e., H_2O-NaCl fluids) exsolved from ascending magmatic fluids through high-permeability pathways, such as volcanic conduits, dikes, and fractures, beneath active and dormant volcanoes can be responsible for the emplacement of magmatic bodies in the shallow crust. Guo and Keppler [58], by performing laboratory-based electrical conductivity measurements, have shown that the electrical conductivity of aqueous fluids generally increases with NaCl concentration. Pommier et al. [54, 55] have also proposed that silicate melts might exhibit higher resistivity compared to NaCl aqueous solutions circulating through the pores and fractures of rocks. Similarly, results from 3D modeling of MT data at the Lastarria volcano in northern Chile demonstrate two low-resistivity areas interpreted as crustal magmatic reservoirs (i.e., magmatic heat sources). One (~5–10 Ωm) located to the south of Lastarria volcano at shallow depths (7–8 km) is fed by a mid-crustal reservoir (~1 Ωm) extending to a depth of 15 km beneath the volcano [40]. The lower crustal low-resistivity zones (50–70 Ωm) in central Mongolia containing continental intraplate volcanism are also associated with areas including small amounts of saline fluids formed by metamorphic dehydration and devolatilization reactions in a thermally perturbed lower crust [34].

Previous broad-band and long-period MT studies in volcanic environments have determined the presence of low-resistivity zones in the mid-to-lower crust and upper mantle. These deep-seated anomalous regions have been generally associated with magma reservoirs, including melt phases, magma pathways, and crustal magma plumbing systems beneath volcanic fields (e.g., [17, 34, 42, 44, 45, 52, 59, 60]). In the East Anatolian Collision Zone (northeastern Turkey), formed by the collision of the Arabian Plate with Eurasia, two distinct low-resistivity zones, which probably

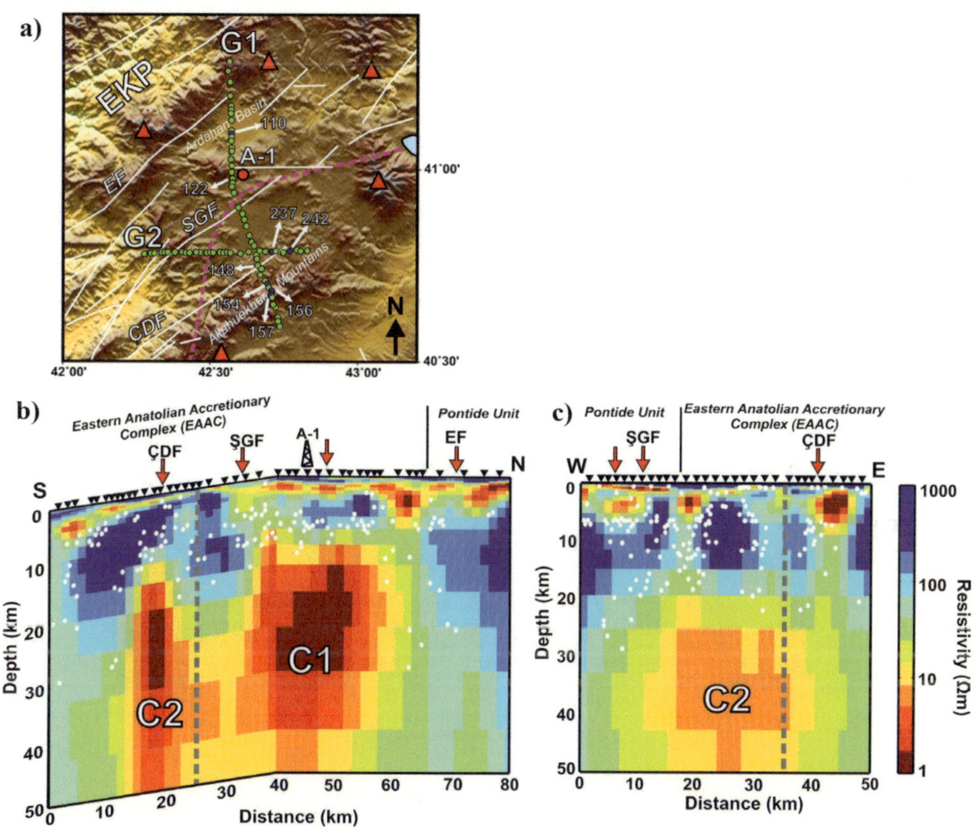

Figure 3.
Topographic map of the Erzurum-Kars Plateau (Northeastern Türkiye) superposed with fault branches (white solid lines) (a). Yellow and blue dots indicate the locations of MT sites. Red triangles mark volcanic centers, and the red dot represents the position of the borehole (A-1). The MT profiles are labeled G1 (a roughly north-south directed profile) and G2 (an east-west directed profile). The resistivity cross sections G1 (b) and G2 (c) are derived from the 3D model. Red arrows show the surface traces of faults. The gray broken line denotes the intersection of the profiles, and the white dots mark the hypocenters of earthquakes. C1 and C2 are mid-to-lower crustal low-resistivity zones (modified from Ref. [17]).

indicate local accumulation of melt at mid-to-lower crustal depths (15–45 km) throughout the Erzurum-Kars Plateau, were identified (**Figure 3**) by Hacıoğlu et al. [17]. These low-resistivity zones have been previously imaged further west of the collision zone [38], suggesting that the electrical characteristics of the collision zone are continuous from west to east beneath the Anatolian Plateau. Additionally, the spatial extent of these zones may indicate the presence of two localized crustal flow channels situated laterally parallel to the orogeny. By combining land and lake-bottom MT measurements, Bertrand et al. [44] characterized a singular mid-crustal low-resistivity zone associated with a complex of relatively mafic magma, consistent with petrologic and other geophysical data. Additionally, discrete low-resistivity anomalies, interpreted as plumes, rise from the margins of the magmatic complex and connect with near-surface geothermal systems.

The 3D resistivity model at the Central Andes, which currently hosts the world's largest zone of silicic partial melt in the form of the Altiplano-Puna Magma (or Mush) Body and the Southern Puna Magma Body, characterized by the relationship between volcanism and plutonism, indicates distinct electrically low-resistivity zones above 10 km below sea level beneath Uturuncu, above the Altiplano-Puna Magma Body. This implies that magma and/or fluids are moving upward through pathways [41]. Intracrustal magma reservoirs beneath Lazufre, fed by ascending melts, have also been attributed to a large volume magmatic reservoir (i.e., the Southern Puna Magma Body) at mid-crustal depths [40, 41]. However, as part of a mature transcrustal magmatic system, the electrical resistivity structure is more complex at Uturuncu. **Figure 4** shows the cross section of the 3D resistivity model with the deep low-resistivity region attributed to the Altiplano-Puna Magma Body (~1 Ωm) and the shallow low-resistivity zone interpreted the presence of a hydrothermal system developed in the region of fossil dacite magma chamber (~3–10 Ωm) at 2.5–4 km below surface [41, 52], implying fluid release and brine condensation above the solidifying pluton

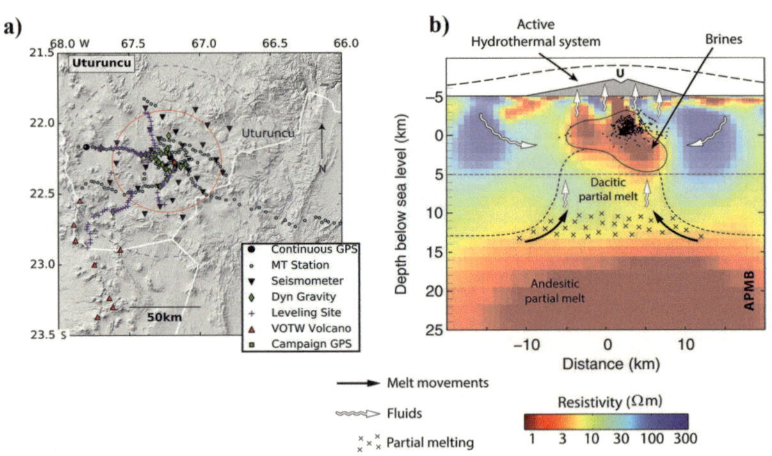

Figure 4.
Location map of the Uturuncu volcano, Bolivia, and distribution of MT sites (a). Interpreted west-east directed 3D resistivity cross section at Uturuncu (b). The deep low-resistivity region is associated with the Altiplano-Puna Magma Body, while the shallow low-resistivity area is explained by the presence of an active hydrothermal system. The label "U" above the resistivity section marks the location of Volcán Uturuncu. The dotted line denotes the region of fluid and/or magma pressurization and depressurization, and the black dots show the earthquake hypocenters. Arrows indicate the inferred flows of silicate melt and aqueous fluids within the crust, and crosses indicate regions of partial melting within the crust above the Altiplano-Puna Magma Body. APMB: Altiplano-Puna Magma Body. Modified from Pritchard et al. [41].

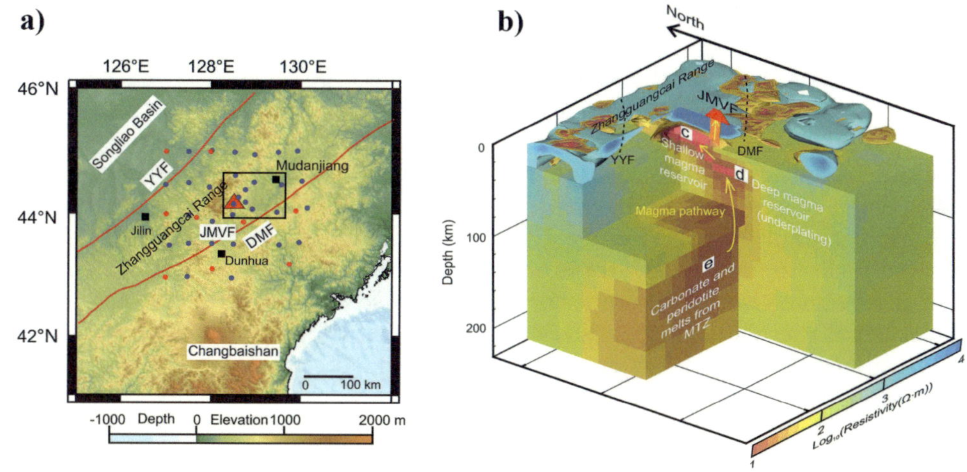

Figure 5.
Topographic map of the Jingpohu Monogenetic Volcanic Field (JMVF) in northeast China, showing the distribution of MT sites (a). Red dots represent long-period MT (LMT) sites, while blue dots represent broad-band (BBMT) sites. The red lines indicate faults in the area. The 3D conceptual model of the JMVF (b), based on the inversion of MT data, illustrates the intraplate plumbing system beneath the volcanic field. Yellow arrows indicate the magma pathways, and bold letters c, d, and e indicate resistivity anomalies. MTZ: Mantle Transition Zone. Modified from Li et al. [22].

(i.e., solidified or partially solidified dacite) [61]. The presence of such a system could also impact ore formation in this area, as brines are known to effectively sequester and concentrate metals such as Cu and Mo (e.g., [62, 63]). Similarly, there are several examples of both plutonic and volcanic systems, such as the Southern Rocky Mountain volcanic field (e.g., [64]), the Yerington batholith, Nevada [65], and the Kohistan arc, Pakistan (e.g., [66]). Granite plutons and batholits, the result of intrusions that have accumulated over hundreds of thousands to millions of years (i.e., fossil systems) [67], have also been studied by MT method (e.g., [68, 69]).

MT studies in volcanic domains show that magma plumbing systems can transport and store magmas at crustal depths (e.g., [22, 35, 42, 45, 70]). Magma upwelling can occur through large-scale faults that serve as conduits for the ascent of magmatic materials. Li et al. [22] proposed that magma transfer occurs through fault-guided dike systems in the Jingpohu intraplate monogenic volcanic field in northeast China based on their MT study. They described a low-resistivity (1–2 Ωm) anomaly system interpreted as a magma plumbing system extending from the asthenosphere to the surface, characterized by magma storage zones at different depths (i.e., shallow and deep crustal magma reservoirs) and a magma pathway related to high-temperature carbonate melts, peridotite, and/or hydrated melts in the asthenosphere (**Figure 5**). At Osorno stratovolcano in the Southern Andes, a complex magmatic system with two reservoirs at different depths (4–8 km and 7–15 km) and compositions (dacitic and andesitic) has been suggested through the MT data and petrological constraints. The eruptions of Osorno volcano have been attributed to the ascent of deep crustal basaltic magma, with the large-scale Liquiñe-Ofqui fault zone enhancing magma ascent as dikes and facilitating the reactivation of associated reservoirs and the volcano [71]. The inversion of long-period and broad-band MT data acquired around Villarrica volcano also highlights the importance of the Andean Transverse Fault in magmatic transport and compositional partitioning in the active volcanic arc of the Andes [35].

4. Mapping hydrothermal systems

Volcanic hydrothermal systems encompass various types of geothermal environments, including hydrothermal convective systems, hot dry rocks, and partially melted zones. An ideal hydrothermal system consists of three key components: a heat source, an impermeable layer, and groundwater. The interaction between the heat source, typically a volcanic or magmatic body, and the groundwater plays a critical role in heat transport and storage. Groundwater circulates through porous and fractured rocks (and/or faults), transferring heat from the heat source to other parts of the geothermal system. The impermeable layer, also known as the clay cap (i.e., hydrothermal alteration zone, argillic zone, or seal), is often formed by prolonged high-temperature fluid-rock interactions above the geothermal reservoir. This layer is predominantly characterized by smectitic clays at lower temperatures (70–200°C) and more resistive illitic and/or chloritic clays at temperatures above 200°C [72]. It prevents the escape of heated water and steam from the reservoirs.

Low-resistivity zones attributed to hydrothermal fluids (e.g., [73, 74]), alteration products (e.g., [75]), and crustal magma reservoirs, including magma chambers,

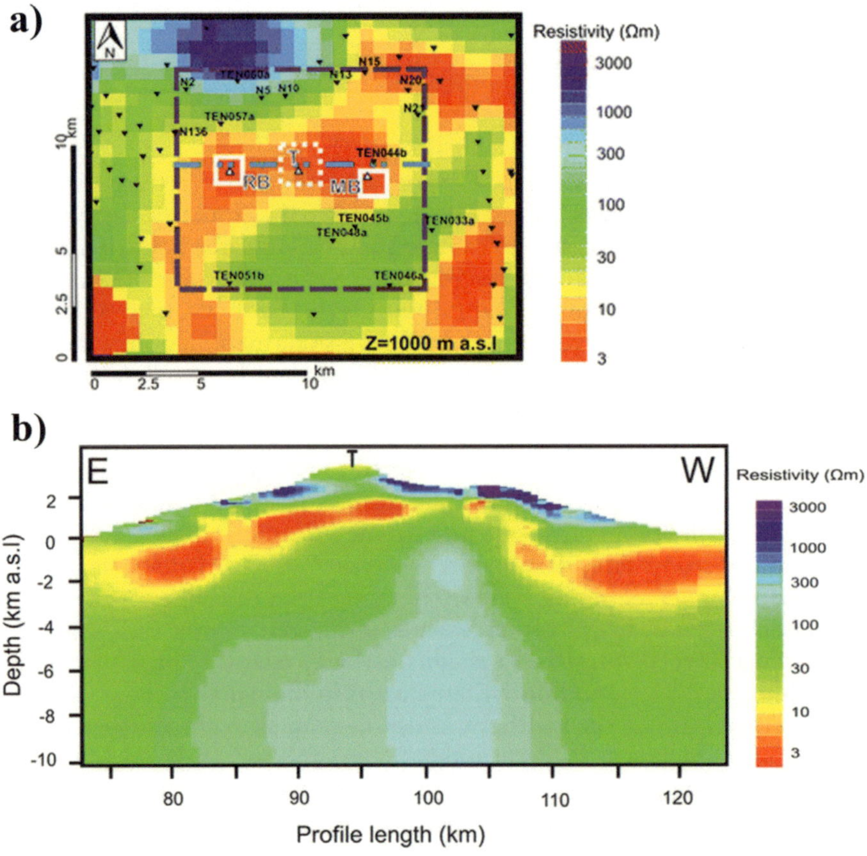

Figure 6.
Plan view of the 3-D resistivity model at 1 km depth above sea level (a). Black triangles represent MT sites. The resistivity cross section derived from the 3D inversion model in the east-west direction, crossing Teide (T) volcano (b). The low-resistivity area (red), interpreted as a smectite clay cap, is associated with the hydrothermal alteration zone (modified from Ref. [24]).

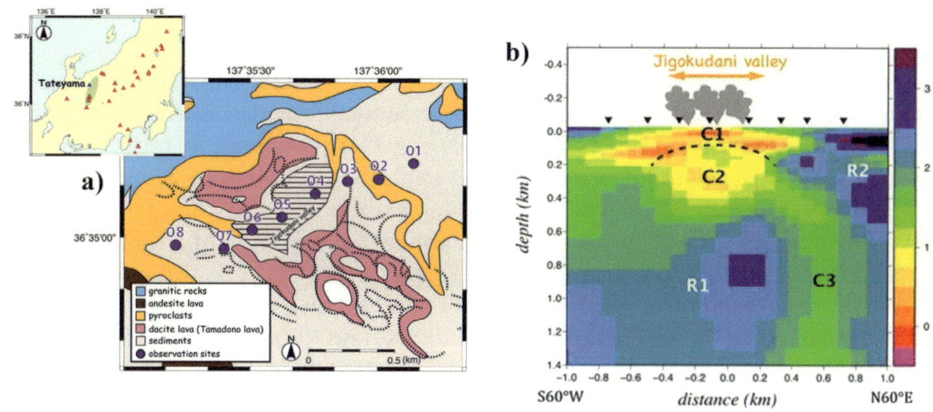

Figure 7.
Simplified geological map of Jigokudani Valley, Japan (a). The hatched area indicates the Jigokudoni Valley, while the dotted lines represent the traces of hydrothermal eruption craters. The inset map displays central Japan, with red triangles marking the locations of active volcanoes. 2D resistivity model across the Jigokudoni Valley, with inverted triangles marking the locations of AMT sites (b). The gray smoke illustration represents the fumarolic zone. C1–C3 represent low-resistivity areas, and R1 and R2 denote regions of higher resistivity (modified from Ref. [79]).

partially melted zones (i.e., crustal mush bodies), deep magma sources, and their magmatic conduits (e.g., [21, 23, 40, 49, 59, 76]) have been identified using MT in various volcanic areas around the world. MT imaging at the south-eastern margin of the Taupo Volcanic Zone (TVZ), in the central North Island of New Zealand, provides evidence of the connections between the near-surface hydrothermal systems and the underlying magmatic systems that lie below the brittle-ductile transition, as indicated by vertical low-resistivity anomalies in the resistivity models [77]. The resistivity distribution beneath Tenerife Volcanic Island, the largest island in the Canarian archipelago (Spain), also presents a model of a typical conventional geothermal system, with a low-resistivity (< 10 Ωm) hydrothermal alteration zone residing above a moderately resistive (20–100 Ωm) reservoir zone (**Figure 6**). Additionally, a medium-high-resistivity area in the 3D resistivity model indicates the presence of slowly solidified magma at depth and a hotter part of the geothermal system [24, 78]. At Jigokudani Valley, Tateyama Volcano, Japan [79], which is a composite volcano located in the Hida Mountains, through an audio-frequency MT (AMT) survey described two low-resistivity layers. The uppermost layer (C1) is associated with the lacustrine sediments of an extinct crater, and a lower low-resistivity zone (C2) contains hydrothermal fluids in the uppermost part and a mix of volcanic gases and hydrothermal fluids in the lower part. They also identified a relatively low-resistivity zone (50–100 Ωm) acting as a conduit for magmatic gases in the resistive granitic basement at depths below 500 m beneath the volcanic edifice (**Figure 7**). Seki et al. [80] also performed 3D modeling with additional sites and obtained results consistent with the previous study. In the Gonghe Basin, on the northeast Tibetan Plateau, which is characterized by the formation of hot dry rocks, Gao et al. [81] attempted to image the 3D subsurface structure to better identify the enhanced geothermal systems in the basin. The electrical resistivity model unveils distinct low resistivity anomalies interpreted as widespread hot dry rocks or hydrothermal reservoirs at depths above ~3 km. Several vertical narrow conductive zones associated with hidden faults in the resistivity model are linked to the channels that transport heat from the heat source in the middle crust (~15–35 km) to the shallow reservoir zone.

5. Imaging caldera structures

Calderas, one of the most enigmatic geological formations identifiable on Earth and other terrestrial planets [82], form through a sequence of volcanic processes that lead to the collapse of the ground surface following a large volcanic eruption, as well as through lateral magma migration. Cashman et al. [83] characterized key features of collapse calderas, including the formation of stably zoned magma chambers through crystal fractionation, the initiation of an eruption by the injection of a vertical and pressurized dike either in an axisymmetric position or at the chamber margin, magma withdrawal beginning from the top of the melt lens and moving downward, and caldera formation by the collapse of an underpressured magma chamber after a portion of the magma has been withdrawn, with the timing of collapse depending on the strength and thickness of the overlying country rock relative to the width of the magma chamber.

Caldera-forming eruptions represent the culmination of a prolonged geological process that involves the generation of magma at depth, its ascent and differentiation, and its eventual eruption at the Earth's surface. However, compared to volcanic eruptions without caldera formation, caldera-forming eruptions are rare since they require very specific stress and thermodynamic conditions for collapse to occur [84]. Recent studies have shown that caldera-forming eruptions can be initiated internally (by the magma chamber) or externally (by the roof), depending on factors such as tectonic conditions, magma volume, and the aspect ratio of the caldera roof (e.g., [85–89]).

Identifying the interior structure of calderas and their associated subvolcanic structures, such as the shape, size, and depth of underlying magmatic bodies, and the structural controls such as boundary or ring faults and their role in caldera collapse, provides significant constraints on the structure and thermodynamic state of the volcanoes. For example, the formation of a collapse caldera mostly depends on the emptying of a horizontally elongated (sill-shaped) magmatic body, leading to the generation of a surface depression with a diameter roughly equal to that of the reservoir [84]. In other words, caldera collapse occurs only after the withdrawal of a critical volume of magma, which depends on the depth and geometry of the reservoir (e.g., [83, 84]). On the other hand, bounding faults control the style of collapse and the structure of the resulting depression [87], and post-caldera eruptions and resurgences are also linked to activity along these faults [90, 91].

Although detailed MT imaging of caldera structures is scarce, recent studies have shown that MT provides significant insights into subsurface structures beneath them (e.g., [18, 92–94]). In the Cappadocian Volcanic Province, Central Anatolia (Türkiye), Hacıoğlu et al. [18] attempted to image a totally hidden caldera complex and presented the effects of transtensional tectonics on the collapse and burial process of the caldera complex using the MT method. Accordingly, the 3D electrical resistivity model revealed a large depression with low-resistivity values and also structural margins. This area is interpreted as a buried caldera complex with a maximum diameter of ~12 km and a depth of ~3 km below sea level, formed by partially overlapping nested caldera structures from subsequent volcanic eruptions and attributed to the probable source area for the Miocene-Pliocene ignimbrite

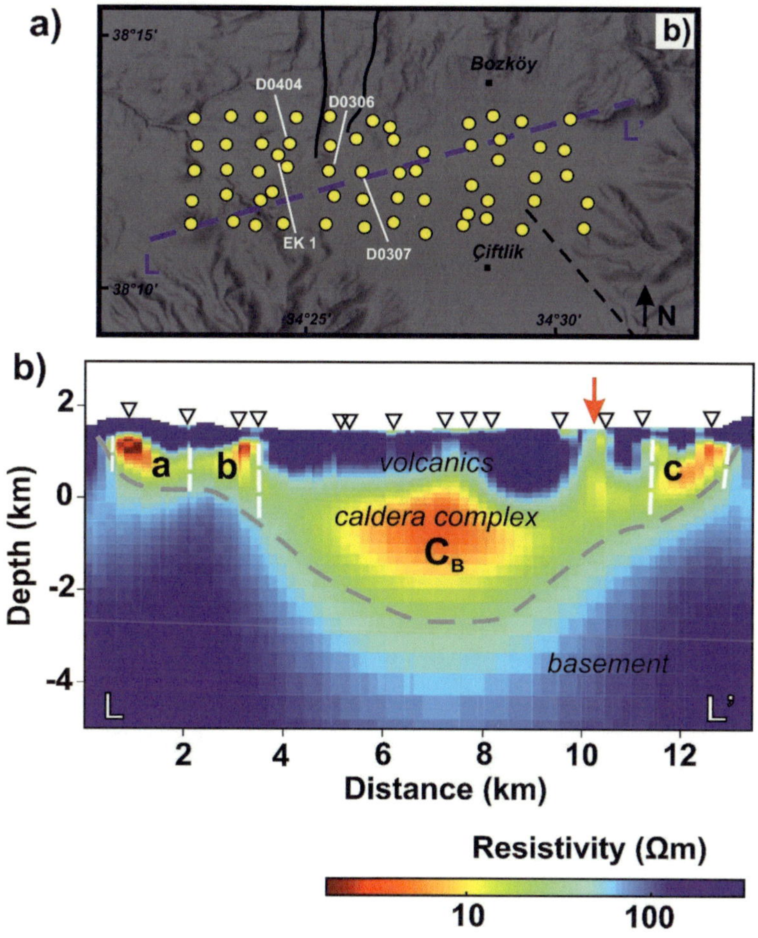

Figure 8.
Topographic map showing the locations of MT sites (yellow dots) and faults (solid and dashed black lines) in the Çiftlik Plain, Cappadocia Volcanic Province, Central Anatolia, Türkiye (a). The blue-dashed line (L-L') on the map denotes the location of the electrical resistivity section (b) derived from the 3D inversion model. White inverted triangles mark the MT sites, and a red arrow indicates the surface trace of a fault in the study area. The gray-dashed line below the broad conductive zone outlines the hidden caldera (CB), and bold letters a, b, and c indicate resistivity anomalies associated with partially overlapped calderas forming a large depression (i.e., caldera complex) (modified from Ref. [18]).

emplacements (**Figure 8**). Thus, a totally hidden caldera was revealed for the first time using MT data. In the San Juan-Silverton caldera complex, Colorado (USA), which is part of the extensive southern Rocky Mountains volcanic field, Rodriguez et al. [94] used MT and AMT data to better understand the subsurface geology and mineral resources. The hydrothermal plumbing associated with moderate-to-low resistivity values in the inversion models acts as flow paths for mineralizing fluids and the formation of metallic ores (**Figure 9**). MT profiles that crossed different parts of the caldera complex, providing images of the caldera ring-fault zones along the caldera margins and the topographic margins, also help understand the caldera processes that developed during intrusive events.

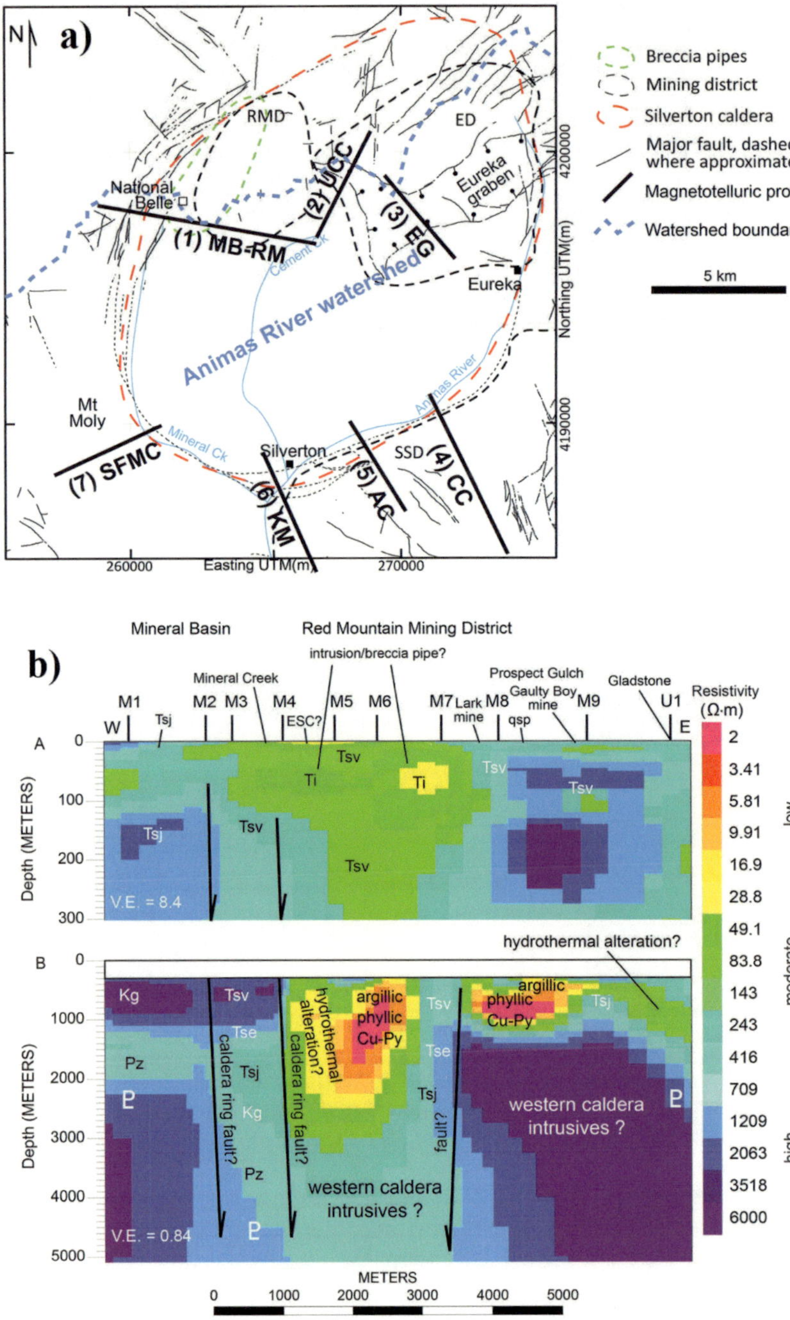

Figure 9.

Structural map of the Silverton caldera showing the caldera, graben structures, major faults, mining districts, breccia pipes, and the MT profiles (a). The resistivity section across the Mineral Basin and Red Mountain mining district (MB-RM: Mineral Basin–Red Mountain) includes interpreted geological units and faults. Black number labels at the top of the profile indicate the projected locations of magnetotelluric (MT) and audiomagnetotelluric (AMT) sites. The top panel (A) shows the AMT model at depths ranging from 0 to 300 m, while the bottom panel (B) shows the MT model at depths ranging from 0 to 5000 m (modified from Ref. [94]). For abbreviations shown on the resistivity models, see Ref. [94].

6. Applications in volcanic hazard assessment and monitoring

Volcanic hazard assessment is critical for mitigating risks associated with volcanic activity. Since the timing, location, style, and magnitude of volcanic eruptions are controlled by complex physical and chemical processes within magmatic systems and their structural components, understanding the characteristics of these systems, particularly the magma plumbing systems that feed volcanic eruptions, is crucial for studying volcanic eruption mechanisms and assessing volcanic risks and hazards. Moreover, the depth, size, and geometry of underlying magmatic bodies beneath active volcanoes strongly affect volcanic hazards. Therefore, understanding the presence of magma reservoirs and their pathways beneath volcanic edifices is crucial for assessing the potential risks and hazards of future eruptions.

Subsurface imaging of magmatic systems is essential for predicting possible eruptions and mitigating potential risks and hazards. The primary application of MT in volcanic hazard assessment is the detection and characterization of magma reservoirs, which allows for better predictions of volcanic behavior, such as potential eruption sites and the magnitude of eruptions (e.g., [24, 45]). MT is also used to map hydrothermal systems that often accompany volcanic activity (see previous sections). In these systems, magma may be located close to the surface, posing a risk for phreatic and phreatomagmatic eruptions, which occur when water interacts with hot rock or magma at depth. Additionally, MT can provide insights into volcanic gas emissions, which are crucial for monitoring and predicting eruptions (e.g., [70, 79, 95]). Furthermore, the method is utilized to monitor the location of subsurface fracture networks created by fluid injection and fluid distribution in EGS (e.g., [96–99]). In these studies, the approximate direction of fluid flow in fracture networks was determined by comparing repeated MT data acquired before and after hydraulic stimulations.

7. Discussion

MT is a powerful tool for imaging subvolcanic structures. By providing a detailed image of subsurface electrical resistivity distribution, it enhances our understanding of volcanic, geothermal, and tectonic processes. Accordingly, in volcanic environments, MT surveys have been successfully applied to map the characteristics of magma storage regions, including magma reservoirs (i.e., partially molten rocks) and magma chambers (i.e., melt lenses), such as sills, dikes, plutons, and batholiths. MT also helps identify pathways of magma ascent, detect hydrothermal alteration zones and collapsed caldera structures, and assess volcanic risks and hazards (see previous sections).

Geophysical imaging is a unique tool for constructing regional images of subsurface structures since it is not possible to directly probe the subsurface on a large scale. Petrological analysis of erupted materials and chemical analysis of volcanic gases can provide insights into the physicochemical state of a magmatic system. However, they are not suitable for constructing a 3D image of the subsurface [84]. On the other hand, geophysical imaging, including MT, has significant limitations and challenges. The interpretation of geophysical data is inherently ambiguous and non-unique, allowing for multiple interpretations of field data. For instance, the low-resistivity values at upper crustal depths, as imaged in electrical resistivity models, may be caused by partial melts, partial melts with coexisting aqueous fluids, or hot geothermal fluids. The MT method in volcanic areas can provide important insights into the subsurface electrical resistivity distribution, which can help identify magma and other fluids. However, distinguishing magma from other low-resistivity liquids in the lithosphere is not always straightforward, and there are challenges. Both magma and other fluids, such as saline aqueous fluids or partial melts, can cause low-resistivity values. Magma typically has higher conductivity due to its high temperature and the presence of dissolved ions, but salinity or the degree of partial melting are also significant factors that may complicate the interpretation. Additionally, high temperatures and pressures in magmatic systems are often attributed to magma, whereas lower temperatures may indicate the presence of brines or hydrothermal fluids. While MT can detect high-temperature zones, distinguishing between the presence of magma and other fluids depends on the geological context. The depth extend of the low-resistivity zones is also important. Low-resistivity areas located at mid-to-lower crustal depths are more likely to be attributed to the presence of magma, given the temperatures required to maintain rock in a molten or partially molten state, whereas aqueous fluids are less likely to exist at these depths. In addition to that, each geophysical method, including MT, has different capabilities for resolving subsurface structures at various spatial resolutions and may have blind spots. For example, long-period MT provides an overview of large-scale subvolcanic structures but faces challenges in distinguishing small-scale features due to decreasing resolution at greater depths. Conversely, short-period EM waves (i.e., the AMT method) can yield high-resolution data at shallow depths (e.g., up to ~2 km depending on near-surface conductive sediments), though they might not capture the larger-scale structures. Similarly, MT imaging may have struggled to detect melt that is not well connected in deep crustal rocks. Therefore, a multi-disciplinary approach is necessary to investigate the structure, composition, and physical state of subvolcanic structures, particularly the magmatic systems, for which MT provides a geophysical image and constraints on their depth and shape beneath volcanic edifices.

Combining multi-disciplinary methods may be a relevant approach to better understanding the interior structure of volcanic systems. However, it is important to note the different spatial scales of distinct methods. For instance, while geophysical methods, including MT combined with petrological barometry, offer a powerful way to identify the depth of magmatic bodies, reconciling information from these methods is challenging due to differences in temporal and spatial scales—namely, micrometers to centimeters for petrological barometry and meters to kilometers for geophysical observations. Thus, linking these two datasets remains extremely challenging [2]. Similarly, geochemical studies can resolve minute changes in chemical parameters in minerals on the submillimeter scale [84], whereas MT imaging beneath volcanoes generally focuses on larger-scale features.

MT may have difficulties detecting the postulated large magma chambers, similar to seismic imaging (e.g., [100]). Instead, previous studies explain the presence of vast quantities of hot igneous rock in the crust of many active and dormant volcanoes, containing only small amounts of melt (e.g., [22, 23, 34, 41]). Here, the limitation arises from the resolution of MT images, similar to other geophysical images that are not capable of identifying the detailed geometry or scale of melt distribution, which is significant for estimating melt fractions in subvolcanic magma reservoirs and to assess the eruption potential of a large magma reservoir. Furthermore, the MT data may not have sufficient resolution to detect small-sized magma pockets [24] due to resolution limits that may be attributed to several factors, such as small resistivity contrasts between magma chambers and host rocks (i.e., located in a low-resistivity area (<10 Ωm)) and low MT site coverage or insufficiently distributed sites above the chambers.

To understand magma storage in the crust, it is critical to determine the depth at which magma is stored, the duration of its storage, and the fundamental processes occurring inside the magma reservoir as it cools and is recharged [2]. Recent MT studies have proposed elevated melt fractions in laterally extended large volumes of magma at Mount St. Helens [101], the Afar volcanic province [102], and Volcán Uturuncu [52], implying significant magma storage may occur at mid- and lower-crustal depths or straddle the Moho, as well as the involvement of MASH processes and hot zones where mantle-derived melts interact with the lithosphere. Evidence from geology, geochemistry, and petrology also shows that substantial magma processing and chemical differentiation occur in the lower and middle crust, supporting the ideas of MASH processes [1, 2].

Inversion and mathematical frameworks can significantly bias geophysical results. Inversion of geophysical data, such as MT data, does not yield a unique solution. Mathematically, it can be demonstrated that an infinite number of resistivity models can fit a set of observed data within a certain statistical tolerance. In other words, the MT inverse problem is non-unique in practice due to data errors, spatially undersampled data (ill-posed problems), a limited frequency band, model parameterization, and physical assumptions [27, 103]. Inverse models also typically incorporate the effects of both the targets and the surrounding rocks, providing a gradual change in resistivity rather than sharp resistivity boundaries (i.e., smooth inversion) (e.g., [104]). Furthermore, MT data can constrain the conductance of structures rather than the individual thicknesses and resistivity values. Therefore, different combinations of thickness and conductivity can adequately fit the observed MT data. Additionally, if the conductance of a layer is less than the total conductance of the overlying layers, it is usually not possible to detect this anomalous layer from the MT data [105]. These challenges can be addressed by incorporating additional data or *a priori* information into the inversion (e.g., [15, 106, 107]) or by performing a joint inversion of multiple datasets (e.g., [108]), which can constrain the range of models produced by the inversion and reduce the uncertainty of subsurface models. Another important aspect when interpreting resistivity models is the sensitivity of data to particular parts of the model. For instance, in magmatic geothermal systems, a low-resistivity clay alteration zone overlying a mid-crustal reservoir may indicate the coupling effects between a magma reservoir and a shallow-seated hydrothermal reservoir. Similarly, a magma reservoir beneath a low-resistivity clay alteration layer may reduce the sensitivity of the MT data to the structure below, leading to an underestimation of the electrical resistivity of the underlying layer. Additionally, it is important to note that MT may face challenges in resolving the lower boundaries of low-resistivity zones, which can lead to the appearance of artificially connected low-resistivity zones at deeper parts of the crust. Therefore, resolution tests are needed to interpret MT inverse models in the presence of shallow low-resistivity layers, such as clay layers or shallow-seated magma reservoirs. It is also worth noting that the presence of a well-covered 3D MT array is crucial for obtaining high-resolution subsurface models.

8. Conclusion

This chapter has presented the role of magnetotellurics (MT) in advancing our understanding of volcanic systems, focusing on its application in imaging subvolcanic structures. The ability to visualize and interpret the internal complexities of volcanoes through MT provides critical insights into the distribution and characteristics of magma reservoirs, hydrothermal systems, and caldera structures. By mapping electrical resistivity variations, the spatial distribution of magma reservoirs, including magma chambers and complex plumbing systems, can be elucidated. This is essential for assessing volcanic hazards, as it offers a clearer picture of the potential volume, location, and behavior of eruptible magma. Furthermore, MT is also capable of imaging low-resistivity zones associated with hydrothermal alteration, fluid movement, and geothermal reservoirs beneath volcanic edifices, which enhances our comprehension of geothermal dynamics and helps identify potential resources. MT also provides insights into the structural complexities and processes involved in caldera collapse. By revealing the depth, shape, and size of underlying magmatic bodies, it aids in understanding the factors leading to caldera formation and the implications for volcanic hazards.

In conclusion, integrating MT with other geophysical methods and geological studies provides a comprehensive approach to understanding volcanic systems. Although MT alone cannot address all aspects of volcanic systems, it contributes to a better understanding of subvolcanic structures. The continued development and application of MT in volcanic domains promise to deepen our knowledge of these systems and their impact on both the environment and human society.

Author details

Özlem Hacıoğlu
Karadeniz Technical University, Trabzon, Türkiye

*Address all correspondence to: ozlem.hacioglu@ktu.edu.tr

References

[1] Spark RSJ, Annen C, Blundy JD, Cashman KV, Rust AC, Jackson MD. Formation and dynamics of magma reservoirs. Philosophical Transactions of the Royal Society A. 2008;**377**:20180019. DOI: 10.1098/rsta.2018.0019

[2] Sparks RSJ, Huppert HE, Kerr RC, McKenzie DP. Tait SR post-cumulus processes in layered intrusions. Geological Magazine. 1985;**122**:555-568. DOI: 10.1017/S0016756800035470

[3] Edmonds M. Architecture and dynamics of magma reservoirs. Philosophical Transactions of the Royal Society A. 2019;**377**:20180298. DOI: 10.1098/rsta.2018.0298

[4] McKenzie DP. The generation and compaction of partially molten rock. Journal of Petrology. 1984;**25**:713-765. DOI: 10.1093/petrology/25.3.713

[5] Bachmann O, Bergantz GW. On the origin of crystal-poor rhyolites: Extracted from batholithic crystal mushes. Journal of Petrology. 2004;**45**:1565-1158. DOI: 10.1093/petrology/egh019

[6] Sisson TW, Bacon CR. Gas-driven filter pressing in magmas. Geology. 1999;**27**:613-616. DOI: 10.1130/0091-7613(1999)027<0613:GDFPIM>2.3.CO;2

[7] Tait SR, Jaupart C. The production of chemically stratified and adcumulate plutonicigneous rocks. Mining Magazine. 1996;**60**:99-114. DOI: 10.1180/minmag.1996.060.398.07

[8] Pistone M et al. Gas-driven filter pressing in magmas: Insights into in-situ melt segregation from crystal mushes. Geology. 2015;**43**:699-702. DOI: 10.1130/g36766.1

[9] Cashman KV, Sparks RSJ, Blundy J. Vertically extensive and unstable crystals mushes: A unifying view of igneous processes associated with volcanoes. Science. 2017;**355**(6331):eaag3055. DOI: 10.1126/science.aag3055

[10] Marsh BD. Dynamics of magmatic systems. Elements. 2006;**2**:287-292. DOI: 10.2113/gselements.2.5.287

[11] Lejeune AM, Richet P. Rheology of crystal-bearing silicate melts: An experimental study at high viscosities. Journal of Geophysical Research - Solid Earth. 1995;**100**:4215-4229. DOI: 10.1029/94JB02985

[12] Pistone M, Caricchi L, Ulmer P, Burlini L, Ardia P, Reusser E, et al. Deformation experiments of bubble-and crystal-bearing magmas: Rheological and microstructural analysis. Journal of Geophysical Research - Solid Earth. 2012;**117**:B05208. DOI: 10.1029/2011jb008986

[13] Pistone M, Caricchi L, Ulmer P, Reusser E, Ardia P. Rheology of volatile-bearing crystal mushes: Mobilization vs. viscous death. Chemical Geology. 2013;**345**:16-39. DOI: 10.1016/j.chemgeo.2013.02.007

[14] Hacıoğlu Ö, Başokur AT, Diner Ç, Meqbel N, Arslan HI, Oğuz K. The effect of active extensional tectonics on the structural controls and heat transport mechanism in the Menderes massif geothermal province: Inferred from three-dimensional electrical resistivity structure of the Kurşunlu geothermal field (Gediz Graben, western Anatolia). Geothermics. 2020;**85**:101708. DOI: 10.1016/j.geothermics.2019.07.006

[15] Hacıoğlu Ö, Başokur AT, Diner Ç. Geothermal potential of the eastern

end of the Gediz basin, western Anatolia, Turkey revealed by three-dimensional inversion of magnetotelluric data. Geothermics. 2021;**91**:102040. DOI: 10.1016/j.geothermics.2020.102040

[16] Cordell D, Unsworth MJ, Díaz D. Imaging the Laguna del Maule volcanic field, Central Chile using magnetotellurics: Evidence for crustal melt regions laterally-offset from surface vents and lava flows. Earth and Planetary Science Letters. 2018;**488**:168-180. DOI: 10.1016/j.epsl.2018.01.007

[17] Hacıoğlu Ö, Başokur AT, Çiftçi ET. Crustal structure of a young collision zone: The Arabia–Eurasia collision in northeastern Turkey investigated by magnetotelluric data. Earth, Planets and Space. 2018;**70**:161. DOI: 10.1186/s40623-018-0932-3

[18] Hacıoğlu Ö, Başokur AT, Meqbel N, Arslan HI, Efeçınar T. Magnetotellurics unveils a hidden caldera complex beneath the Cappadocia Volcanic Province, Central Anatolia. Türkiye Journal of Volcanology and Geothermal Research. 2023;**442**:107877. DOI: 10.1016/j.jvolgeores.2023.107877

[19] Cheng Y, Pang Z, Di Q, Chen X, Kong Y. Three-dimensional resistivity structure in the hydrothermal system beneath Ganzi Basin, eastern margin of Tibetan plateau. Geothermics. 2021;**93**:102062. DOI: 10.1016/j.geothermics.2021.102062

[20] Yldız A, Kozak M, Başaran C, Özyıldırım Ö. Integrated approach for geothermal exploration: Case study from Salar area (Afyonkarahisar, Turkey). Geothermics. 2024;**122**:103068. DOI: 10.1016/j.geothermics.2024.103068

[21] Hill GJ, Bibby HM, Ogawa Y, Wallin ER, Bennie SL, Caldwell TG, et al. Structure of the Tongariro volcanic system: Insights from magnetotelluric imaging. Earth and Planetary Science Letters. 2015;**432**:115-125. DOI: 10.1016/j.epsl.2015.10.003

[22] Li Y, Weng A, Xu W, Zou Z, Tang Y, Zhou Z, et al. Translithospheric magma plumbing system of intraplate volcanoes as revealed by electrical resistivity imaging. Geology. 2021;**49**:1337-1342. DOI: 10.1130/G49032.1

[23] Başokur AT, Koçyiğit A, Hacıoğlu Ö, Arslan HI, Meqbel N. Magnetotelluric imaging of the shallow-seated magma reservoir beneath the Karadağ stratovolcano, Central Anatolia. Turkey Journal of Volcanology and Geothermal Research. 2022;**427**:107567. DOI: 10.1016/j.jvolgeores.2022.107567

[24] Piña-Varas P, Ledo J, Queralt P, Marcuello A, Perez N. On the detectability of Teide volcano magma chambers (Tenerife, Canary Islands) with magnetotelluric data. Earth, Planets and Space. 2018;**70**:14. DOI: 10.1186/s40623-018-0783-y

[25] Chave A, Jones A. The Magnetotelluric Method: Theory and Practice. Cambridge: Cambridge University Press; 2012. DOI: 10.1017/CBO9781139020138

[26] Jiracek GR, Haak V, Olsen KH. Practical magnetotelluric in a continental rift environment. In: Olsen KH, editor. Continental Rifts: Evolution, Structure, and Tectonics. Amsterdam: Elsevier; 1995. pp. 103-129. DOI: 10.1016/S0419-0254(06)80011-X

[27] Simpson F, Bahr K. Practical Magnetotellurics. Cambridge: Cambridge University Press; 2005. DOI: 10.1017/CBO9780511614095

[28] Tikhonov AN. On determining electrical characteristics of the deep

layers of the Earth's crust. Doklady. 1950;**73**:281-285

[29] Cagniard L. Basic theory of the magnetotelluric method in geophysical prospecting. Geophysics. 1953;**18**:605-635. DOI: 10.1190/1.1437915

[30] Hohmann GW. Three-dimensional EM modeling. Geophysical Surveys. 1983;**6**:27-53. DOI: 10.1007/BF01453994

[31] Kaufman AA, Keller GV. The Magnetotelluric Sounding Method, Methods in Geochemistry and Geophysics. Amsterdam: Elsevier Scientific Publishing Company; 1981

[32] Berdichevsky MN, Dmitriev VI. Models and Methods of Magnetotellurics. Berlin: Springer; 2008

[33] Ye T, Huang Q, Chen X, Zhang H, Chen J, Zhao L, et al. Magma chamber and crustal channel flow structures in the Tengchong volcano area from 3-D MT inversion at the intracontinental block boundary southeast of the Tibetan plateau. Journal of Geophysical Research - Solid Earth. 2018;**123**:11112-11126. DOI: 10.1029/2018JB015936

[34] Comeau MJ, Becken M, Grayver AV, Käufl JS, Kuvshinov AV. The geophysical signature of a continental intraplate volcanic system: From surface to mantle source. Earth and Planetary Science Letters. 2022;**578**:117307. DOI: 10.1016/j.epsl.2021.117307

[35] Pavez M, Brasse H, Kapinos G, Díaz D, Lara LE, Schill E. Magma storage and transfer in the Villarrica volcanic chain, South Chile: MT insights into volcanotectonic interactions. Journal of Volcanology and Geothermal Research. 2023;**439**:107832. DOI: 10.1016/j.jvolgeores.2023.107832

[36] Unsworth M, Comeau MJ, Diaz D, Brasse H, Heit B, Favetto A, et al. Crustal structure of the Lazufre volcanic complex and the southern Puna from 3-D inversion of magnetotelluric data: Implications for surface uplift and evidence for melt storage and hydrothermal fluids. Geosphere. 2023;**19**(5):1210-1230. DOI: 10.1130/GES02506.1

[37] Komori S, Takakura S, Mitsuhata Y, Yokota T, Uchida T, Makino M, et al. Three-dimensional resistivity structure in Toya caldera region, Southwest Hokkaido, Japan — Constraints on magmatic and geothermal activities. Geophysics. 2024;**89**:B31-B50. DOI: 10.1190/geo2022-0558.1

[38] Türkoğlu E, Unsworth M, Çağlar İ, Tuncer V, Avşar Ü. Lithospheric structure of the Arabia–Eurasia collision zone in eastern Anatolia: Magnetotelluric evidence for widespread weakening by fluids? Geology. 2008;**36**:619-622. DOI: 10.1130/G24683A

[39] Laumonier M, Gaillard F, Sifre D. The effect of pressure and water concentration on the electrical conductivity of dacitic melts: Implication for magnetotelluric imaging in subduction areas. Chemical Geology. 2015;**418**:66-76. DOI: 10.1016/j.chemgeo.2014.09.019

[40] Díaz D, Heise W, Zamudio F. Three-dimensional resistivity image of the magmatic system beneath Lastarria volcano and evidence for magmatic intrusion in the backarc (northern Chile). Geophysical Research Letters. 2015;**42**:5212-5218. DOI: 10.1002/2015GL064426

[41] Pritchard ME, de Silva SL, Michelfelder G, Zandt G, McNutt SR, Gottsmann J, et al. Synthesis: PLUTONS: Investigating the relationship between pluton growth and volcanism in the Central Andes. Geosphere.

2018;**14**(3):954-982. DOI: 10.1130/GES01578.1

[42] Liu Y, Du J, Hu Z, Hu D, Luo Y, Chen C. Crustal magma plumbing system beneath the quaternary volcanic area (northern Hainan Island, China) revealed by magnetotelluric data. Journal of Volcanology and Geothermal Research. 2021;**419**:107362. DOI: 10.1016/j.jvolgeores.2021.107362

[43] Yang B, Lin W, Hu X, Fang H, Qiu G, Wang G. The magma system beneath Changbaishan- Tianchi volcano, China/North Korea: Constraints from three-dimensional magnetotelluric imaging. Journal of Volcanology and Geothermal Research. 2021;**419**:107385. DOI: 10.1016/j.jvolgeores.2021.107385

[44] Bertrand EA, Kannberg P, Caldwell TG, Heise W, Constable S, Scott B, et al. Inferring the magmatic roots of volcano-geothermal systems in the Rotorua Caldera and Okataina volcanic Centre from magnetotelluric models. Journal of Volcanology and Geothermal Research. 2022;**431**:107645. DOI: 10.1016/j.jvolgeores.2022.107645

[45] Dong Z, Xiao Q, Deng Y, Han B, Tang J, Wang L, et al. Preliminary magnetotelluric investigation of crustal magma plumbing system beneath the Wulanhada volcanic field, northern China: Implications for the magma reservoir and pathway. Journal of Volcanology and Geothermal Research. 2023;**443**:107938. DOI: 10.1016/j.jvolgeores.2023.107938

[46] Çoban H, Karsli O, Caran Ş, Yilmaz K. Sediment-derived melt-related metasomatised mantle wedge as a source of post-subduction quaternary adakitic porphyries associated with absarokiteshoshonite from the Karadağ stratovolcano (Karaman, Central Anatolia, Turkey). Journal of Asian Earth Sciences. 2020;**196**:104380. DOI: 10.1016/j.jseaes.2020.104380

[47] Devine JD, Murphy MD, Rutherford MJ, Barclay J, Sparks RSJ, Young SR, et al. Petrologic evidence for pre-eruptive pressure-temperature conditions, and recent reheating, of andesitic magma erupting at the Soufriere Hills volcano, Montserrat, W.I. Geophysical Research Letters. 1998;**25**:3669-3672. DOI: 10.1029/98GL01330

[48] Devine JD, Rutherford MJ, Norton GE, Young SR. Magma storage region processes inferred from geochemistry of Fe–Ti oxides in andesitic magma, Soufrière Hills volcano, Montserrat, W.I. Journal of Petrology. 2003;**44**:1375-1400. DOI: 10.1093/petrology/44.8.1375

[49] Guevara L, Apaza FD, Favetto A. Conductivity distribution beneath Socompa volcano, northwestern Argentina, from 3-D magnetotelluric characterization. Journal of Volcanology and Geothermal Research. 2023;**441**:107889. DOI: 10.1016/j.jvolgeores.2023.107889

[50] Mancini R, Díaz D, Brasse H, Godoy B, Hernández MJ. Conductivity distribution beneath the San Pedro-Linzor volcanic chain, North Chile, using 3-D magnetotelluric modeling. JGR: Solid Earth. 2019;**124**:4386-4398. DOI: 10.1029/2018JB016114

[51] Comeau MJ, Unsworth MJ, Ticona F, Sunagua M. Magnetotelluric images of magma distribution beneath Volcán Uturuncu, Bolivia: Implications for magma dynamics. Geology. 2015;**43**(3):243-246. DOI: 10.1130/G36258.1

[52] Comeau MJ, Unsworth MJ, Cordell D. New constraints on the magma distribution and composition beneath

Volcán Uturuncu and the southern Bolivian Altiplano from magnetotelluric data. Geosphere. 2016;**12**(5):1391-1421. DOI: 10.1130/GES01277.1

[53] Roberts JJ, Tyburczy JA. Partial-melt electrical conductivity: Influence of melt composition. Journal of Geophysical Research. 1999;**104**:7055-7065. DOI: 10.1029/1998JB900111

[54] Pommier A, Gaillard F, Pichavant M, Scaillet B. Laboratory measurements of electrical conductivities of hydrous and dry Mount Vesuvius melts under pressure. Journal of Geophysical Research. 2008;**113**:B05205. DOI: 10.1029/2007JB005269

[55] Pommier A, Tarits P, Hautot S, Pichavant M, Scaillet B, Gaillard F. A new petrological and geophysical investigation of the present-day plumbing system of Mount Vesuvius. Geochemistry, Geophysics, Geosystems. 2010;**11**(7):Q07013. DOI: 10.1029/2010GC003059

[56] Pommier A, Le-Trong E. SIGMELTS: A web portal for electrical conductivity calculations in geosciences. Computational Geosciences. 2011;**37**:1450-1459. DOI: 10.1016/j.cageo.2011.01.002

[57] Afanasyev A, Blundy J, Melnik O, Sparks S. Formation of magmatic brine lenses via focussed fluid-flow beneath volcanoes. Earth and Planetary Science Letters. 2018;**486**:119-128. DOI: 10.1016/j.epsl.2018.01.013

[58] Guo H, Keppler H. Electrical conductivity of NaCl-bearing aqueous fluids to 900 °C and 5 GPa. Journal of Geophysical Research - Solid Earth. 2019;**124**:1397-1411. DOI: 10.1029/2018JB016658

[59] Harangi S, Novák A, Kiss B, Seghedi I, Lukács R, Szarka L, et al. Combined magnetotelluric and petrologic constrains for the nature of the magma storage system beneath the Late Pleistocene Ciomadul volcano (SE Carpathians). Journal of Volcanology and Geothermal Research. 2015;**290**:82-96. DOI: 10.1016/j.jvolgeores.2014.12.006

[60] Ślęzak K, Díaz D, Vargas JA, Cordell D, Reyes-Cordova F, Segovia MJ. Magnetotelluric image of the Chilean subduction zone in the Salar de Atacama region (23°-24° S): Insights into factors controlling the distribution of volcanic arc magmatism. Physics of the Earth and Planetary Interiors. 2021;**318**:106765. DOI: 10.1016/j.pepi.2021.106765

[61] Weis P. The dynamic interplay between saline fluid flow and rock permeability in magmatic-hydrothermal systems. Geofluids. 2015;**15**:350-371. DOI: 10.1111/gfl.12100

[62] Tattitch BC, Candela PA, Piccoli PM, Bodnar RJ. Copper partitioning between felsic melt and H_2O-CO_2 bearing saline fluids. Geochimica et Cosmochimica Acta. 2015;**148**:81-99. DOI: 10.1016/j.gca.2014.08.025

[63] Tattitch B, Blundy J. Partitioning of Cu and Mo between felsic melts and saline magmatic fluids: Influence of salinity, fO_2 and fS_2. Applied Earth Science. 2017;**126**(2):99. DOI: 10.1080/03717453.2017.1306300

[64] Lipman PW, Bachmann O. Ignimbrites to batholiths: Integrating perspectives from geological, geophysical, and geochronological data. Geosphere. 2015;**11**(3):705-743. DOI: 10.1130/GES01091.1

[65] Dilles JH, Wright JE. The chronology of early Mesozoic arc magmatism in the Yerington district of western Nevada and its regional implications. Geological Society of

America Bulletin. 1988;**100**:644-652. DOI: 10.1130/0016-7606(1988)100<0644:TCOEMA>2.3.CO;2

[66] Jagoutz O, Schmidt MW. The formation and bulk composition of modern juvenile continental crust: The Kohistan arc. Chemical Geology. 2012;**298**:79-96. DOI: 10.1016/j.chemgeo.2011.10.022

[67] Lipman PW. The roots of ash-flow calderas in North America: Windows into the tops of granitic batholiths. Journal of Geophysical Research. 1984;**89**:8801-8841. DOI: 10.1029/JB089iB10p08801

[68] Roy KK, Rao CK, Chattopadhyay A. Magnetotelluric survey across Singhbhum granite batholith. Proceedings of the Indian Academy of Sciences – (Earth and Planetary Sciences). 1989;**98**:147-165. DOI: 10.1007/BF02863240

[69] Piña-Varas P, Soto R, Clariana P, Ayala C, Rubio FM, Ledo JJ, et al. High-resolution scan of the Pyrenean crustal structurecombining magnetotelluric and gravity data. Tectonophysics. 2023;**864**:230022. DOI: 10.1016/j.tecto.2023.230022

[70] Gresse M, Uyeshima M, Koyama T, Hase H, Aizawa K, Yamaya Y, et al. Hydrothermal and magmatic system of a volcanic island inferred from magnetotellurics, seismicity, self-potential, and thermal image: An example of Miyakejima (Japan). Journal of Geophysical Research: Solid Earth. 2021;**126**:e2021JB022034. DOI: 10.1029/2021JB022034

[71] Díaz D, Zúniga F, Castruccio A. The interaction between active crustal faults and volcanism: A case study of the Liquie-Ofqui fault zone and Osorno volcano, southern Andes, using magnetotellurics. Journal of Volcanology and Geothermal Research. 2020;**393**:106806. DOI: 10.1016/j.jvolgeores.2020.106806

[72] Ussher G, Harvey C, Johnstone R, Anderson E. Understanding the resistivity observed in geothermal systems. In: Proceedings World Geothermal Congress; 28 May-10 June 2000; Kyushu-Tohoko, Japan. 2000. pp. 1915-1920

[73] Ingham MR, Bibby HM, Heise W, Jones KA, Cairns P, Dravitzki S, et al. A magnetotelluric study of Mount Ruapehu volcano, New Zealand. Geophysical Journal International. 2009;**179**:887-904. DOI: 10.1111/j.1365-246X.2009.04317.x

[74] Matsunaga Y, Kanda W, Takakur S, Koyama T, Saito Z, Sek K, et al. Magmatic hydrothermal system inferred from the resistivity structure of Kusatsu-Shirane volcano. Journal of Volcanology and Geothermal Research. 2020;**390**:106742. DOI: 10.1016/j.jvolgeores.2019.106742

[75] Matsushima N, Utsugi M, Takakura S, Yamasaki T, Hata M, Takeshi H, et al. Magmatic–hydrothermal system of Aso volcano, Japan, inferred from electrical resistivity structures. Earth Planet Space. 2020;**72**:57. DOI: 10.1186/s40623-020-01180-8

[76] Aizawa K, Koyama T, Hasel H, Uyeshima M, Kanda W, Utsugi M, et al. Three-dimensional resistivity structure and magma plumbing system of the Kirishima volcanoes as inferred from broadband magnetotelluric data. Journal of Geophysical Research - Solid Earth. 2014;**119**:198-215. DOI: 10.1002/2013JB010682

[77] Bertrand A, Caldwell TG, Hill GJ, Wallin EL, Bennie SL, Cozens SA, et al. Magnetotelluric imaging of upper-crustal convection plumes beneath the Taupo

volcanic zone, New Zealand. Geophysical Research Letters. 2012;**39**:L02304. DOI: 10.1029/2011GL050177

[78] Piña-Varas P, Ledo J, Queralt P, et al. 3-D magnetotelluric exploration of Tenerife geothermal system (Canary Islands, Spain). Surveys in Geophysics. 2014;**35**:1045-1064. DOI: 10.1007/s10712-014-9280-4

[79] Seki K, Kanda W, Ogawa Y, Tanbo T, Kobayashi T, Hino Y, et al. Imaging the hydrothermal system beneath the Jigokudani valley, Tateyama volcano, Japan: Implications for structures controlling repeated phreatic eruptions from an audio-frequency magnetotelluri survey. Earth Planet Space. 2015;**67**:6. DOI: 10.1186/s40623-014-0169-8

[80] Seki K, Kanda W, Tanbo T, Ohba T, Ogawa Y, Takakura S, et al. Resistivity structure and geochemistry of the Jigokudani Valley hydrothermal system, Mt. Tateyama. Japan. Journal of Volcanology and Geothermal Research. 2016;**325**:15-26. DOI: 10.1016/j.jvolgeores.2016.06.010

[81] Gao J, Zhang H, Zhang S, Chen X, Cheng Z, Jia X, et al. Three-dimensional magnetotelluric imaging of the geothermal system beneath the Gonghe Basin, Northeast Tibetan plateau. Geothermics. 2018;**76**:15-25. DOI: 10.1016/j.geothermics.2018.06.009

[82] Cole JW, Milner DM, Spinks KD. Calderas and caldera structures: A review. Earth-Science Reviews. 2005;**69**:1-26. DOI: 10.1016/j.earscirev.2004.06.004

[83] Cashman KV, Giordano G. Calderas and magma reservoirs. Journal of Volcanology and Geothermal Research. 2014;**288**:28-45. DOI: 10.1016/j.jvolgeores.2014.09.007

[84] Martí J, Geyer A, Folch A, Gottsmann J. A review on collapse caldera modelling. In: Caldera Volcanism: Analysis, Modelling and Response. Amsterdam: Elsevier; 2008. pp. 233-283. DOI: 10.1016/s1871-644x(07)00006-x

[85] Martí J, Folch A, Macedonio G, Neri A. Pressure evolution during caldera forming eruptions. Earth and Planetary Science Letters. 2000;**175**:275-287. DOI: 10.1016/S0012-821X(99)00296-4

[86] Jellinek AM, De Paolo GJ. A model for the origin of large silicic magma chambers: Precursors of caldera-forming eruptions. Bulletin of Volcanology. 2003;**65**:363-381. DOI: 10.1007/s00445-003-0277-y

[87] Gudmundsson A. Magma-chamber geometry, fluid transport, local stresses, and rock behaviour during collapse-caldera formation. In: Gottsmann J, Marti J, editors. Caldera Volcanism: Analysis, Modelling and Response. Amsterdam: Elsevier; 2008. pp. 313-349. DOI: 10.1016/S1871-644X(07)00008-3

[88] Gudmundsson A. Magma chambers: Formation, local stresses, excess pressures, and compartments. Journal of Volcanology and Geothermal Research. 2012;**237-238**:19-41. DOI: 10.1016/j.jvolgeores.2012.05.015

[89] Gregg PM, de Silva SL, Grosfils EB, Parmigiani JP. Catastrophic caldera-forming eruptions: Thermomechanics and implications for eruption triggering and maximum caldera dimensions on earth. Journal of Volcanology and Geothermal Research. 2012;**241-242**:1-12. DOI: 10.1016/j.jvolgeores.2012.06.009

[90] Saunders SJ. The shallow plumbing system of Rabaul caldera: A partially intruded ring fault? Bulletin of Volcanology. 2001;**63**:406-420. DOI: 10.1007/s004450100159

[91] Saunders SJ. The possible contribution of circumferential fault intrusion to caldera resurgence. Bulletin of Volcanology. 2004;**67**:57-71. DOI: 10.1007/s00445-004-0360-z

[92] Di Giuseppe MG, Troiano A, Carlino S. Magnetotelluric imaging of the resurgent caldera on the island of Ischia (southern Italy): Inferences for its structure and activity. Bulletin of Volcanology. 2017;**79**:85. DOI: 10.1007/s00445-017-1170-4

[93] Martínez-Moreno FJ, Monteiro Santos FA, Madeira J, Pous J, Bernardo I, Soares A, et al. Investigating collapse structures in oceanic islands using magnetotelluric surveys: The case of Fogo Island in Cape Verde. Journal of Volcanology and Geothermal Research. 2018;**357**:152-162. DOI: 10.1016/j.jvolgeores.2018.04.028

[94] Rodriguez RD, Yager DB, Anderson ED, Hoogenboom BE, Runkel RL, Smith BD, et al. Deep resistivity geophysics of the San Juan–Silverton caldera complex, San Juan County, Colorado (USA). Geosphere. 2024;**20**(3):910-934. DOI: 10.1130/GES02550.1

[95] Seki K, Kanda W, Mannen K, Takakura S, Koyama T, Noguchi R, et al. Imaging the source region of the 2015 phreatic eruption at Owakudani, Hakone volcano, Japan, using high-density audio-frequency magnetotellurics. Geophysical Research Letters. 2021;**48**:e2020GL091568. DOI: 10.1029/2020GL091568

[96] Wannamaker PE, Rose PE, Doerner WM, Berard BC, McCulloch J, Nurse K. Magnetotelluric surveying and monitoring at the Coso geothermal area, California, in support of the enhanced geothermal systems concept: Survey parameters and initial results. In: Proceedings of 29th Workshop on Geothermal Reservoir Engineering, 26-28 January 2004. Stanford, CA: Stanford University; 2004. SGP-TR-175

[97] Peacock JR, Thiel S, Reid P, Heinson G. Magnetotelluric monitoring of a fluid injection: Example from an enhanced geothermal system. Geophysical Research Letters. 2012;**39**(17):3-7. DOI: 10.1029/2012GL053080

[98] Peacock JR, Thiel S, Heinson GS, Reid P. Time-lapse magnetotelluric monitoring of an enhanced geothermal system. Geophysics. 2013;**78**(3):B121-B130. DOI: 10.1190/geo2012-0275.1

[99] Didana YL, Heinson G, Thiel S, Krieger L. Magnetotelluric monitoring of permeability enhancement at enhanced geothermal system project. Geothermics. 2017;**66**:23-38. DOI: 10.1016/j.geothermics.2016.11.005

[100] Farrell J, Smith RB, Husen S, Diehl T. Tomography from 26 years of seismicity revealing that the spatial extent of the Yellowstone crustal magma reservoir extends well beyond the Yellowstone caldera. Geophysical Research Letters. 2014;**41**:3068-3073. DOI: 10.1002/2014GL059588

[101] Bedrosian PA, Peacock JR, Bowles-Martinez E, Schultz A, Hill GJ. Crustal inheritance and a top-down control on arc magmatism at mount St. Helens. Nature Geoscience. 2018;**11**:865-870. DOI: 10.1038/s41561-018-0217-2

[102] Desissa M, Johnson N, Whaler K, Hautot S, Fisseha S, Dawes G. A mantle magma reservoir beneath an incipient mid-ocean ridge in Afar, Ethiopia. Nature Geoscience. 2013;**6**:861-865. DOI: 10.1038/ngeo1925

[103] Parker RL. The inverse problem of electromagnetic induction: Existence and construction of solutions based on incomplete data. Journal of Geophysical Research - Solid Earth. 1980;**85**:4421-4428. DOI: 10.1029/JB085iB08p04421

[104] Miensopust MP, Queralt P, Jones AG. The 3D MT modellers (2013) Magnetotelluric 3-D inversion—A review of two successful workshops on forward and inversion code testing and comparison. Geophysical Journal International. 2013;**193**(3):1216-1238. DOI: 10.1093/gji/ggt066

[105] Jones AG. Imaging the continental upper mantle using electromagnetic methods. Lithos. 1999;**48**:57-80. DOI: 10.1016/S0024-4937(99)00022-5

[106] Heise W, Caldwell TG, Bibby HM, Bennie SL. Three-dimensional electrical resistivity image of magma beneath an active continental rift, Taupo volcanic zone, New Zealand. Geophysical Research Letters. 2010;**37**(10):L10301. DOI: /10.1029/2010GL043110

[107] Jegen M, Avdeeva A, Berndt C, Franz G, Heincke B, Holz S, et al. 3-D magnetotelluric image of offshore magmatism at the Walvis ridge and rift basin. Tectonophysics. 2016;**683**:98-108. DOI: 10.1016/j.tecto.2016.06.016

[108] Moorkamp M, Heincke B, Jegen M, Roberts AW, Hobbs RW. A framework for 3-D joint inversion of MT, gravity and seismic refraction data. Geophysical Journal International. 2011;**184**(1):477-493. DOI: 10.1111/j.1365-246X.2010.04856.x

Chapter 6

The Contribution of Underwater Systems for the Study of the Emerged Volcanoes and Seamounts of Southern Italy

Tiziana Sgroi and Paolo Favali

Abstract

This chapter examines the signals collected by multidisciplinary seafloor observatories and subsea instruments in Southern Tyrrhenian region. It provides an overview of the systems, emphasizing the key characteristics of the sensors. Various types of signals (such as seismological, magnetic, deformation, and oceanographic data), detected underwater and related to onshore and offshore volcanic structures, are analysed and compared, where possible, with data from land-based stations. The focus is on several case studies from Southern Italy, a region of high geological activity, and tectonic and volcanic features. This area includes two of the world's most active volcanoes, Mt. Etna and Stromboli, as well as the Marsili Seamount, the largest volcanic seamount in the Tyrrhenian Sea bathyal plain. Although extensive research has been conducted on Mt. Etna and Stromboli, the functioning of their deepest and submerged parts, as well as the nature and evolution of their activity, remains incomplete. This is also particularly true for volcanic seamounts like Marsili. Seafloor instrumentation, including Ocean Bottom Seismometers and Hydrophones (OBSHs), and multiparametric seafloor observatories play a crucial role in monitoring volcanic activity contributing significantly to understanding of volcanic behaviour and its temporal evolution. Additionally, swath bathymetry data aid in visualizing tectonic and volcanic structures.

Keywords: emerged volcanoes and seamounts, underwater observatories and instruments, time series, signal analysis, southern Italy

1. Introduction

Covering approximately 70% of the Earth's surface, the ocean plays a crucial role in regulating the global climate and acting as a major component of the Earth's geo- and ecosystem. It absorbs more than 90% of the excess heat resulting from greenhouse gas emissions, leading to changes in ocean temperature, acidity, and stratification. These changes impact marine biodiversity and the functioning of marine ecosystems. Additionally, anthropogenic factors, such as the impacts of pollution, eutrophication, intensive coastal development, and extensive agricultural practices, contribute significant stress to marine and coastal environments, alongside the overexploitation of marine resources.

Understanding the ocean's complex interplay of physical, biological, chemical, and geological systems presents a significant challenge for the early twenty-first century. The ocean's role in regulating carbon dioxide (CO_2) exchange and storage is critical for controlling global climate and supporting life. Moreover, the ocean provides numerous resources essential to human society.

Geological hazards predominantly originated in oceanic areas, with approximately 70% of the Earth's volcanic activity occurring on the ocean floor. These hazards include significant seismic events and submarine landslides, which are primary sources of tsunamis. Such events can cause catastrophic damage and have substantial socio-economic impacts. Historical examples of catastrophic events that occurred in the studied area include the 28 December 1908 Messina Straits Earthquake (M_w 7.1 [1]), which generated tsunamis and caused widespread devastation. A comprehensive scientific approach necessitates coordinated and complementary monitoring and observation programmes to grasp the ocean's interrelated processes through an interdisciplinary lens [2].

Observing natural systems is often complicated by temporal variability, which can be challenging to characterize accurately. When combined with spatial variability, this complexity significantly impedes our understanding of system structure and function, particularly in distinguishing natural from anthropogenic changes over time. Eulerian observatories, which are fixed platforms relative to the Earth's crust, are equipped with sensors for making physical, chemical, and environmental *in situ* measurements and are critical for sustained ocean observations [3]. These observations are vital for understanding complex processes and require a network of long-term, comprehensive observations [4]. Multidisciplinary standardized reference data facilitate interdisciplinary analysis required to improve geo-hazard warning and effectively evaluate the influences of climate change and anthropogenic impacts [5].

Eulerian observatories generally fall into two main categories: stand-alone seafloor monitoring modules, such as Ocean Bottom Seismometers and Hydrophones (OBSHs), and seafloor observatory. OBSHs are typically deployed and recovered via free-fall landing and pop-up procedures or by Remotely Operated Vehicles (ROVs). In contrast, more complex multiparametric platforms, known as 'seafloor and water column observatories', offer extended lifetimes and enhanced performance. These observatories provide valuable insights into sub-seafloor, seafloor, and water column interactions, offering comprehensive data on geosphere, biosphere, and hydrosphere interactions and their temporal variability [6].

Globally, numerous initiatives have established networks of ocean observatories and research infrastructures [7]. In Europe, research infrastructures in the marine domain are focused on enhancing cooperation, coordination, and integration to support healthy and productive oceans, seas, coastal and inland waters (see Dañobeitia et al. [8] for details and references). These efforts aim to mitigate the increasing socio-economic impacts by improving knowledge and information.

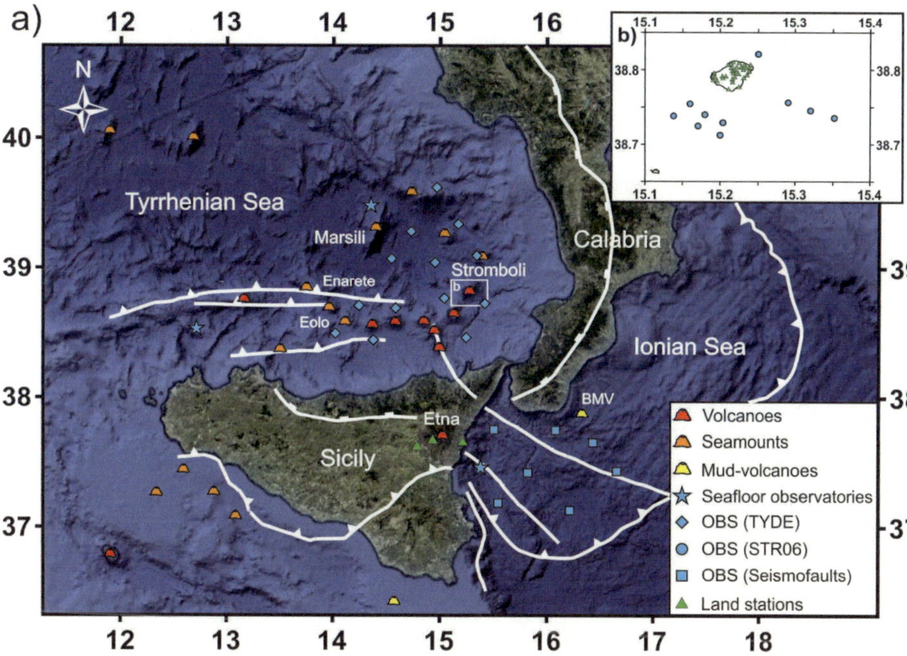

Figure 1.
(a) Shaded-relief morphological map (latitude N and longitude E) of Southern Italy (based on Google Earth Pro [9]) indicating subaerial volcanoes, seamounts, and mud volcanoes (BMV-Bortoluzzi Mud Volcano), and the position of OBSHs, seafloor observatories, and land stations (see legend). The major tectonic structures are also reported in white (lines faults, lines with rectangles outer extensional front and with triangles outer thrust front) (modified after Sgroi et al., Polonia et al., and Lavecchia et al. [10–12]); (b) zoom on the area of Stromboli volcano where the Stromboli (STR06) experiment was performed (see Section 3).

This chapter explores the contributions of underwater monitoring systems, both multiparametric observatories and individual modules, to our understanding of emerged volcanoes and seamounts. These systems have significantly advanced the study of volcanic activity in Southern Italy, enabling comprehensive monitoring and data collection from both submerged and emerged volcanic structures. By comparing signals collected by these underwater systems with those from land-based stations, a clearer understanding of volcanic behavior within the regional tectonic context can be achieved. The focus of this study is on Southern Italy, particularly the main volcanoes (Mt. Etna, Stromboli, and Marsili Seamount), the Aeolian Arc, and mud volcanoes. **Figure 1** shows the Southern Italy area covered by the activities of the experiments and projects described in this chapter.

2. Geological setting

The Mediterranean region is characterized by the convergence of the Eurasian and African tectonic plates, resulting in both collisional and extensional processes along their margins (e.g., Dewey et al. [13]). Volcanic activity has occurred within the tectonic framework since the Oligocene. During the Pliocene-Pleistocene, the direction of extension shifted from east-west (E-W) to southeast-northwest (SE-NW) (e.g., Argnani and Savelli [14]). The complex geodynamic environment of the

Southern Tyrrhenian Sea is primarily governed by the sinking and rollback of the Ionian subducting slab (e.g., Argnani [15]). This tectonic activity led to the formation of two back-arc sub-basins within the Tyrrhenian Sea: the Vavilov (4.3–3.6 Ma) and the Marsili (<2 Ma) [16, 17].

The Marsili Basin includes the Marsili Seamount, the largest European underwater volcanic structure, associated with the northwestward subduction of the Ionian lithosphere below the Calabrian Arc. The Southern Tyrrhenian Sea region hosts two of the world's most active volcanoes, Stromboli and Mt. Etna, as well as several seamounts.

The Marsili Basin has a roughly elliptical shape with a crustal thickness of 7–8 km and a lithosphere thickness of around 30 km (kilometers) [18, 19]. Satellite gravity data modeling indicates that the crust within the Marsili Basin is of oceanic origin. The central portion of the basin features a flat seafloor at a water depth of 3500 metres, interrupted by the Marsili Seamount, which rises to a minimum water depth of 500 m. The volcanic structure extends 60 km in length and 20 km in width, is oriented predominantly north-northeast-south-southwest (NNE-SSW), and is characterized by steep flanks with slopes averaging about 20° [16, 20].

The Aeolian Arc includes seven islands, listed from west to east as follows: Alicudi, Filicudi, Salina, Vulcano, Lipari, Panarea, and Stromboli. It also encompasses several seamounts: Sisifo, Enarete, and Eolo in the western sector; and Lametini, Alcione, and Glabro in the eastern sector. This arc surrounds the Marsili Basin and extends northwards to include the Palinuro Seamount [20]. Volcanic activity within this region is influenced by a strike-slip/extensional tectonic regime, with both Mt. Etna and the central Aeolian Islands exhibiting such tectonic features [21]. Volcanic activity has migrated from west to east along the Aeolian Islands, currently affecting the central-eastern part of the arc. The western islands have not experienced volcanic activity in the past 30–40 ka [22]. Significant hydrothermal activity is observed offshore the Aeolian Islands, with the most active site located off the coast of Panarea [23].

Belonging to the Aeolian Arc, Stromboli rises approximately 3000 metres from the seafloor and stands 924 metres above sea level. This volcano has exhibited continuous eruptive activity for over 2000 years, and possibly as long as 5000 years [24]. The current eruptive behavior is characterized by intermittent explosive events, where well-collimated gas jets laden with molten lava fragments produce brief eruptions lasting 5–15 seconds, occurring at a typical rate of 3–10 events per hour. This persistent volcanic activity is occasionally interrupted by lava effusions or major explosions [25].

Mud volcanoes are ubiquitous both on land and in marine environments. They are formed due to various processes, typically involving overpressured clay- and fluid-rich subsurface levels that drive the ascent of mixed and liquefied clay, water, and gases (e.g., CO_2, methane (CH_4), and nitrogen (N)). This results in the formation of a conical or truncated-conical landform [26].

The Western Ionian Sea coastal region is densely populated and supports numerous social, commercial, and industrial activities, rendering it highly exposed to natural hazards, such as volcanic eruptions and earthquakes and, in some cases, tsunamis. The active volcano in this region is Mt. Etna situated in the forearc of the Calabrian Arc subduction system. Mt. Etna is a large basaltic composite volcano covering a broad sub-rounded surface of 1178 km^2, with a maximum diameter of approximately 45 km, an elevation of about 3350 metres, and in activity since 3.5 ka [27]. Its eastern flank extends beneath the sea surface to a depth of 2000 metres below sea level.

Mt. Etna has exhibited nearly continuous volcanic activity for the past 600 ka [28], causing significant destruction in Eastern Sicily. Despite extensive and continuous monitoring, the dynamics of its deepest and submerged parts, as well as the nature of its activity or its evolutionary origin, remain incompletely understood.

3. Technology used and description of the experiments and projects

This section provides a brief overview of the technology used and the experiments and projects carried out.

3.1 Stand-alone seafloor monitoring modules

Stand-alone seafloor monitoring modules, such as OBSHs (**Figure 2a**), typically possess the following key characteristics:

- Limited autonomy, depending on the sensors and the sampling rates used, which may allow operation for up to 1–2 years.
- Local data storage capability.
- Acoustic communications for controlling the start of the acquisition.

Figure 2.
(a) An example of OBSH owned by GEOMAR; (b) GEOSTAR seafloor observatory on the deck of R/V URANIA during the deployment in Tyrrhenian Sea; (c) relay buoy to allow near-real-time communication between seafloor observatory and inland; and (d) SN1 observatory during the operation, offshore Eastern Sicily to connect it to the electro-optical cable by ROV.

- Uncertainty in the orientation of the three-component sensors.
- Free-fall landing and pop-up recovery procedures utilizing acoustic release of the ballast.
- Alternative deployment and recovery options using ROVs.

Table 1 summarizes the experiments conducted and the primary technical specifications of the OBSHs utilized (extracted from References [29–35]).

3.2 Multiparametric platform

Multiparametric seafloor and water column observatories are more advanced platforms compared to stand-alone seafloor monitoring modules. These observatories offer extended operational lifetimes, enhanced performance, and the capability to detect signals over longer time scale. They are powerful tools for monitoring the sub-seafloor, seafloor, and water column, providing valuable insights into interactions among geosphere, biosphere, and hydrosphere, as well as their temporal evolution and variability. The key features of these observatories include:

- Multiple payload capabilities.
- Autonomy.
- Communication capabilities.
- Remote reconfiguration potential.
- Accurate positioning.
- Data acquisition procedures compatible with those used in land-based observatories.

A seafloor observatory is an unmanned station designed for long-term operations, supporting various instrumented packages across multiple scientific disciplines (**Figure 2b**). These platforms can be configured in different ways: autonomous, acoustically linked with an infrastructure (a moored buoy, **Figure 2c**; or another observatory), or cabled (**Figure 2d**) [6].

In recent years, integrating sensors into undersea telecommunications cables has become increasingly vital for addressing societal threats, such as climate change, sea level rise, ocean warming, earthquakes, and tsunamis. This approach helps bridge critical gaps in fundamental data needed to model, understand, and respond effectively to these challenges. The SMART (Science Monitoring And Reliable Telecommunications) Subsea Cables initiative leverages the growing demand for connectivity through underwater telecommunication cables [36]. Over the past 2 years, numerous global projects have been launched, ranging from early planning stages to implementation [37].

Table 2 summarizes the main information of the multiparametric platforms and their use in projects. **Table 3** lists the sensors used for each case. The content of these tables is extracted from Gasparoni et al. [38], which contains additional detailed information.

DOI: http://dx.doi.org/10.5772/intechopen.1007585

Area	Duration	OBSH	Seismometer	Hydrophone	Recorder/ Digitizer	Sampling rate (Hz)	Clock
Southern Tyrrhenian Sea (TYDE)	6 months (2000–2001)	14	PMD (0.016–32 Hz); Spahr Webb-SCRIPPS (0.02–100 Hz)	OAS E-2PD (0.5–50 Hz)	SEND MLS (21–16 bits)	50	GPS synch. on board
Marsili Seamount (Prometheus)	11 days (2006)	01	Trilium 3C broadband 40s	OAS E-2PD	SEND MLS (16 bits)	200	GPS synch. on board
Stromboli (STR06)	10 days (2006)	10	3C SM6 B-Coil (4.5 Hz)	HTI-1 (1–30 kHz)	GEOPRO SEDIS V (24 bits)	Variable (500, 250, 125, 62.5)	TCXO 0.01 PPM, 0–50 C
Ionian Sea (Seismofaults)	1 year (2017–2018)	08	Trilium 3C broadband 120 s	0.001—5 kHz	Güralp CD24	100	TCXO

Extracted from References [29–35].

Table 1.
List of the experiments with the OBSH main technical specifications.

n°	Project	Area	Duration	Platform
1	GEOSTAR-2 (1999–2001)	Southern Tyrrhenian Sea (1950 m)	Sept 2000–Apr 2001	GEOSTAR
2	ORION-GEOSTAR-3 (2002–2005)	Southern Tyrrhenian Sea (3350 m)	Dec 2003– Apr 2004	GEOSTAR, SN3
3	ORION-GEOSTAR-3 (2002–2005)	Southern Tyrrhenian Sea (3350 m)	Jun 2004–Apr 2005	GEOSTAR, SN3
4	GNDT (2000–2003)	Ionian Sea (2105 m)	Oct 2002–May 2003	SN1
5	INGV-INFN (2005–2008)	Ionian Sea (2105 m)	Jan 2005–May 2008	SN1 (cabled)
6	MUR-EMSO (2010 on)	Ionian Sea (2105 m)	Jun 2012–May 2013 Dec 2023 on	SN1 (cabled)

Extracted from Gasparoni et al. [38].

Table 2.
Main information on the multiparametric platforms and their use in projects.

Ser. No.	Platform	Sensors	Type	Sampling rate
1	GEOSTAR	3C broadband seismometer	Güralp CMG-1 T	100 Hz
		Scalar magnetometer	GEM Syst. GSM-19 L	1 sample/min
		Vectorial magnetometer	INGV/Tecnomare	1 sample/10 s
		Gravimeter	CNR-IFSI	1 Hz
		ADCP	RDI 300 kHz	1 prof/min
		CTD	Seabird SBE-16	1 sample/h
		Turbidity	Chelsea Alphatracka II	1 sample/h
		Current meter	FSI 3D-ACM	2 Hz
		Automatic water sampler	McLane RAS-48-500	500 mL/10 days
		Precision tilt meter	Applied Geomech. 716	1 sample/h
		Compass	Falmouth Ostar	1 Hz
2–3	GEOSTAR	Same of n. 1	Same of n. 1	Same of n. 1
	SN3	pH/Eh	Tecnomare with AMT	1 sample/6 h
		Hydrophone	OAS E-2PD	100 Hz
		3C broadband seismometer	PMD/EEntec EP300-DT	100 Hz
4	SN1	3C broadband seismometer	Güralp CMG-1 T	100 Hz
		Gravimeter	CNR-IFSI	1 Hz
		Low-frequency hydrophone	OAS E-2PD	100 Hz
		Current meter	FSI 3D-ACM	2 Hz
		CTD	Seabird SBE37-SM	1 sample/h
5	SN1	Same of n. 4	Same of n. 4	Same of n. 4
		Scalar magnetometer	Mar. Magnetics Sentinel 3000	1 sample/h
6	SN1	Same of n. 5	Same of n. 5	Same of n. 5
		Current meter (new)	Nobska MAVS-3	2 Hz
		Vectorial magnetometer	INGV	0.5 Hz
		Low-frequency hydrophone	SMID DT-405D(V)1	2 kHz
		DPG	SCRIPPS	100 Hz
		ADCP	RDI Sentinel 600 kHz	I prof/h
		4 hydrophones (bio-acoust.)	SMID TR-401(V)1	96 kHz
		Compass	Falmouth Ostar	1 Hz
		Absolute pressure gauge	Paroscientific 8CB4000-I	40–60 samples /min
		Inertial unit	ILMRK20-AHRS150-02-100	200 Hz

Extracted from Gasparoni et al. [38].

Table 3.
List of the sensors used for each case indicated in **Table 2.**

3.3 Description of the experiments and projects

This section provides a brief overview of the experiments and projects.

TYDE (*TY*rrhenian *D*eep-sea *E*xperiment, see the first row of **Table 1**).

The TYDE experiment, funded by the European Commission (EC) under Fourth Framework Program (FP4) within the Access to EC Research Infrastructures (2000–2003), conducted a long-term, six-month experiment in Southern Tyrrhenian Sea to contribute to the geodynamic understanding of this crucial region of the Mediterranean. Fourteen OBSHs, provided by GEOMAR and Hamburg University, were deployed [29, 30].

PROMETHEUS (*PRO*gram of *M*editerranean *E*xploration for *TH*ermal *E*nergy *U*se, see the second row of **Table 1**).

This project, funded by a private company, focused on an integrated study of the Marsili Seamount [39]. The primary objective was to characterize the seamount using swath bathymetry, chirp sub-bottom profiling, magnetometric and gravimetric surveys, water and seabed samplings. A seismic monitoring was also carried out through an OBSH deployed in different points of the seamount, including its flat summit. The sampling rate of 200 Hz per channel was selected to detect signals indicative of hydrothermal activity [35].

Stromboli experiment (*STR06*, see the third row of **Table 1**).

The STR06 geophysical experiment, funded by the Italian Civil Protection Department, aimed to acquire onshore and offshore seismic data through a combined land-based and marine seismic network. These data were used to reconstruct the seismic tomography of the Stromboli volcano and investigate its deep structures, including the location of magma chambers. Wide-angle refraction seismics was conducted around the volcano using a 4 GI-GUN tuned array. Data were recorded by the permanent seismic network of the *Istituto Nazionale di Geofisica e Vulcanologia* (INGV), along with 20 temporary land stations and 10 OBSHs deployed on the southeast (SE), southwest (SW), and northeast (NE) submerged flanks of the volcano following a detailed morpho-bathymetric analysis [31, 32].

Seismofaults experiment (see the fourth row of **Table 1**).

This experiment involved the deployment of eight OBSHs in the Ionian Sea for 1 year. It provided a detailed view of the seismogenic structures in the region, which are potential sources of medium- to high-magnitude earthquakes. The integration of offshore seismic data with land-based observations significantly improved the detection and identification of active tectonic structures in the area [33, 34].

GEOSTAR-2 (Second phase: Deep-Sea Scientific Mission, see the row 1 of **Table 2**).

GEOSTAR-2 was a project funded by the EC within the FP4-MAST3 program. The multiparametric seafloor observatory prototype, developed during the first phase, and named GEOSTAR (GEophysical and Oceanographic STation for Abyssal Research), completed its inaugural deep-sea mission, which lasted 7 months at a water depth of approximately 2000 metres in the Tyrrhenian Sea. Communication systems were enhanced using three independent approaches: floats developed in the first phase, an acoustic link between the underwater system and a surface unit; and a surface buoy equipped with an acoustic underwater link as well as both radio and satellite links to land [40].

ORION-GEOSTAR-3 (*O*cean *R*esearch by *I*ntegrated *O*bservatory *N*etworks, see rows 2 and 3 of **Table 2**).

ORION-GEOSTAR-3 was a project funded by the EC within the Fifth Framework Program (FP5). This project marked the transition from a single-node to a multinode system, deploying two multiparametric observatories (a main node and a satellite one). A deep-water network was established, enabling underwater acoustic communication between the satellite and the main node, which then communicated acoustically with a relay surface buoy linked to land via radio and satellite. ORION-GEOSTAR-3 implemented one of the first near-real-time deep-sea networks, which operated for nearly 2 years at depths exceeding 3300 metres in the Tyrrhenian abyssal plain, located at the northwestern base of the Marsili Seamount [41].

GNDT-SN1 (Design of Monitoring and Warning Networks in Marine Areas close to High-Seismic-Risk Zone: First Implementation in Eastern Sicily).

GNDT-SN1was an Italian project funded by *National Earthquake Defence Group* (GNDT) (see rows 4 and 5 of **Table 2**). The project developed SN1, a benthic, GEOSTAR-type multidisciplinary observatory [38], and conducted a stand-alone experiment off the coast of Eastern Sicily at a water depth of 2000 metres for 7 months during 2002–2003. In 2005, thanks to a specific agreement between *Istituto Nazionale di Fisica Nucleare* (INFN) and INGV, SN1 was redeployed at the same location and connected to an electro-optical cable, making it the first real-time European benthic observatory and one of the few worldwide at that time. SN1 is currently one of the operational nodes of the research infrastructure EMSO (European Multidisciplinary Seafloor and Water-Column Observatory, www.emso.eu). The maintenance costs of this cabled infrastructure are covered by the MUR-EMSO (*Ministry of Education, Universities and Research*) funds (see row 6 of **Table 2**), which is part of Italy's contribution through the *Ordinary Fund for Research Institutions* (FOE), allocated annually to INGV as Italy's representative entity in EMSO.

The following section describes some of the scientific results obtained using data recorded by infrastructures in the experiment and projects previously described.

4. Case studies

A brief description of the main results of studies performed with data recorded by underwater platforms is reported here.

4.1 Stromboli

Stromboli was the site chosen for several experiments with the use of OBSH modules.

The OBSH deployed near Stromboli (OBS5, 16 km SSE at a water depth of 1500 m) during the TYDE experiment [29] was the first underwater instrument able to detect low-frequency signals, which were identified as long-period (LP) events and volcanic tremor. **Figure 3** illustrates an example of LP events and tremor, their spectrograms and power spectral density (PSD). A comparison between the low-frequency seismicity recorded at OBS5 and explosion quakes (EQs) at a land station located on the summit of the volcano suggests that these recordings cannot be attributed to the same source. EQs manifest as impulsive events with short duration and are characterized by higher frequency content compared to the LP events, representing the signal generated by the explosion of gas bubbles near the surface of the volcano. As low-frequency seismicity is considered a key indicator of fluid migration within the volcanic edifice, with the source mechanism typically attributed to fluid resonance in volcanic conduits and fractures. Analyzing data recorded by OBS5, Sgroi et al. [25] identified the

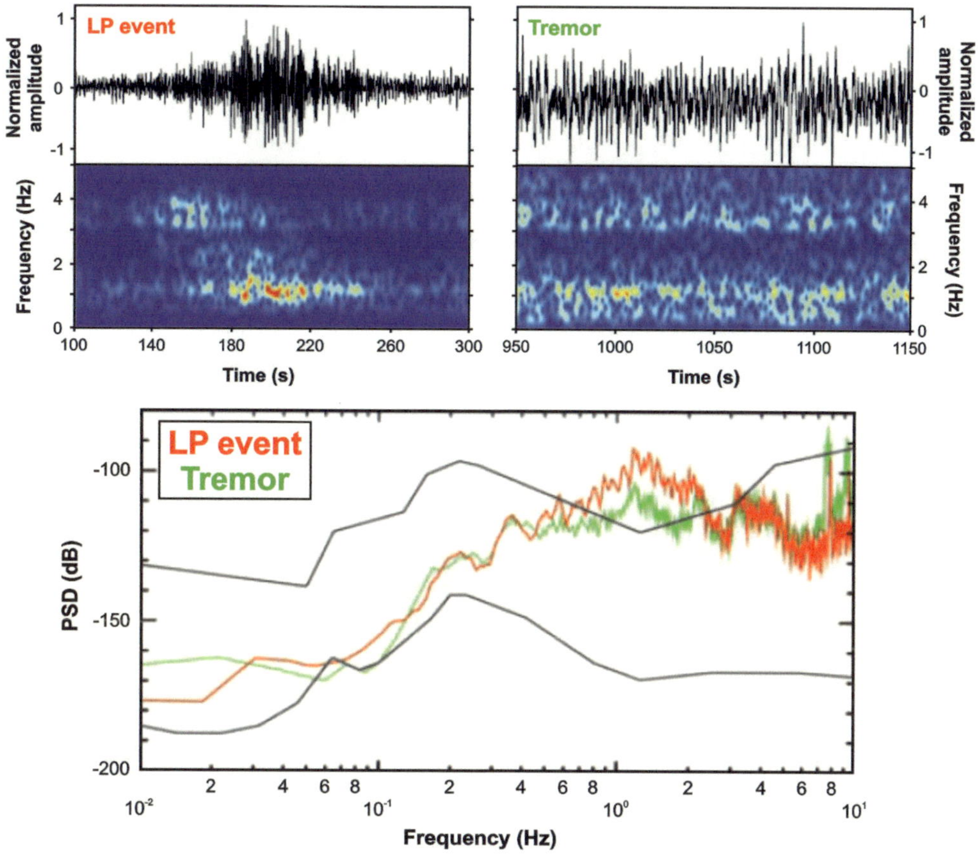

Figure 3.
Vertical component seismograms and spectrograms computed on 200-s window of LP event (top left) and tremor (top right) recorded by OBS5. In the spectrograms, the warm colors (red, yellow, and green) define the dominant spectral amplitudes, while cooler colors (light to dark blue) define lower amplitudes and background. Bottom graph shows a comparison between PSD of LP event (red trace) and tremor (green trace) recorded by OBS5 (modified after Sgroi et al. [25]). Spectral content is similar for both seismic signals, but spectral amplitude of LP event is about 10 dB greater than tremor signal in the range of 1–2 Hz. Black solid lines represent the HNM (High Noise Model) and LNM (Low Noise model) curves according to Peterson [42].

source of the LP events with a magmatic reservoir located approximately at 3-km depth. On the contrary, the source of the EQs (recorded only by land station) was associated with a shallower reservoir at a depth of around 400 m. The distinct features of the seismic signals recorded by both the land station and OBS5 hypothesize the presence of two separate seismic sources [43].

This result was confirmed by the construction of a high-resolution three-dimensional (3D) velocity model of the volcanic edifice, using both onshore and offshore seismic sensors deployed during the STR06 experiment [31, 32]. The joint inversion of artificial and natural seismic sources data revealed a high-velocity seismic body, leading to the identification of a magma chamber at 1–3-km depth [44].

Moreover, another 3D velocity (V_P) model of the Southern Tyrrhenian Sea was developed for the first time by integrating both land and marine (OBSHs and seafloor observatories) [45]. This approach has provided new insights into the tectonic and volcanic structures of the region. The results of the 3D inversion were more accurate due to the inclusion of the bathymetry, which exhibits significant variation (ranging from approximately −3500 m in the Marsili Basin to +3350 m at Mt. Etna). This tomography model identified two main low-velocity anomalies under the central Aeolian Islands and Mt. Etna related to active volcanism. These two volumes are separated and located at different depths (above 16 km for Aeolian, between 24 and 28 km for Mt. Etna).

4.2 Submerged flanks of Stromboli, Marsili, and other seamounts

Detailed mapping and sampling of submerged volcanic structures have revealed complex eruption histories. Comparing submerged and emerged volcanic systems significantly enhances our understanding of volcanic processes. **Figure 4** illustrates the 3D bathymetry of the submerged flanks of Stromboli, the Marsili Seamount, and two seamounts, Eolo and Enarete, located west of the Aeolian Arc (see **Figure 1**).

The SSE submerged flank of Stromboli displays an irregular seafloor morphology, with several outcrops of volcanic basement, separated by NE-SW trending lineaments. These features likely result from the channelized flow of volcanoclastic deposits towards the slope and bathyal plain (**Figure 4a**). In contrast, the SW submerged flank has a more regular morphology, marked by volcanic blocks deposited after mass wasting along the slope or outcrops of volcanic basement (**Figure 4b**) [31]. The Eolo Seamount is characterized by large scars along its southern and eastern flanks, shaping the current morphology of the volcanic edifice, while at the Enarete Seamount the collapse of the eastern flank is distinctly evident. These findings are interpreted as later manifestations of volcanic activity, developed during Mid-Late Pleistocene [46].

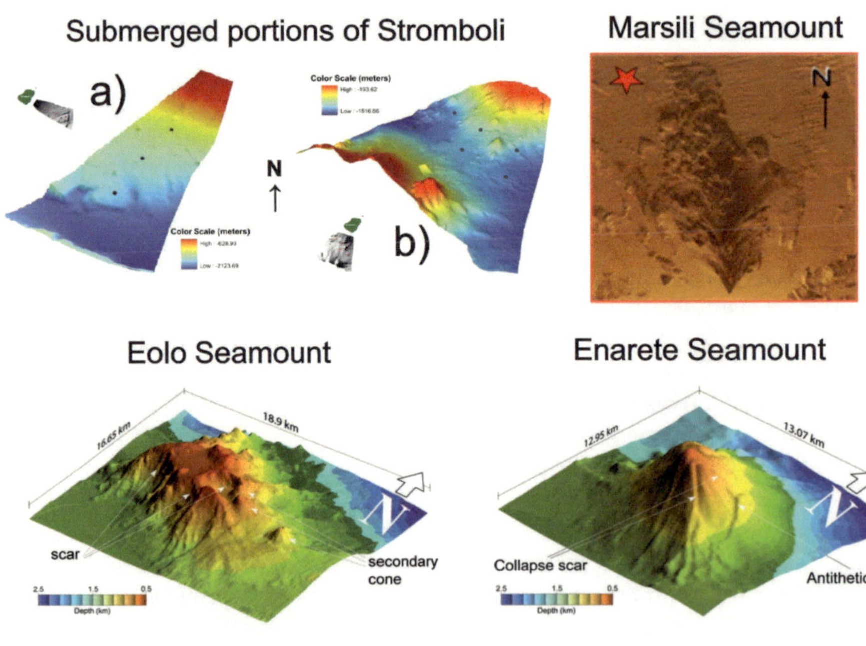

Figure 4.
3D bathymetry of Stromboli ('a' SSE and 'b' SW submerged flanks, modified after Marsella et al. [31]); Marsili Seamount, the red star indicates the area of deployment of the two observatories during the ORION-GEOSTAR-3 project (modified after Marani and Gamberi [20]); Eolo and Enarete Seamounts (modified after Bortoluzzi et al. [46]).

4.3 Marsili seamount

Accurate bathymetric maps reconstructed around the Marsili Seamount revealed that the volcanic structure is complex, with a broad, gently sloping summit and steep flanks [20]. The Marsili Basin exhibits a high heat flow, averaging 250 mWm^{-2}, and reaching 500 mWm^{-2} in the uppermost and central portion of the seamount [47, 48].

Detailed magnetic and gravimetric surveys were conducted across the entire seamount area [39], which overlapped with the positive heat-flow anomalies (**Figure 5a,b**). During 2003–2005, long-term geophysical and oceanographic monitoring was carried out using two observatories deployed at depths exceeding 3300 m in the NW vicinity of the seamount [41].

An analysis of the first 8 months of data collected by the seismometer at the main node revealed the occurrence of over 700 earthquakes, including 130 teleseisms (M > 4.8), 218 regional events (M > 2.3), and 185 localized local events (M > 1.9), along with 209 non-localized low-magnitude earthquakes. Among the non-localized local events, 160 were detected exclusively by the underwater observatory, indicating

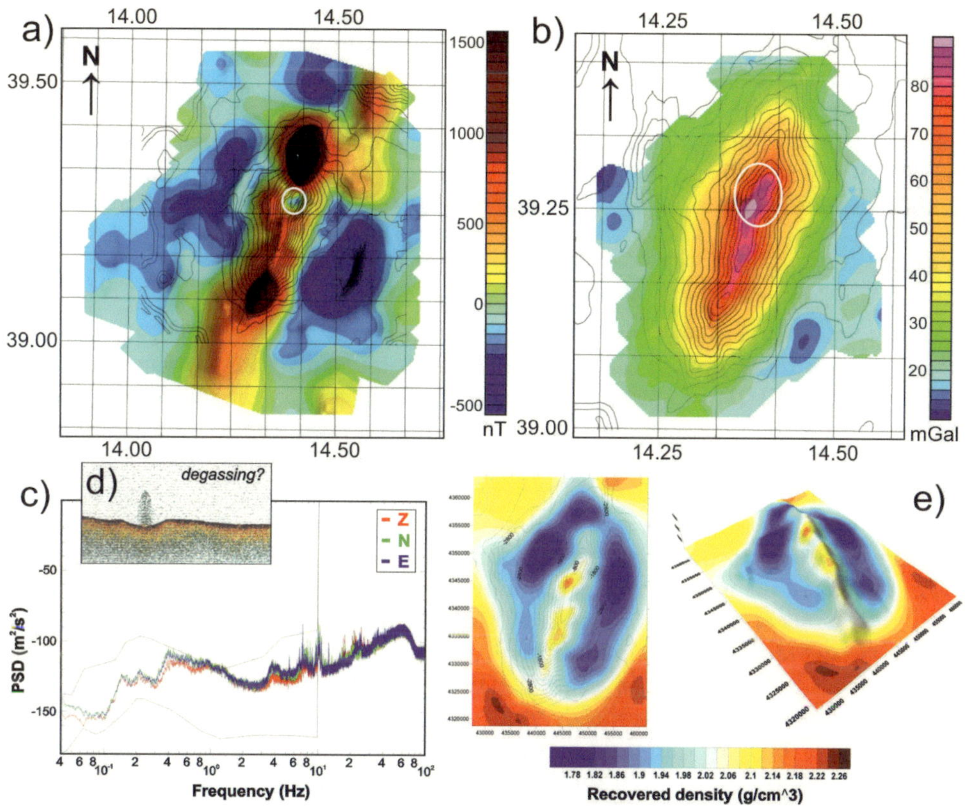

Figure 5.
Marsili Seamount: (a) magnetic anomaly field reduced to the magnetic pole, (b) gravimetric survey ('a' and 'b', modified after Caso et al. [49]). Low magnetization and high density are recognizable in correspondence of the central summit zone of Marsili (white circles); (c) PSD of OBSH three components (red, Z; green, N-S; and blue, E-W) computed on one-hour long seismic signal recorded on the summit of the seamount (18 July 2006, h 12:00), the LNM and HNM lines according to Peterson [42]; (d) 80-m-high episode of fluid venting in the proximity of the Marsili Seamount crest recorded with chirp investigations (courtesy of Institute of Marine Sciences-Italian National Research Council (ISMAR-CNR)); and (e) recovered density model of the seamount (modified after Caso et al. [49]).

a degree of local activity (data extracted from Ciafardini [50]). The broadband seismometer recorded notable short-duration seismic signals at frequencies exceeding 10 Hz, which could be attributed to phase transitions between/by liquid-supercritical fluid states of seawater, consistent with hydrothermal circulation. Additionally, gravimetric measurements showed fluctuations in the gravity signal that coincided with increases in high-frequency seismic noise, suggesting the presence of magmatic hydrothermal fluids [51]. The vectorial magnetometer time series indicated a decrease in the ratio of vertical to horizontal magnetic field components, corresponding to an increase in electrical conductivity. This variation can be interpreted as a sign of deep circulation of thermally mineralized fluids [41].

The presence of an active hydrothermal system is also indicated by seismic activity recorded by the OBSH on the flat crest of the seamount [52]. The signals are characterized by higher frequencies compared to normal seismic noise, within the 4–60 Hz frequency range, and an unexpectedly high number of non-tectonic seismic events (approximately 800 within a few days). These events can be categorized into two main groups: 720 with frequencies between 4 and 10 Hz, and 80 events with the 20–60 Hz range [35]. The first group is likely associated with hydrothermal fluid circulation processes, while the second may be generated by phase changes in supercritical fluid. **Figure 5c** illustrates an example of the PSD of the seismic noise, with visible peaks in the 20–60 Hz range. Additional evidence of fluid venting was detected through chirp sonar investigations (**Figure 5d**). The observed anomalous gravity field can only be explained by assuming an average density of the volcanic structure of approximately 2 g/cm^3 (**Figure 5e**). The collective evidence strongly supports the hypothesis that the Marsili Seamount constitutes a geothermal field of significant potential [53].

4.4 Mud volcanoes

Morphologically similar to volcanoes [54], the mud volcanoes are mainly connected to the tectonic activity and are characterized by the expulsion of mud, brine, and gases, such as methane (CH_4) and carbon dioxide (CO_2). Their activity can range from slow, continuous emission to more vigorous, episodic bursts. These eruptions form conical mounds or, in some cases, expansive flat areas covered in mud. The surrounding tectonic activity may influence the behavior of the mud volcano, potentially triggering more significant eruptions. Among many examples [55], two notable instances of mud volcanoes have been discovered offshore Sicily, south of the island and within the Ionian Sea.

The first example was discovered offshore south of Sicily and studied by Holland et al. [56]. The second, named Bortoluzzi Mud Volcano (BMV) in honor of Giovanni Bortoluzzi, was recently identified through a bathymetric survey in the Ionian Sea during the Seismofaults experiment [33]. This survey revealed significant changes in seafloor topography and the discovery of a circular structure [57]. The BMV has offered valuable insights into fluid and gas dynamics of the Earth's crust, particularly in the Calabrian Arc subduction zone. **Figure 6** presents a high-resolution bathymetric map, three chirp profiles, a shaded-relief map, and the seismicity data from 1990 to 2019 across the BMV. Notably, a cluster of hypocentres is observed approximately 40 km northeast of the BMV.

4.5 Mt. Etna

The Western Ionian Sea is an area of great interest from tectonic and volcanic points of view. It is particularly suited to multidisciplinary studies due to its high

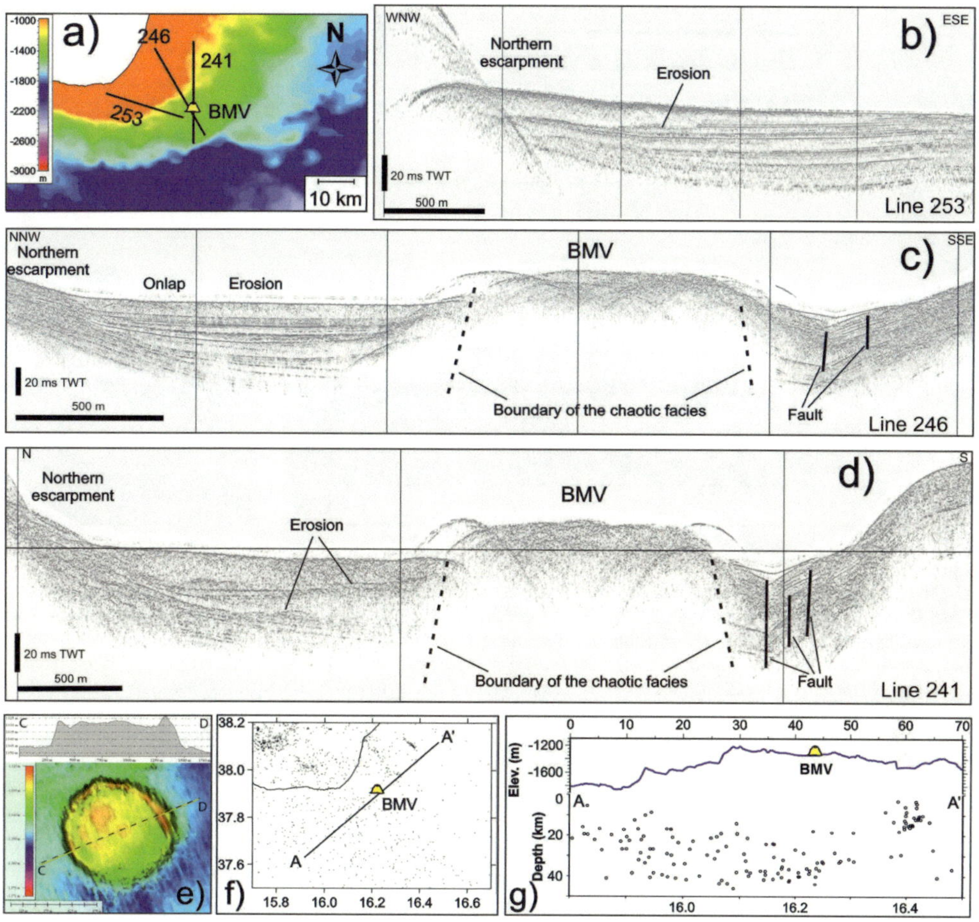

Figure 6.
(a) High-resolution bathymetric map (modified after Bosman et al. [58]), data were collected during the Seismofaults experiment [33, 34], map of the 3-single-channel chirp profiles (numbered 253, 246, and 241); (b), (c), and (d) chirp profile lines shown in 'a'; (e) shaded-relief map of the BMV and location of bathymetric section (C-D) (modified after Bosman et al. [58]); (f) map and SW-NE section (A-A' profile) of the seismic events that occurred around the BMV from 1990 to 2019; (g) plot of the seismic events within a distance of about 10 km from the transect (earthquake data from Sgroi et al. [59] and Barberi et al. [60]).

seismic activity, including some of the strongest earthquakes (M7+) that have generated in some cases intense tsunami waves [61, 62]. Moreover, this region encompasses Mt. Etna, one of the most active volcanoes in the Mediterranean, whose deeper and submerged activity is not fully understood.

The SN1 multidisciplinary observatory was deployed, with the aim to focus on its tectonic and volcanic complexity. Long-term multiparametric analysis of the SN1 time series data is revealed to be crucial for identifying volcanic processes and their variations [43, 63, 64].

During the stand-alone deployment in 2002–2003, SN1 well recorded several hundred seismic events (teleseismic, regional, and local). Notably, 485 local events detected by SN1 were not reported in standard seismological bulletins [65]. During this period, a significant flank eruption was characterized by moderate lava emission, intense explosive activity, and abundant ash emission. This eruption began 2 hours after a seismic crisis and provided a good opportunity to study low-frequency

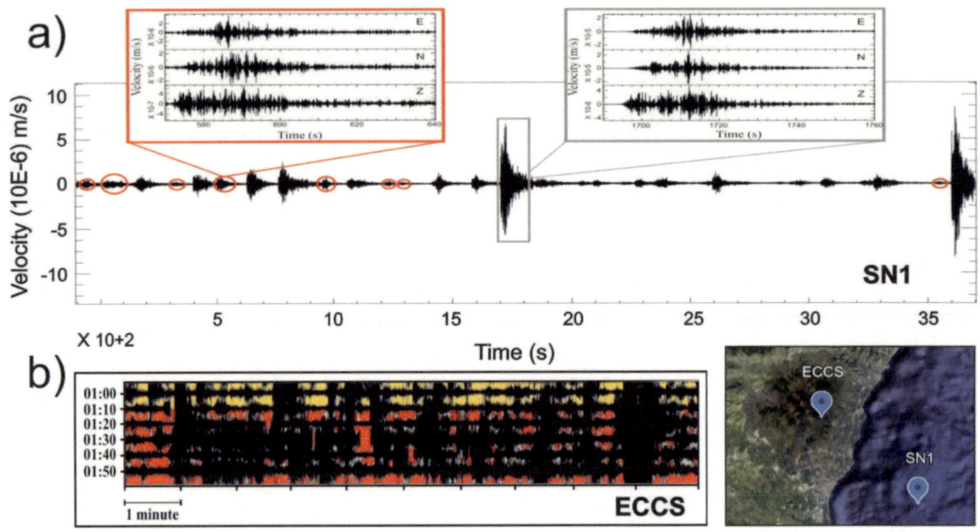

Figure 7.
(a) One-hour vertical component seismogram (Butterworth 4–12 Hz bandpass) recorded by SN1 during the onset of Mt. Etna eruption (27 October 2002) and more than 40 earthquakes were recorded by SN1 together with land stations. The events recorded by the seafloor seismometer, but not reported in seismic bulletins are circled in red (an example is shown in the top-left inset). The top-right inset represents an Etnean event (M_D 3.5, [66]) from the swarm accompanying the eruption onset, recorded by land and offshore stations (modified after Sgroi et al. [65]); (b) ECCS land-station seismogram recorded at the same time of 'a'. Due to the high amplitude of volcanic tremor, some low-magnitude earthquakes are not clearly visible on land-station seismograms. The bottom-right map shows the two seismic stations ECCS and SN1 located at about 7.5 and 50 km from Mt. Etna summit.

seismic signals recorded by SN1 during explosive volcanic activity. **Figure 7** illustrates an example of a one-hour seismic event recorded by SN1 during the onset of the Mt. Etna eruption, with unreported events circled in red. For comparison, the same hour of data from the ECCS land-station seismogram is also shown (see the caption of **Figure 7** for the comments).

The comparison between seismic data recorded during both an eruption and a cycle of lava fountain at Etna is shown in **Figure 8**. This figure presents the strain release and the number of earthquakes per day (a), and the root mean square (RMS; b) computed on the vertical component of the seismic signal from October 2002 to January 2003 (the eruption onset and end are marked with red arrows), while **Figure 8c,d** present the same type of plot but from 1 January to 30 April 2013 (red asterisks marking the occurrence of lava fountains).

In 2013, Mt. Etna experienced another significant eruption phase characterized by 13 intense episodes of lava fountains. Also, an intense volcanic ash fallout occurred and it was completely documented by SN1's oceanographic sensor. These volcanic events were clearly observed by SN1, deployed at approximately 50 km from the volcano summit. The intensity and depth of the lava fountains generated signals that were more effectively detected by SN1 than by land-based stations at similar distances, such as the ESLN station, which is the closest land-based station [63].

Due to its location, SN1 has recorded various signal types associated with instabilities in the Western Ionian linked to both Mt. Etna's volcanism and tectonic structures, potential sources of destructive earthquakes [63, 65]. For example, the high-frequency, impulsive short-duration events (SDEs) continuously detected by SN1 are likely related to gas movements caused by Mt. Etna's volcanism.

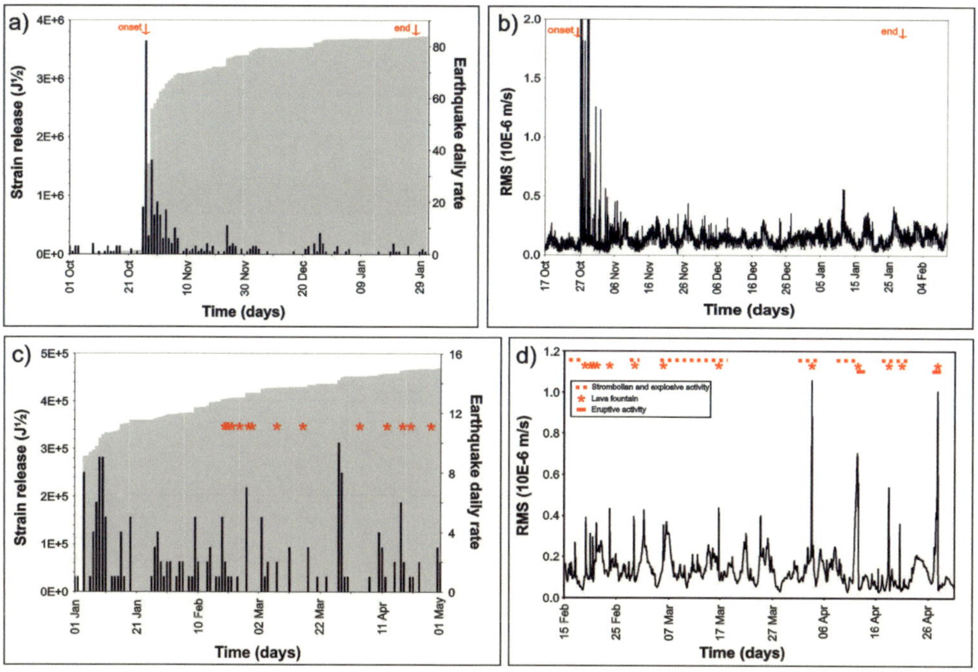

Figure 8.
(a) Cumulative strain release (grey area, earthquake data from Alparone et al. [67]) and histogram of the number of earthquakes per day (1 October 2002–2031 January 2003), the onset and end of the eruption are marked with red arrows; (b) RMS computed on the vertical component of the SN1 seismic signal filtered 0.8–8 Hz; (c) cumulative strain release (gray area, earthquake data from Alparone et al. [68]) and histogram of the number of earthquakes per day (1 January–30 April 2013), the lava fountain episodes are marked with red asterisks; (d) RMS computed on the vertical component of the SN1 seismic signal filtered 0.8–8 Hz.

An analysis performed on continuous background signal allowed for the differentiation of volcanic tremor from oceanographic noise, highlighting the distinct characteristics of these signals [69]. **Figure 9** highlights that the predominant contribution to SN1's signals occur in the 0.01–0.1 Hz frequency band, where the signal is well correlated with sea wave height data (blue plot). In the 0.1–0.8 Hz band, there is an increase in background noise amplitude observed at both SN1 and land stations. After the EPOZ land station near the coast became operational on 20 March 2013, it was found to be more sensitive to sea wave impacts compared to SN1 in the 0.1–0.8 and 0.8–8 Hz bands. Despite this, both SN1 and EPOZ land station effectively detect sea wave motion during lava fountain events (indicated with black asterisks), with SN1 showing clear peaks associated with increased tremor activity. The 0.8–8 Hz frequency band captures volcanic tremor signals, with notable peaks corresponding to lava fountain episodes.

The joint use of the seismic signals recorded by SN1 and land stations improves the accuracy of the source location. The deep geometry of Mt. Etna plumbing system remains inadequately understood using only land-based stations. Integrating seafloor measurements with the existing dense on-land network is essential to advance this understanding. Sharp et al. [70] proposed the existence of a deep source beneath Mt. Etna, at approximately a 20-km depth based on deep sounding techniques. This finding was corroborated by a 3D tomography study that revealed a significant P-wave low-velocity anomaly at the crust-mantle interface beneath Mt. Etna [45] and associated to deep (about 30 km) tectonic activity in the

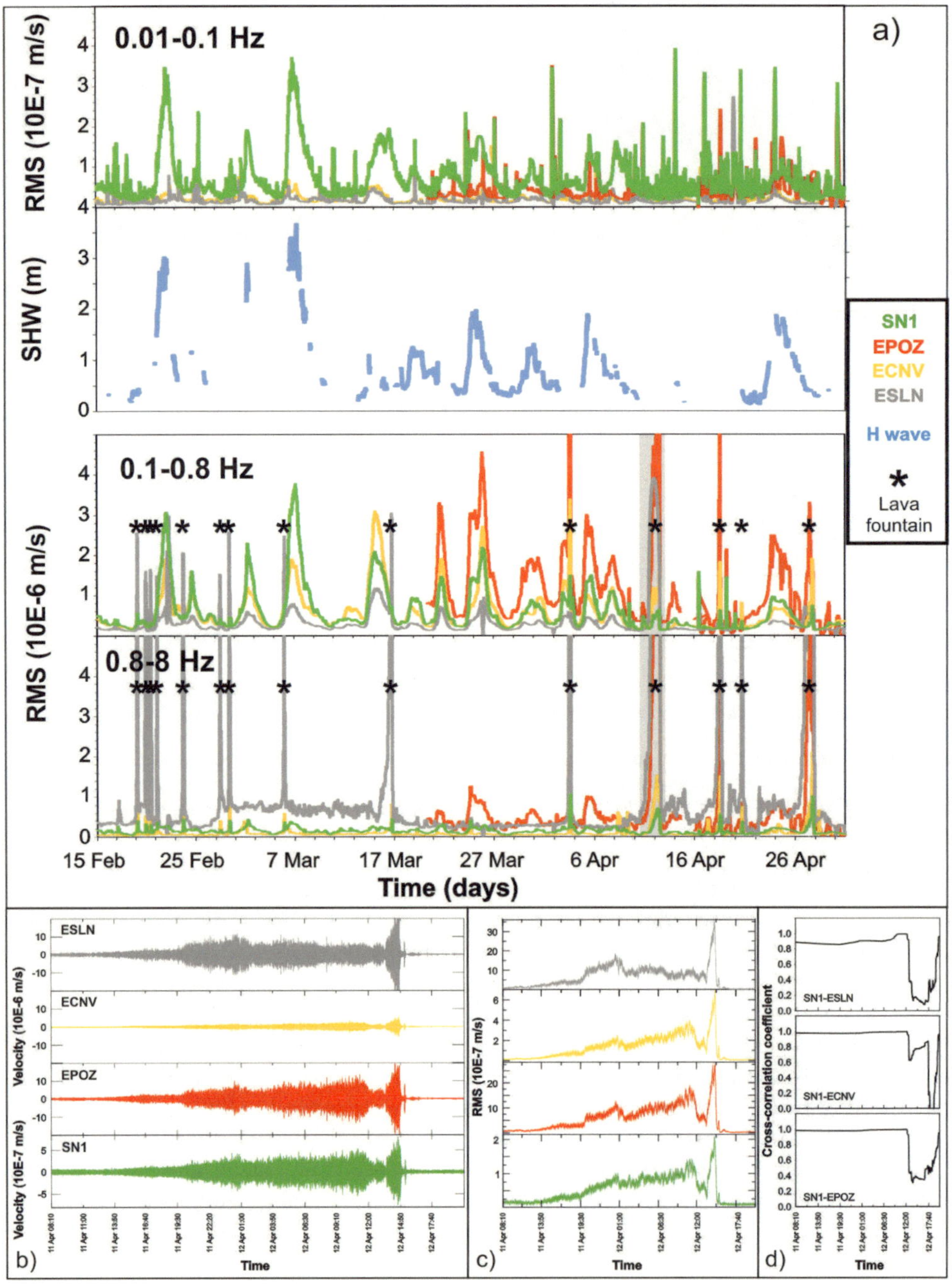

Figure 9.

(a) RMS on the seismic signal recorded by SN1 and 3-land seismic stations (15 February–30 April 2013) in three different frequency bands (0.01–0.1 Hz; 0.1–0.8 Hz; and 0.8–8 Hz), the light gray rectangle encloses an example of lava fountain episode (see 'd'); (b) vertical component of background seismic signal (filtered 0.8–8 Hz) recorded by 3-land stations and SN1 during a lava fountain episode that occurred from 11 April to 12 April 2013; (c) RMS computed on the previous background seismic signal (filtered 0.8–8 Hz); (d) correlation between signals recorded by SN1 and each of the land stations, during the different phases of the lava fountain episode. Maximum correspondence among SN1 and land-station signals is observed at the maximum phase of explosive activity.

northwestern sector of the volcano. Michon et al. [71] compared volcanic systems at Piton de la Fournaise, Hawaii, and Mt. Etna, concluding that the regional stress field plays a crucial role in controlling magma transfer along deep plumbing systems.

5. Conclusions

Southern Italy is a hotspot for volcanic activity, encompassing both emerged volcanoes, like Mt. Etna and the Aeolian Islands, particularly Stromboli, as well as submerged volcanic structures, such as the Marsili Seamount.

The integration of advanced underwater systems has revolutionized volcanic studies. Key contributions of these systems include:

i. *Real-time monitoring:* Using geophysical, geochemical, and oceanographic sensors, real-time monitoring can detect volcanic tremor, gas emissions, and anomalies in environmental parameters, offering early warnings of potential eruptions.

ii. *Geophysical surveys:* These surveys provide data on the magnetic, gravimetric, and seismic characteristics of both emerged and underwater volcanoes, helping to elucidate their internal structures and magmatic processes.

iii. *Seafloor samplings:* The collection of rocks, sediments, and gas samples from underwater volcanoes aids in understanding their composition, age, and eruption history. This information is crucial for comparing submerged and emerged volcanic systems.

High-resolution swath bathymetry offers detailed maps of seafloor topography, essential for determining the most suitable sites for the deployment of underwater systems and for identifying volcanic structures, like cones, craters, and lava flows.

Data recorded on the seafloor have led to several significant insights into the volcanic activity and geodynamics of Southern Italy, and long-term sea-based observations are indispensable for directly studying volcanic processes.

Continued development in underwater exploration technologies, as well as long-term sea-based observations, is essential for directly studying volcanic processes. These advancements have enhanced our understanding of complex volcanic activity and are crucial for safeguarding populations in this volcanically active region.

Acknowledgements

The authors wish to deeply thank all the colleagues who contributed to the success of the European and Italian experiments and projects mentioned here. They come from the following institutions: GEOMAR, Hamburg University, CNR, Florence University, INGV, 'Sapienza' Rome University, IFREMER, CNRS, Tecnomare-ENI, TFH, TUB, SERCEL, INAF, IPGP, OGS, Roma-3 University, Chieti University, Catania University, Messina University, Palermo University, and INFN.

Conflict of interest

The authors declare no conflict of interest.

Author details

Tiziana Sgroi and Paolo Favali*
Istituto Nazionale di Geofisica e Vulcanologia (INGV), Roma, Italy

*Address all correspondence to: paolo.favali@ingv.it

References

[1] Pino NA, Piatanesi A, Valensise G, Boschi E. The 28 December 1908 Messina Straits earthquake (M_w 7.1): A great earthquake throughout a century of seismology. Seismological Research Letters. 2009;**80**(2):243-259. DOI: 10.1785/gssrl.80.2.243

[2] Favali P, Dañobeitia JJ, Howe BM, Ruhl HA. The Discovery of the Unknown Planet: The Ocean. Lausanne: Frontiers Media SA; 2023. 265 p. DOI: 10.3389/978-2-8325-3029-0

[3] Lampitt RS, Favali P, Barnes CR, Church MJ, Cronin MF, Hill KL, et al. In situ sustained Eulerian observatories. In: Hall J, Harrison DE, Stammer D, editors. Proceedings of the OceanObs'09: Sustained Ocean. Observations and Information for Society Conference 2; 21-25 September 2009. Venice, Italy. ESA Publication WPP-306; 2010. DOI: 10.5270/OceanObs09.pp.27

[4] Weller RA, Baker DJ, Glackin MM, Roberts SJ, Schmitt RW, Twigg ES, et al. The challenge of sustaining ocean observations. Frontiers in Marine Science. 2019;**6**:105. DOI: 10.3389/fmars.2019.00105

[5] Ruhl HA, André M, Beranzoli L, Çağatay MN, Colaço A, Cannat M, et al. Societal need for improved understanding of climate change, anthropogenic impacts, and geo-hazard warning drive development of ocean observatories in European seas. Progress in Oceanography. 2011;**91**:1-33. DOI: 10.1016/j.pocean.2011.05.001

[6] Favali P, Beranzoli L. Seafloor Observatory Science: a Review. Annals of Geophysics 2006;**49**(2-3):515-567. DOI: 10.4401/ag-3125

[7] Favali P, Beranzoli L, De Santis A, editors. Seafloor Observatories: A New Vision of the Earth from the Abyss, Springer-Praxis Books in Geophysical Sciences. Berlin Heidelberg: Springer-Verlag; 2015. 676 p. DOI:10.1007/978-3-642-11374-1

[8] Dañobeitia JJ, Pouliquen S, Pade N, Arvanitidis C, Sanders R, Stanica A, et al. The role of the marine research infrastructures in the European marine observation landscape: Present and future perspectives. Frontiers in Marine Science. 2023;**10**:1047251. DOI: 10.3389/fmars.2023.1047251

[9] Google Earth Pro. Google Earth LLC. 2022. Available from: https://earth.google.com/web/

[10] Sgroi T, De Nardis R, Lavecchia G. Crustal structure and seismotectonics of Central Sicily (southern Italy): New constraints from instrumental seismicity. Geophysical Journal International. 2012;**189**(3):1237-1252. DOI: 10.1111/j.1365-246X.2012. 05392.x

[11] Polonia A, Torelli L, Artoni A, Carlini M, Faccenna C, Ferranti L, et al. The Ionian and Alfeo-Etna fault zones: New segments of an evolving plate boundary in the Central Mediterranean Sea? Tectonophysics. 2016;**675**:69-90. DOI: 10.1016/j.tecto.2016.03.016

[12] Lavecchia G, Bello S, Andrenacci C, Cirillo D, Pietrolungo F, Talone D, et al. QUIN 2.0- new release of the QUaternary fault strain INdicators database from the southern Apennines of Italy. Scientific Data. 2024;**11**(1):189. DOI:. DOI: 10.1038/s41597-024-03008-6

[13] Dewey J, Helman M, Knott S, Turco E, Hutton D. Kinematics of the

western Mediterranean. The Geological Society of London—Special Publications. 1989;**45**:265-283. DOI: 10.1144/GSL.SP.1989.045.01.15

[14] Argnani A, Savelli C. Cenozoic volcanism and tectonics in the southern Tyrrhenian Sea: Space-time distribution and geodynamic significance. Journal of Geodynamics. 1999;**27**:409-432. DOI: 10.1016/S0264-3707(98)00025-8

[15] Argnani A. Evolution of the southern Tyrrhenian slab tear and active tectonics along the western edge of the Tyrrhenian subducted slab. The Geological Society of London—Special Publications. 2009;**311**:193-212. DOI: 10.1144/SP311

[16] Marani M, Trua T. Thermal constriction and slab tearing at the origin of a super-inflated spreading ridge: Marsili volcano (Tyrrhenian Sea). Journal of Geophysical Research. 2002;**107**(B9):2188. DOI: 10.1029/2001JB000285

[17] Cocchi L, Caratori Tontini F, Muccini F, Marani MP, Bortoluzzi G, Carmisciano C. Chronology of the transition from a spreading ridge to an accretional seamount in the Marsili back-arc basin (Tyrrhenian Sea). Terra Nova. 2009;**21**(5):369-374. DOI: 10.1111/j.1365-3121.2009.00891.x

[18] Sartori R, Torelli L, Zitellini N, Carrara G, Magaldi M, Mussoni P. Crustal features along a W–E Tyrrhenian transect from Sardinia to Campania margins (Central Mediterranean). Tectonophysics. 2004;**383**(3-4):171-192. DOI: 10.1016/j.tecto.2004.02.008

[19] Scarascia S, Lozej A, Cassinis R. Crustal structures of the Ligurian, Tyrrhenian and Ionian seas and adjacent onshore areas interpreted from wide—Angle seismic profiles. Bollettino di Geofisica Teorica ed Applicata. 1994;**36**(141-144):5-20

[20] Marani MP, Gamberi, F. Distribution and nature of submarine volcanic landforms in the Tyrrhenian Sea: the arc vs the back-arc. In: Marani MP, Gamberi F, Bonatti E, editors. From Seafloor to Deep Mantle: Architecture of the Tyrrhenian Back-arc Basin. Memorie Descrittive Carta Geologica d'Italia. Vol. 64. Roma: System Cart; 2004. pp. 109-126

[21] Pondrelli S, Piromallo C, Serpelloni E. Convergence vs. retreat in southern Tyrrhenian Sea: Insights from kinematics. Geophysical Research Letters. 2004;**31**(6):L06611. DOI: 10.1029/2003GL019223

[22] Lucchi F, Tranne CA, Rossi PL. New chronostratigraphic constraints for the eruptive history of Alicudi volcano (Aeolian archipelago, Italy). Geoacta. 2008;7(7389):115-131

[23] Gugliandolo C, Italiano F, Maugeri T. The submarine hydrothermal system of Panarea (southern Italy): Biogeochemical processes at the thermal fluids-sea bottom interface. Annals of Geophysics. 2006;**49**(2-3):783-792. DOI: 10.4401/ag-3139

[24] Allard P, Carbonnelle J, Métrich N, Loyer H, Zettwoog P. Sulphur output and magma degassing budget of Stromboli volcano. Nature. 1994;**368**:326-330

[25] Sgroi T, Montuori C, Agrusta R, Favali P. Low-frequency seismic signals recorded by OBS at Stromboli volcano (southern Tyrrhenian Sea). Geophysical Research Letters. 2009;**36**:L04305. DOI: 10.1029/2008GL036477

[26] Etiope G, Milkov AV. A new estimate of global methane flux from onshore and shallow submarine mud volcanoes to the atmosphere. Environmental Geology. 2004;**46**:997-1002. DOI: 10.1007/s00254-004-1085-1

[27] Global Volcanism Program. Mt. Etna (211060). [Database] Volcanoes of the World (v. 5.2.3; 20 Sep 2024). Distributed by Smithsonian Institution, compiled by Venzke E. 2024. DOI: 10.5479/si.GVP.VOTW5-2024.5.2

[28] Branca S, Coltelli M, Groppelli G. Geological evolution of a complex basaltic stratovolcano: Mount Etna, Italy. Italian Journal of Geosciences. 2011;**30**(3):306-317. DOI: 10.3301/IJG.2011.13

[29] Dahm T, Thorwart M, Flueh ER, Braun T, Herber H, Favali P, et al. Ocean bottom seismometers deployed in Tyrrhenian Sea. Eos, Transactions American Geophysical Union. 2002;**83**(29):309-315

[30] Sgroi T, Braun T, Dahm T, Frugoni F. An improved seismicity picture of the southern Tyrrhenian area by the use of OBS and land-based networks: The TYDE experiment. Annals of Geophysics. 2006;**49**(2-3):801-817. DOI: 10.4401/ag-3130

[31] Marsella E, Favali P, Castellano M, Aiello G, Bortoluzzi G, Di Fiore V, et al. The Stromboli Geophysical Experiment. Preliminary Report on Wide Angle Refraction Seismics and Morphobathymetry of Stromboli Island (Southern Tyrrhenian Sea, Italy) Based on Integrated Offshore-Onshore Data Acquisition (Cruise STR06 R/V URANIA). Bologna; 2007. CNR-ISMAR Technical Report 102. 84 p. Available from: www.projects.bo.ismar.cnr.it/MEDITERRNEAN/STROMBOLI/STR06 REP

[32] Castellano M, Augusti V, De Cesare W, Favali P, Frugoni F, Montuori C, et al. Seismic tomography experiment at Italy's Stromboli Volcano. Eos, Transactions American Geophysical Union. 2008;**89**(30):269-276

[33] Billi A, Cuffaro M, Beranzoli L, Bigi S, Bosman A, Caruso C, et al. The SEISMOFAULTS project: First surveys and preliminary results for the Ionian Sea area, southern Italy. Annals of Geophysics. 2020;**63**(3):SE326. DOI: 10.4401/ag-8171

[34] Sgroi T, Polonia A, Beranzoli L, Billi A, Bosman A, Costanza A, et al. One year of seismicity recorded through ocean bottom seismometers illuminates active tectonic structures in the Ionian Sea (Central Mediterranean). Frontiers in Earth Science. 2021;**9**:661311. DOI: 10.3389/feart.2021.661311

[35] D'Alessandro A, D'Anna G, Luzio D, Mangano G. The INGV's new OBS/H: Analysis of the signals recorded at the Marsili submarine volcano. Journal of Volcanology and Geothermal Research. 2009;**183**(1-2):17-29. DOI: 10.1016/j.jvolgeores.2009.02.008

[36] Howe BM, Arbic BK, Aucan J, Barnes CR, Bayliff N, Becker N, et al. SMART cables for observing the global ocean: Science and implementation. Frontiers in Marine Science. 2019;**6**:424. DOI: 10.3389/fmars.2019.00424

[37] Howe BM, Angove M, Aucan J, Barnes CR, Barros JS, Bayliff N, et al. SMART subsea cables for observing the earth and ocean, mitigating environmental hazards, and supporting the blue economy. Frontiers in Earth Science. 2022;**9**:775544. DOI: 10.3389/feart.2021.775544

[38] Gasparoni F, Furlan F, Bruni F, Zanon F, Favali P, Beranzoli L, Marinaro G, De Santis A, Gerber HW. GEOSTAR-class observatories 1995-2012: A technical overview. In: Favali P, Beranzoli L, De Santis A, editors. Seafloor Observatories: A New Vision of the Earth from the Abyss, Springer-Praxis Books in Geophysical Sciences. Heidelberg:

Springer-Verlag Berlin; 2015. pp. 229-304. DOI:10.1007/978-3-642-11374-1_11

[39] Paltrinieri D, Viezzoli C, Signanini P, Di Sabatino B, Caso C, Iezzi B, et al. Joint Research Project "PROMETHEUS" an Integrated Study of the Marsili Seamount, Tyrrhenian Sea. Bologna, Italy: CNR-ISMAR Interim Report MRS06; 2006. 46 p. Available from: www.projects.bo.ismar.cnr.it/MARSILI

[40] Beranzoli L, Favali P, Smriglio G, editors. Science Technology Synergy for Research in the Marine Environment: Challenges for the XXI Century. In: Developments in Marine Technology. Vol. 12. Amsterdam: Elsevier; 2002. 268 p

[41] Beranzoli L, Ciafardini A, Cianchini G, De Caro M, De Santis A, Favali P, Frugoni F, Marinaro G, Monna S, Montuori C, Qamili E, Sgroi T, Vitale S. A first insight into the Marsili volcanic seamount (Tyrrhenian Sea, Italy): Results from ORION-GEOSTAR3 experiment. In: Favali P, Beranzoli L, De Santis A, editors. Seafloor Observatories: A New Vision of the Earth from the Abyss, Springer-Praxis Books in Geophysical Sciences. Berlin Heidelberg: Springer-Verlag; 2015. pp. 623-641. DOI:10.1007/978-3-642-11374-1_24

[42] Peterson JR. Observations and Modelling of Seismic Background Noise. Open-File Report 93-322. Albuquerque, New Mexico: USGS; 1993. 94 p. DOI: 10.3133/ofr93322

[43] Giovanetti G, Monna S, Lo Bue N, Embriaco D, Frugoni F, Marinaro G, et al. Observing volcanoes from the seafloor in the Central Mediterranean area. Remote Sensing. 2016;**8**:298. DOI: 10.3390/rs8040298

[44] Patanè D, Barberi G, De Gori P, Cocina O, Zuccarello L, Garcia-Yeguas A, et al. The shallow magma chamber of Stromboli Volcano (Italy). GeophysIcal Research Letters. 2017;**44**:1-8. DOI: 10.1002/2017GL073008

[45] Monna S, Sgroi T, Dahm T. New insights on volcanic and tectonic structures of the southern Tyrrhenian (Italy) from marine and land seismic data. Geochemistry, Geophysics, Geosystems. 2013;**14**:3703-3719. DOI: 10.1002/ggge.20227

[46] Bortoluzzi G, Ligi M, Romagnoli C, Cocchi L, Casalbore D, Sgroi T, et al. Interactions between volcanism and tectonics in the western Aeolian sector, southern Tyrrhenian Sea. Geophysical Journal International. 2010;**183**(1):64-78. DOI: 10.1111/j.1365-246X.2010.04729.x

[47] Della Vedova B, Bellani S, Pellis G, Squarci P. Deep temperatures and surface heat flow distribution. In: Vai GB, Martini IP, editors. Anatomy of an Orogen: The Apennines and Adjacent Mediterranean Basin. Great Britain, UK: Kluwer Academic Publishers; 2001. pp. 65-79. DOI: 10.1007/978-94-015-9829-3_7

[48] Verzhbitskii EV. Heat flow and matter composition of the lithosphere of the World Ocean. Oceanology. 2007;**47**:564-570. DOI: 10.1134/S000143700704013 3

[49] Caso C, Signanini P, De Santis A, Favali P, Iezzi G, Marani MP, et al. Submarine geothermal systems in southern Tyrrhenian Sea as future energy resource: The example of Marsili seamount. In: Proceedings of the World Geothermal Congress 2010; 25-30 April 2010. Bali, Indonesia; 2010. pp. 1-9

[50] Ciafardini A. Study of the S-wave attenuation of Mediterranean earthquakes recorded by a broadband seismometer installed in a seafloor observatory (Marsili Basin, Southern

Tyrrhenian Sea) [PhD thesis]. Chieti: "G. d'Annunzio" Univerity; 2006

[51] Battaglia M, Gottsmann J, Carbone D, Fernàndez J. 4D volcano gravimetry. Geophysics. 2008;**73**:WA3-WA18. DOI: 10.1190/1.2977792

[52] Caso C. Volcano Marsili exploration for geothermal purposes: Seismic noise analysis [PhD thesis]. Chieti: "G. d'Annunzio" University; 2007

[53] Paltrinieri D, Favali P, Italiano F, Signanini P, Caso C, Armani FB. The Marsili seamount offshore geothermal reservoir: A big challenge for an energy transition model. Energies. 2022;**15**(5):1900. DOI: 10.3390/en15051900

[54] Borgia A, Aubert M, Merle O, van Wyk de Vries B. What is a volcano? In: Cañón-Tapia E, Szakács A, editors. The Geological Society of America (Special Publications). Vol. 470. 2010. DOI: 10.1130/2010.2470(01)

[55] Etiope G, Caracausi A, Favara R, Italiano F, Baciu C. Methane emission from the mud volcanoes of Sicily (Italy). Geophysical Research Letters. 2002;**56**(8):1215. DOI: 10.1029/2001GL014340

[56] Holland CW, Etiope G, Milkov AV, Michelozzi E, Favali P. Mud volcanoes discovered offshore Sicily. Marine Geology. 2003;**199**:1-6. DOI: 10.1016/S0025-3227(03)00125-7

[57] Cuffaro M, Billi A, Bigi S, Bosman A, Caruso CG, Conti A, et al. The Bortoluzzi mud volcano (Ionian Sea, Italy) and its potential for tracking the seismic cycle of active faults. Solid Earth. 2019;**10**:741-763. DOI: 10.5194/se-10-741-2019

[58] Bosman A, Cuffaro M, Conti A, Gasperini L, Petracchini L, Sgroi T. High-Resolution Multibeam Bathymetry along the Sicily and Calabria continental margin (Bortoluzzi Mud Volcano, Ionian Sea, Italy). GFZ Data Services. 2019. DOI: 10.5880/FIDGEO.2019.014

[59] Sgroi T, Polonia A, Barberi G, Billi A, Gasperini L. New seismological data from the Calabrian arc reveal arc-orthogonal extension across the subduction zone. Scientific Reports. 2021;**11**:473. DOI: 10.1038/s41598-020-79719-8

[60] Barberi G, Sgroi T, Marchetti A. The new earthquake Locations in the Western Ionian Sea (WISeLoc_1990-2019). [Data set]: INGV, Roma; 2024. DOI: 10.13127/wiseloc/1990-2019

[61] Boschi E, Guidoboni E, Ferrari G, Mariotti D, Valensise G, Gasperini P. Catalogue of strong Italian earthquakes from 461 B.C. to 1997. Annals of Geophysics. 2000;**43**(4). DOI: 10.4401/ag-3668

[62] Tinti S, Maramai A, Graziani L. The new catalogue of Italian tsunamis. Natural Hazards. 2004;**33**:439-465. DOI: 10.1023/B:NHAZ.0000048469.51059.65

[63] Sgroi T, Monna S, Embriaco D, Giovanetti G, Marinaro G, Favali P. Geo-hazards in the western Ionian Sea: Insights from non-earthquake signals recorded by the NEMO-SN1 seafloor observatory. Oceanography. 2014;**27**(2):154-166. DOI: 10.5670/oceanog.2014.51

[64] Monna S, Falcone G, Beranzoli L, Chierici F, Cianchini G, De Caro M, et al. Underwater geophysical monitoring for European multidisciplinary seafloor and water-column observatories. Journal of Marine Systems. 2014;**130**:12-30. DOI: 10.1016/j.jmarsys.2013.09.010

[65] Sgroi T, Beranzoli L, Di Grazia M, Ursino A, Favali P. New observations of local seismicity by the SN-1 seafloor observatory in the Ionian Sea, off-shore eastern Sicily (Italy). Geophysical Journal International. 2007;**169**(2):490-501. DOI: 10.1111/j.1365-246X.2007.03348.x

[66] ISIDe Working Group. Italian Seismological Instrumental and Parametric Database (ISIDe) [Data Set]. INGV; 2007. DOI: 10.13127/ISIDE

[67] Alparone SC, Maiolino V, Mostaccio A, Scaltrito A, Ursino A, Barberi G, et al. Mt. Etna Seismic Catalog 2000-2010 [Data set]. INGV; 2015. DOI: 10.13127/ETNASC/2000_2010

[68] Alparone SC, Barberi G, Di Grazia G, Giampiccolo E, Maiolino V, Mostaccio A, et al. Mt. Etna Seismic Catalog 2011-2013 (EtnaSC_2011_2013) (Version 1) [Data Set]. INGV; 2020. DOI: 10.13127/ETNASC/2011_2013

[69] Sgroi T, Di Grazia G, Favali P. Volcanic tremor of Mt. Etna (Italy) recorded by NEMO-SN1 seafloor observatory: A new perspective on volcanic eruptions monitoring. Geosciences. 2019;**9**:115. DOI: 10.3390/geosciences903011

[70] Sharp A, Davis P, Gray F. A low velocity zone beneath Mount Etna and magma storage. Nature. 1980;**287**:587-591. DOI: 10.1038/287587a0

[71] Michon L, Ferrazzini V, Di Muro A, Villeneuve N, Famin V. Rift zones and magma plumbing system of piton de la Fournaise volcano: How do they differ from Hawaii and Etna? Journal of Volcanology and Geothermal Research. 2015;**303**:112-129. DOI: 10.1016/j.jvolgeores.2015.07.031

Chapter 7

Statistical Analysis of Seismic and Tsunami Waves Generated by the 2022 Tonga Volcano's Eruption

Leonid Chernogor and Yiyang Luo

Abstract

On January 15, 2022, a catastrophic explosive eruption occurred at the Hunga-Tonga-Hunga-Ha'apai (HTHH) volcano in the Kingdom of Tonga, marking a significant geological event. The eruption, located at 20°54′S, 175°38′W, involved 3–5 powerful explosions, with the most severe at 04:15 UTC, and was accompanied by an earthquake of magnitude $M_s \approx 5.8$. The eruption's source energy was estimated between 6.5 to 23 TJ and 19 to 69 TJ, while the energy of the associated air waves ranged from 16.7 to 75.3 TJ and 2 ± 0.8 PJ. Classified as a supervolcano, the total energy release was approximately 17.6 ± 2.7 Mt TNT or 73.6 ± 11.3 TJ, with an eruption heat energy of about 3.9×10^{18} J or 932 Mt TNT. This eruption caused extensive damage to multiple Earth subsystems, including the lithosphere, ocean, atmosphere, ionosphere, and magnetosphere. This paper examines the statistical characteristics of seismic and tsunami waves resulting from the 2022 HTHH eruption, focusing on Rayleigh waves, slow seismic waves potentially related to Lamb waves, underwater long waves, and tsunami waves.

Keywords: Hunga-Tonga-Hunga-Ha'apai volcano, explosive volcanic eruption, seismic waves, underwater long waves, tsunami waves, Rayleigh waves, lamb waves, statistical analysis

1. Introduction

On January 15, 2022, a catastrophic explosive eruption occurred at the Hunga-Tonga-Hunga-Ha'apai volcano in the Kingdom of Tonga (hereinafter referred to as the "Tonga eruption"), located at 20°54′S, 175°38′W. The event featured five significant explosions, with the most severe occurring at 04:15 UTC, accompanied by an earthquake with magnitude $M_s \approx 5.8$ [1]. The estimated energy of the eruption source ranged between 6.5 to 23 TJ and 19 to 69 TJ in various reports [2–6]. The energy of the associated air waves ranged from 16.7 to 75.3 TJ and 2 ± 0.8 PJ [7–10]. Our findings indicate the total energy did not exceed 17.6 ± 2.7 Mt TNT or 73.6 ± 11.3 TJ, with a thermal energy release of approximately 3.9×10^{18} J or 932 Mt TNT, corresponding to an average power output of around 91 TW [11].

Classified as a supervolcano, the Tonga eruption caused widespread disruptions across Earth's subsystems, including the lithosphere, oceans, atmosphere, ionosphere, and magnetosphere [12]. Traditionally, a supervolcano is defined by a volcanic explosivity index (VEI) of 8, indicating an eruption that expels over 1000 cubic kilometers (240 cubic miles) of material. Such eruptions typically occur at hotspots, like Yellowstone, or subduction zones, such as the Toba eruption 75,000 years ago. These supervolcanoes form when magma from the mantle accumulates in the crust, unable to escape, until the pressure triggers a massive eruption. The Tonga eruption, however, suggests a novel classification approach based on the total energy released, which challenges the traditional focus on eruptive volume alone [6]. This event underscores the need to consider both energy output and sediment dispersal in defining supervolcanoes. Comprehensive modeling of these effects is supported by multi-instrumental observations [13].

The Tonga eruption generated various seismic effects, as documented in [2, 14, 15], and tsunamis, discussed in [9, 16–21]. Atmospheric processes were examined in [3, 6–8, 22–29], while ionospheric disturbances are detailed in [8, 23, 30–37], and geomagnetic effects are presented in [38–41]. These publications observed and analyzed seismic waves, tsunamis, atmospheric waves, ionospheric disturbances, and magnetic field oscillations.

This study focuses on the statistical characteristics of seismic waves and tsunamis generated by the Tonga eruption. Key results have been presented at international conferences [42]. The objective is to provide statistical assessments of the mechanical properties of Rayleigh waves, slow seismic waves, and tsunamis resulting from the eruption. The paper is structured as follows: it begins with a brief description of the data sources used, followed by statistical analyses of earthquake parameters, Rayleigh waves, slow seismic waves, and tsunamis, and concludes with a discussion and summary.

2. Data sources

For the statistical analysis, we utilized seismograms sourced from the Global Seismographic Network (GSN), which is managed by the Incorporated Research Institutions for Seismology (IRIS) Consortium [43]. This network provides a comprehensive set of seismic data from around the world, essential for analyzing earthquake activities. Additionally, tsunami data were obtained from the Federation of Digital Seismograph Networks (FDSN) [44] and GeoNet Data Exploration [45], which offer critical information on tsunami characteristics and propagation.

Our analysis encompassed several key methodologies. We generated correlation fields, including scatter diagrams, to visually assess the relationships between different seismic and tsunami parameters. Regression equations were formulated to quantify these relationships and derive predictive models. Statistical parameters were calculated to understand the variations and trends in earthquake, seismic wave, and tsunami characteristics associated with the Tonga volcano's eruption.

To visualize the spatial distribution of key geographic features related to the Tonga eruption, **Figure 1** was created by using ArcGIS Pro software. The map utilizes the Papua Palau Azimuthal Equidistant projection coordinate system, centered on the Pacific Ocean. This projection is particularly effective for accurately representing distances and directions from a central point, making it well-suited for covering the extensive area of the Pacific Ocean relevant to our study.

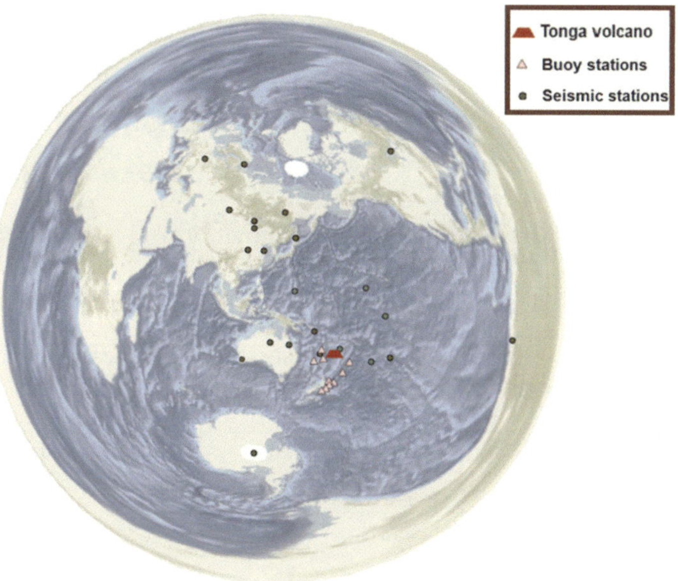

Figure 1.
Spatial distribution of key monitoring stations and the Tonga volcano: visualization using the Papua Palau azimuthal equidistant projection.

Furthermore, **Figure 1** illustrates the location of essential stations, including buoy stations for tsunami detection, seismic stations for earthquake and seismic wave data, and the Tonga volcanoes. By showing their geographic coordinates (GC) and relative positions, the map provides a clear view of the spatial relationships that impact seismic wave and tsunami propagation. This visualization highlights the interconnectedness of the seismic network and the distribution of monitoring stations around the Tonga volcanic region.

More detailed explanations and descriptions of the recorded data, station information, and related parameters are provided in the following sections.

3. Statistical analysis of the earthquakes

In this section, we examine and analysis the statistical characteristics of the earthquakes triggered by the Tonga volcano eruption.

The Tonga volcano eruption triggered a moderate earthquake with an approximate magnitude of $M_s \approx 5.8$, as reported by the US Geological Survey [1]. Our analysis of the seismic impact is detailed in **Table 1**, which includes data from 11 monitoring stations. Observations were made across a range of distances from the volcano, spanning from 0.76 to 11.20 million meters (Mm). We observed a decline in the amplitude (A) of surface vibrations, decreasing from 1.50×10^{-4} to 1.33×10^{-6} meters, and a reduction in the velocity (V) of surface propagation, ranging from 6×10^{-6} to 4.26×10^{-8} meters per second. The oscillation period (T) showed relatively minor variations, ranging from 20 to 31.2 seconds. **Figure 2** illustrates the correlation between amplitude of surface vibrations and distance, with observational data fitted using four distinct regression models (see **Figure 2a–d** respectively):

$$A = 8.588 \cdot 10^{-5} r^{-2.025}, \quad R^2 = 0.9911, \boldsymbol{\sigma} = 4.33 \cdot 10^{-6}\text{m} \tag{1}$$

$$A = e^{-6.799 - 3.065r + 0.585r^2 - 0.0342r^3}, \quad R^2 = 0.9962, \boldsymbol{\sigma} = 3.20 \cdot 10^{-6}\text{m}, \tag{2}$$

$$A = 5.821 \cdot 10^{-5} r^{-2.941} + 1.637 \cdot 10^{-5} r^{-0.6465}, \quad R^2 = 0.9951, \boldsymbol{\sigma} = 3.63 \cdot 10^{-6}\text{m}, \tag{3}$$

$$A = 11.07 \cdot 10^{-5} r^{-2.146} e^{-0.5259r + 0.2094r^2 - 0.0141r^3}, \quad R^2 = 0.9969, \boldsymbol{\sigma} = 3.13 \cdot 10^{-6}\text{m}, \tag{4}$$

where R^2 represents the coefficient of determination and σ denotes the root mean square (RMS) error.

Station code	Geographic coordinates latitude (°N), longitude (°E)	Distance (r), Mm	Amplitude (A), 10^{6} m	Period (T), s	Velocity (V), 10^{8} m/s	Magnitude (M_s)
MSVF	−17.74, 178.05	0.76	150	25	600	5.45
DZM	−22.07, 166.44	1.89	21.7	22	99	5.60
PPTF	−17.59, −149.57	2.73	4.7	21.9	21.5	5.10
HNR	−9.43, 159.95	2.92	11.7	23	51	5.36
CTAO	−20.09, 146.25	4.00	11.4	20	62	5.67
WRAB	−19.93, 134.35	5.23	7.89	20	39.5	5.67
YSS	46.95, 142.76	8.61	7.66	25.2	36.4	5.65
YAK	62.03, 129.68	10.39	3.57	29.6	12.1	5.63
ATAH	−7.14, −78.40	10.46	1.42	20.5	7.1	5.30
ULN	47.87, 107.05	10.82	2.64	24.4	10.8	5.63
TLY	51.68, 103.64	11.20	1.33	31.2	4.26	5.65

Table 1.
Key parameters of seismic stations and earthquake magnitude.

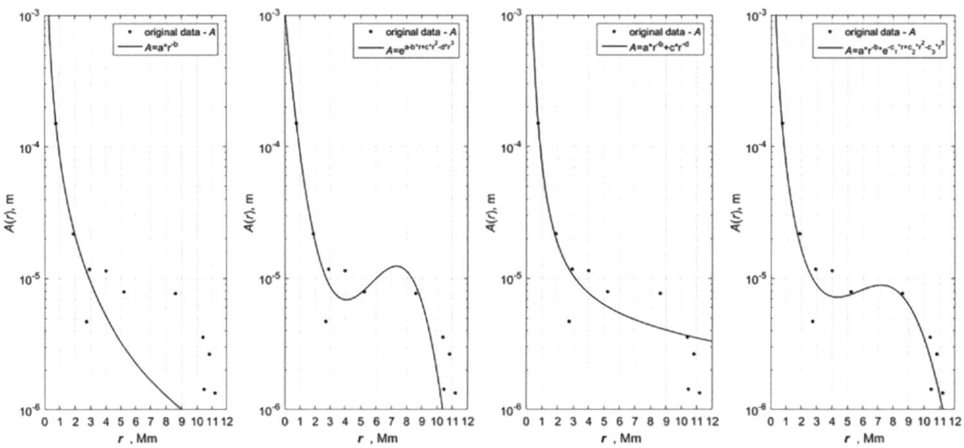

(a) (b) (c) (d)

Figure 2.
Dependence of seismic amplitude attenuation with distance.

According to Eqs. (1) and (2), the attenuation depth, which represents the distance at which the amplitude decreases by a factor of Euler's number (e), is estimated to be approximately 1.4 times the original distance for Eq. (1), and $r \approx 2.5$ Mm for Eq. (2) results in an attenuation depth of approximately e.

Regarding the period of seismic vibrations, a suitable approximation is T = const, specifically T = 25.6 ± 5.6 seconds. These period values exhibited variations within a range of approximately 20%. In investigating the relationship between the surface propagation speed and distance, we evaluated the following four approximations (illustrated in **Figure 3a–d**, respectively):

$$V = 3.573 \cdot 10^{-6} r^{-1.879}, \quad R^2 = 0.9866, \sigma = 2.21 \cdot 10^{-7} \text{m/s}, \tag{5}$$

$$V = e^{-10.07 - 3.001r + 0.5981r^2 - 0.03634r^3}, \quad R^2 = 0.9943, \sigma = 1.56 \cdot 10^{-7} \text{m/s}, \tag{6}$$

$$V = 2.041 \cdot 10^{-6} r^{-3.1} + 9.906 \cdot 10^{-7} r^{-0.7638}, \quad R^2 = 0.9916, \sigma = 1.89 \cdot 10^{-7} \text{m/s}, \tag{7}$$

$$V = 4.457 \cdot 10^{-6} r^{-2.062} e^{-0.5089r + 0.2163^2 - 0.01515r^3}, \quad R^2 = 0.9949, \sigma = 1.59 \cdot 10^{-7} \text{m/s}. \tag{8}$$

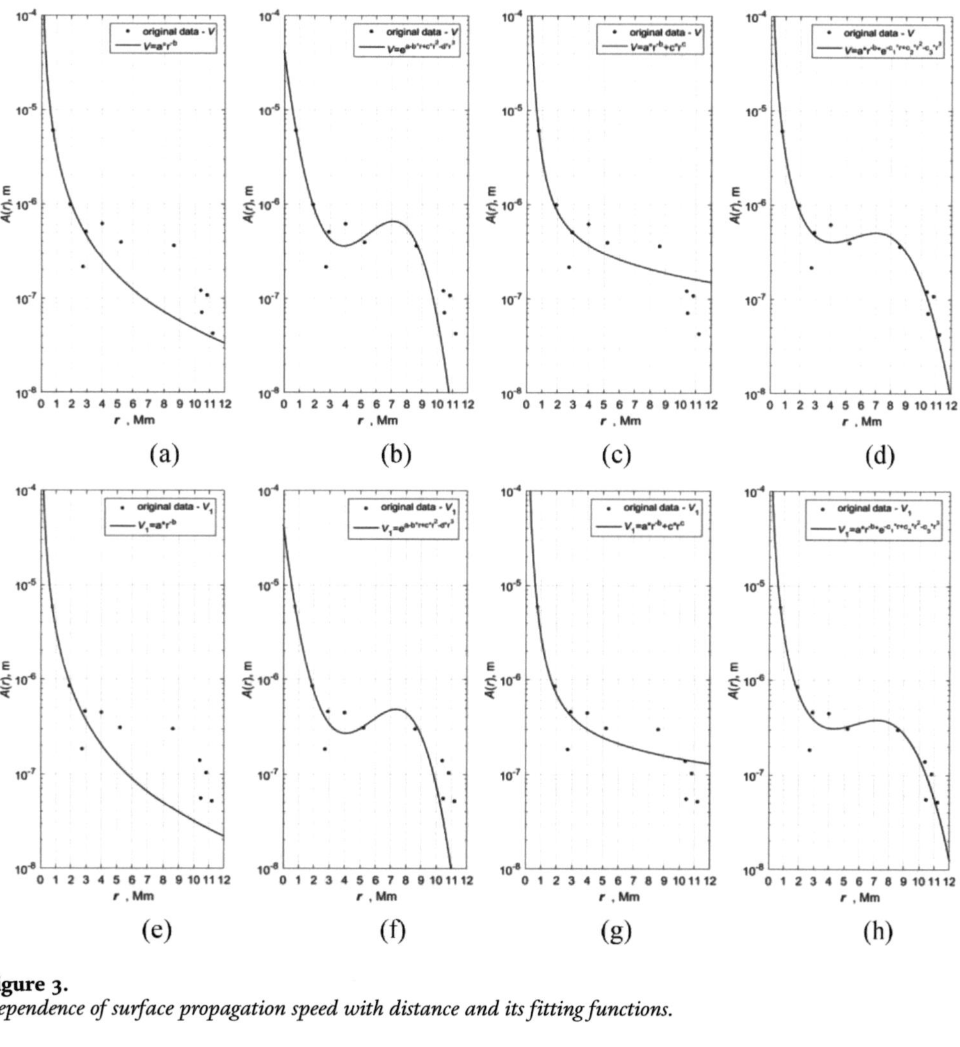

Figure 3.
Dependence of surface propagation speed with distance and its fitting functions.

When estimating V from the values of A, considering $T = T_1 = 25.6$ seconds, we obtained adjusted speed values denoted as $V_1 = A/T_1$. The relationships between V_1 and r are illustrated in **Figure 3e–h**. The corresponding approximating equations are expressed as follows:

$$V_1 = 3.355 \cdot 10^{-6} r^{-2.025}, \quad R^2 = 0.9911, \sigma = 1.69 \cdot 10^{-7} \mathrm{m/s}, \tag{9}$$

$$V_1 = e^{-10.04 - 3.065r + 0.5849r^2 - 0.03419r^3}, \quad R^2 = 0.9962, \sigma = 1.25 \cdot 10^{-7} \mathrm{m/s}, \tag{10}$$

$$V_1 = 2.274 \cdot 10^{-6} r^{-2.941} + 6.395 \cdot 10^{-7} r^{-0.6465}, \quad R^2 = 0.9951, \sigma = 1.42 \cdot 10^{-7} \mathrm{m/s}, \tag{11}$$

$$V_1 = 4.324 \cdot 10^{-6} r^{-2.146} e^{-0.5259r + 0.2094r^2 - 0.0141r^3}, \quad R^2 = 0.9969, \sigma = 1.22 \cdot 10^{-7} \mathrm{m/s}. \tag{12}$$

Based on the estimation and analysis presented above, several key results have emerged:

1. Regarding the earthquake's magnitude, calculated using data from the 11 mentioned stations (refer to **Table 1**) at varying distances from the volcano, ranging from 0.76 Mm to 11.20 Mm, the estimate is $M_s = 5.385 \pm 0.285$, indicating magnitude values fluctuated within a range of 5.3%. Notably, the average value differs from the USGS data by approximately 7.1%.

2. The amplitude of surface vibrations shows a strong dependency on distance, as described by Eqs. (1)–(4). These approximations demonstrate high reliability with R^2 values ranging from 0.9911 to 0.9969 and relatively low standard deviations (σ) between 3.13×10^{-6} and 4.33×10^{-6} meters. Notably, amplitudes tend to stabilize at distances greater than 10 Mm. Among the equations, Eq. (1) provides the most consistent results, with a clear physical interpretation, while Eqs. (2) and (3) should be considered with caution due to their lack of physical clarity.

3. The relationship between surface vibration speed and distance is approximated by Eqs. (5)–(8). Eq. (5) offers the simplest and most intuitive model, whereas more complex relationships in Eqs. (6)–(8) result in higher R^2 values and lower σ, though they lack clear physical explanations. A significant observation is the transition in speed from $V(r) \sim r^{-3.1}$ at distances less than 1.4 Mm to $V(r) \sim r^{-0.7638}$ at distances greater than 1.4 Mm.

4. Using an average vibration period $T_1 \approx 25.6 \pm 5.6$ seconds for the seismic waves, as opposed to variable periods at different stations, provides improved R^2 values and reduced σ values, indicating more reliable estimates of vibration speed.

4. Statistical analysis of the Rayleigh waves

Every earthquake generates seismic waves, including Rayleigh waves. In this section, we examine and analysis the statistical characteristics of the Rayleigh Waves triggered by the Tonga volcano eruption.

Table 2 summarizes the principal parameters of these waves for seven stations. It is observed that as the distance r increases by an order of magnitude, the time delay (Δt) also increases approximately by an order of magnitude.

Station code	Station name/region	Geographic coordinates latitude (°N), longitude (°E)	Distance (r), Mm	Time delay (Δt), s	Velocity (V), km/s
MSVF	Nonsavu, Fiji	−17.74, 178.05	0.76	216	3.52
DZM	Mont Dzumac, New Caledonia	−22.07, 166.44	1.89	540	3.50
PPTF	Pamatai, Papeete, Tahiti, French Polynesia	−17.59, −149.57	2.73	678	4.03
CTAO	Charters Towers, Queensland, Australia	−20.09, 146.25	4.00	1104	3.62
WRAB	Tennant Creek, Northern Territory, Australia	−19.93, 134.36	5.23	1482	3.53
NWAO	Narrogin (SRO), Western Australia, Australia	−32.93, 117.23	6.73	1800	3.74
YSS	Yuzhno-Sakhalinsk, Sakhalinskaya Oblast', Russia	46.96, 142.76	8.61	2202	3.91

Table 2.
Key parameters of Rayleigh waves.

The dependence of time delay on distance is depicted in **Figure 4a** and **b**. Two approximations of $\Delta t(r)$ are considered:

$$\Delta t = 265.4r, \quad R^2 = 0.9931, \boldsymbol{\sigma} = 59.95\text{s}, \tag{13}$$

$$\Delta t = 39.92 + 258.5r, \quad R^2 = 0.9940, \boldsymbol{\sigma} = 61.01\text{s}. \tag{14}$$

Among these two regression models, Eq. (13) is preferred due to its slightly lower R^2 and σ compared to Eq. (14). Eq. (14) lacks physical justification because, as $r \to 0$, it predicts an initial delay (Δt_0) of approximately 39.92 s, which is inconsistent with the simultaneous generation of the seismic wave with the earthquake. Additionally, the value of t_0 is smaller than σ = 61.01 s, making Eq. (13) more reasonable.

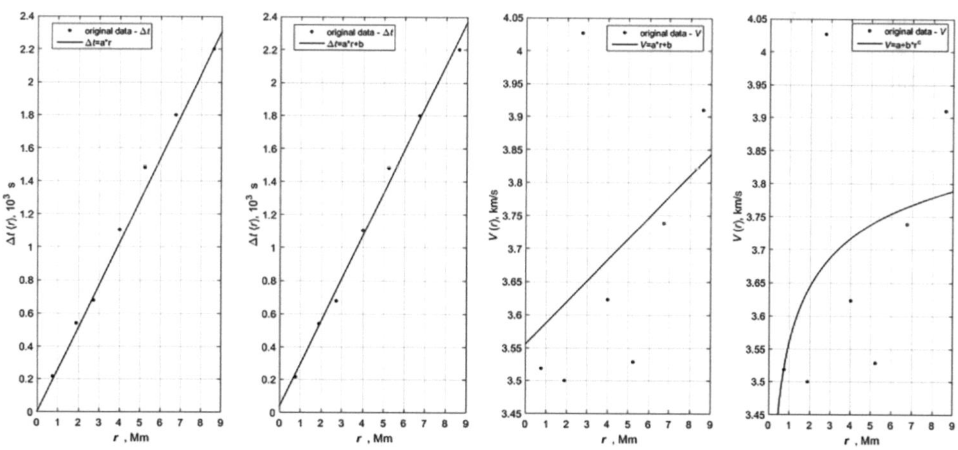

(a) (b) (c) (d)

Figure 4.
Dependence of the time delay of Rayleigh wave arrival at a seismic station (a and b) and wave propagation velocity on distance (c and d).

Furthermore, based on Eq. (13), the average wave speed $\overline{V} = \left(\frac{d\Delta t}{dr}\right)^{-1}$ is calculated to be 3.77 km/s.

The relationship between wave speed and distance, as shown in **Figure 4c** and **d**, is expressed using the following equations:

$$V = 3.556 + 0.03189r, \quad R^2 = 0.1809, \sigma = 0.21\text{km/s}, \tag{15}$$

$$V = 4.151 - 0.5914r^{-0.2236}, \quad R^2 = 0.1892, \sigma = 0.23\text{km/s}. \tag{16}$$

For the power-law dependence $V(r)$, the associated values are approximately $R^2 \approx 0.19$ and $\sigma \approx 200$ to 230 m/s (refer to **Figure 3d**). Consequently, it is reasonable to assume that the wave speed remains constant with V = 3.76 ± 0.26 km/s, and considering the value of period $T \approx 25.6 \pm 5.6$ s as discussed in the previous section, it suggests that the seismic wave is likely a Rayleigh wave.

5. Statistical analysis of the slow seismic waves

The seismogram analysis revealed the generation of a seismic wave with a significantly lower speed compared to the Rayleigh wave, differing by an order of magnitude [42]. In this section, we examine and analysis the statistical characteristics of these slow seismic waves triggered by the Tonga volcano eruption.

Key parameters of these distinct waves are summarized in **Table 3**. The relationship between the time delay and distance is approximated by the following equations, as shown in **Figure 5a**:

Station code	Station name/Region	Geographic coordinates latitude (°N), longitude (°E)	Distance (r), Mm	Delay Time (Δt), s	Velocity (V), m/s
MSVF	Nonsavu, Fiji	−17.74, 178.05	0.76	2220	342.34
RAR	Rarotonga, Cook Islands	−21.21, −159.77	1.62	5400	300.00
XMAS	Kiritimati, Kiribati	2.04, −157.45	3.18	10,300	308.74
JOHN	Johnston Island, Johnston Island, United States	16.73, −169.53	4.20	13,440	312.50
GUMO	Guam, Mariana Islands	13.59, 144.87	5.78	18,600	310.75
QSPA	South Pole Quiet Zone Earth Science Observatory, Antarctica	−89.93, 145.00	7.73	25,560	302.43
TSN	Qingdao, Shandong, China	36.07, 120.32	9.24	28,440	324.89
XAN	Xi'an, Shaanxi, China	34.04, 108.92	10.05	32,940	305.10
TLY	Talaya, Irkutskaya Oblast', Russia	51.68, 103.64	11.20	36,960	303.03
E46A	Sault Ste. Marie, Michigan, USA	46.37, −84.31	11.74	38,400	305.73
KURK	Kurchatov, Kazakhstan	50.72, 78.62	12.89	43,000	299.77
LVZ	Lovozero Karel'skaya, Russia	67.90, 34.65	14.37	47,040	305.48
BEL	Belsk, Poland	51.84, 20.79	16.28	53,160	306.25

Note: The pink-highlighted fields indicate significant deviations from neighboring stations

Table 3.
Key parameters of slow seismic waves generated by Lamb waves.

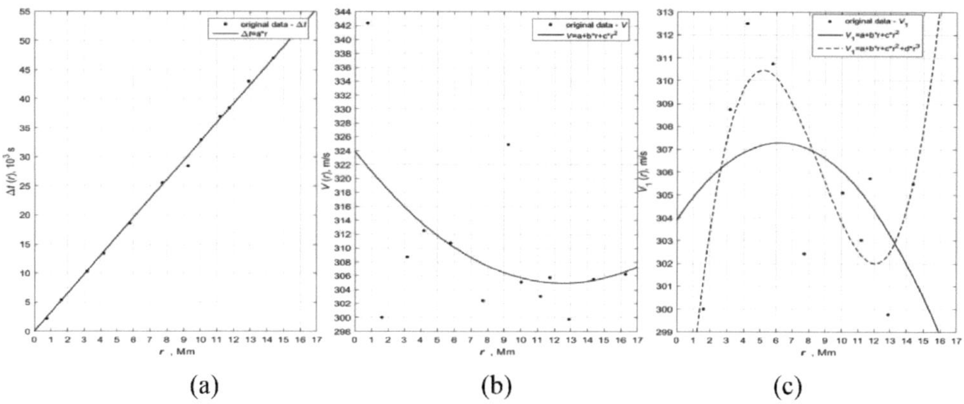

Figure 5.
Dependence of slow seismic wave arrival time delay (a), wave propagation speed (b) and corrected wave propagation speed (c) on distance.

$$\Delta t = 3268r, \quad R^2 = 0.9987, \sigma = 601.5\text{s}. \tag{17}$$

When excluding stations MSVF and TSN (denoted by subscript 1),

$$\Delta t_1 = 3288r, \quad R^2 = 0.9996, \sigma = 306.4\text{s}. \tag{18}$$

From Eqs. (17) and (18), the average wave speeds were calculated to be 306 m/s and 304 m/s, respectively. When assuming a uniform wave speed across all propagation paths, the speed estimates are $V = 321.06 \pm 21.29$ m/s and $V_1 = 306.14 \pm 9.37$ m/s. The error in speed estimation remains within a margin of 6.5%.

Under the assumption of a variable wave speed ($V \neq$ constant), the following approximations are observed (as shown in **Figure 5b** and **c**):

$$V = 323.9 - 3.022r + 0.1205r^2, \quad R^2 = 0.2453, \sigma = 11.2\text{m/s}, \tag{19}$$

$$V_1 = 303.9 + 1.089r - 0.08768r^2, \qquad R^2 = 0.1848, \sigma = 4.4\text{m/s}. \tag{20}$$

For V_1, the following approximation provides a better fit:

$$V_1 = 288.9 + 9.721r - 1.342r^2 + 0.05191r^3, \quad R^2 = 0.6295, \sigma = 3.2\text{m/s}. \tag{21}$$

Now, let us further examine the key parameters of slow seismic waves with two distinct phases, P1 and P2, as detailed in **Table 4**. These oscillations exhibit periods of ~4–6 minutes for P1 and ~ 8–11 minutes for P2. The dependencies of $\Delta t(r)$ for waves with these phases are depicted in **Figure 6** and are consistent with the following approximating equations:

$$\Delta t_{P1} = 239 + 3201r, \quad R^2 = 0.9918, \sigma = 674.4\text{s}, \tag{22}$$

$$\Delta t_{P2} = 559 + 4238r, \quad R^2 = 0.9912, \sigma = 921.5\text{s}. \tag{23}$$

Alternative approximating relations are as follows:

$$\Delta t_{P1} = 3258r, \quad R^2 = 0.9913, \sigma = 599.1\text{s}, \tag{24}$$

Station code	Geographic coordinates latitude (°N), longitude (°E)	Distance (r), Mm	Type/ phase	Delay time (Δt), s	Velocity (V), m/s
MSVF	−17.74, 178.05	0.76	P1	2220	342.34
			P2	2876	264.26
RAR	−21.21, −159.77	1.62	P1	5487	295.24
			P2	7811	207.40
JOHN	16.73, −169.53	4.20	P1	13,328	315.13
			P2	18,433	227.85
XMAS	2.04, −157.45	3.18	P1	11,403	278.87
			P2	15,116	210.37
GUMO	13.59, 144.87	5.78	P1	18,495	312.52
			P2	24,415	236.74

Table 4.
Key parameters of delay time Δt and apparent velocity V for the slow seismic waves with phases P1 and P2.

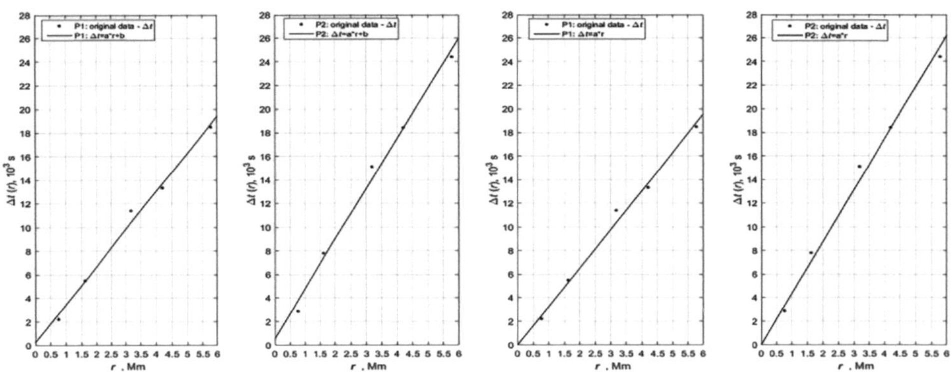

(a) (b) (c) (d)

Figure 6.
Dependence of the time delay of the arrival of the slow seismic waves with phases P1 (a and c) and P2 (b and d) at a seismic station on distance: (a) –Δt = 239 + 3201r; (b) –Δt = 559 + 4238r; (c) –Δt = 3258r; (d) –Δt = 4373r.

$$\Delta t_{P2} = 4373r, \quad R^2 = 0.9899, \sigma = 857.0\text{s}. \tag{25}$$

For Eqs. (22) and (23), the average speeds $\overline{V_{P1}}$ and $\overline{V_{P2}}$ are estimated to be 312 m/s and 236 m/s, respectively. Similarly, for Eqs. (24) and (25), the average speeds are approximately 307 m/s and 229 m/s. The variation between different regression models is approximately 1.6% and 3.0% for phases P1 and P2, respectively.

Table 4 and **Figure** 7 reveal notable fluctuations in speed. Assuming that speed is independent of distance, we obtain the following values: V_{P1} = 310.61 ± 31.73 m/s and V_{P2} = 235.83 ± 28.48 m/s. The velocities of waves with P1 and P2 phases exhibited fluctuations within 10% and 12%, respectively. Notably, these velocities slightly differ from the previously calculated $\overline{V_{P1}}$ and $\overline{V_{P2}}$.

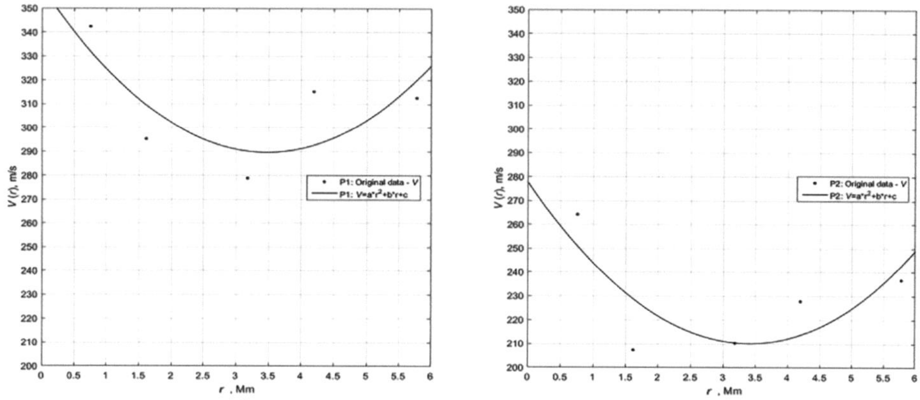

(a) (b)

Figure 7.
Dependence of the propagation velocity of the slow seismic waves with phases P1 (a) and P2 (b) on distance.

If the speeds of the waves with P1 and P2 phases depend on the distance, the following approximations are valid:

$$V_{P1} = 358.7 - 39.68r + 5.707r^2, \quad R^2 = 0.5581, \sigma = 22.34\text{m/s}, \tag{26}$$

$$V_{P2} = 277.8 - 39.66r + 5.813r^2, \quad R^2 = 0.5899, \sigma = 20.84\text{m/s}. \tag{27}$$

Next, we examine the statistical characteristics of seismic waves with P1 and P2 phases. For P1 wave with periods of 250–350 s, Eqs. (22) and (24) describe the time delay as a function of distance. For P2 waves with periods of 500–700 s, Eqs. (23) and (25) are used. Eqs. (22) and (23) show slightly higher R^2 values and σ compared to Eqs. (24) and (25). However, for Eqs. (22) and (23), $\Delta t_0 \approx 0$ as $r \to 0$, which lacks physical explanation, though Δt_0 is within the error limits ($\Delta t_0 < \sigma$). Thus, Eqs. (24) and (25) are preferred, with velocities close to 307 m/s and 229 m/s, respectively.

6. Statistical analysis of the tsunami waves

The Tonga volcano's explosion also generated a globally observed tsunami. In this section, we examine and analyze the statistical characteristics of the tsunami waves triggered by the Tonga volcano eruption.

At distances of 0.37 Mm and 2.53 Mm from the volcano, the maximum tsunami amplitudes were approximately 21 cm and 2.9 cm, respectively (see **Table 5**). By measuring the delay time (Δt) at a known distance (r), the average tsunami speed can be estimated as $V_{ts} = r/\Delta t$. The values of r, Δt, and V_{ts} are provided in **Table 5**.

To compare the calculated maximum tsunami height (h_{cal}) with the observed height (h), we use the relation proposed by Heidarzadeh et al. [17]:

$$M_s = \log h_{cal} + \log \Delta + 5.8 = \log h_{cal} + \log \frac{r}{R_E} + 5.8, \tag{28}$$

where R_E is the radius of the Earth. According to the USGS, $M_s \approx 5.8$. Here, r is in Mm, and h_{cal} is in cm. The results of estimating h_{cal} are also presented in **Table 5**.

Station code	Geographic coordinates latitude (°N), longitude (°E)	Distance (r), Mm	Amplitude (h), cm	Calculable amplitude (h_{cal}), cm	Delay time (Δt), s	Velocity (V_{ts}), m/s
NZG	−23.35, −173.40	0.37	20.77	30.03	1635	227.86
NZF	−29.68, −175.01	1.02	8.36	10.99	4215	241.38
NZK	−24.31, 169.50	1.61	5.43	6.92	5895	273.32
NZE	−36.05, −177.71	1.74	6.31	6.41	6315	275.68
NZD	−36.10, 178.60	1.83	3.13	6.10	10,695	170.96
NZL	−19.29, 166.81	1.87	4.68	5.95	9495	196.88
NZC	−38.20, −179.80	2.01	4.69	5.56	7215	278.72
NZJ	−26.67, 163.95	2.21	8.15	5.04	7753	285.51
NZB	−40.60, 179.10	2.29	2.10	4.86	8115	282.58
NZA	−42.37, 176.91	2.53	2.90	4.40	8895	284.97

Note: The pink-highlighted fields indicate values and stations that are significantly different from their neighboring stations

Table 5.
Key parameters related to tsunami.

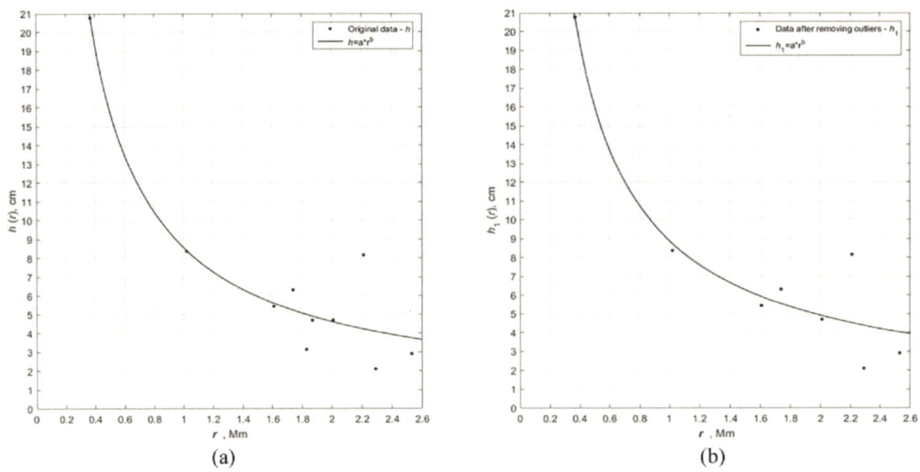

Figure 8.
Dependence of the maximum tsunami-wave amplitude on distance at DART stations: (a) original data including all stations; (b) data after removing outliers, specifically excluding DART stations NZD and NZL.

Table 5 reveals notable differences in Δt and h for stations NZD and NZL compared to their neighboring stations. This justifies the creation of separate graphs for $h(r)$, $\Delta t(r)$, and $V_{ts}(r)$ for all stations, as well as graphs that exclude the NZD and NZL stations, labeled as $h_1(r)$, $\Delta t_1(r)$, and $V_{ts1}(r)$. The maximum tsunami height ($h_{(1)}$) shows a clear dependency on distance (r), with relationships depicted in **Figure 8** and the corresponding approximating equations as follows:

$$h = 8.556r^{-0.8915}, \quad R^2 = 0.9048, \sigma = 1.762\text{cm}, \tag{29}$$

$$h_1 = 8.877r^{-0.8531}, \quad R^2 = 0.9141, \sigma = 1.857\text{cm}. \tag{30}$$

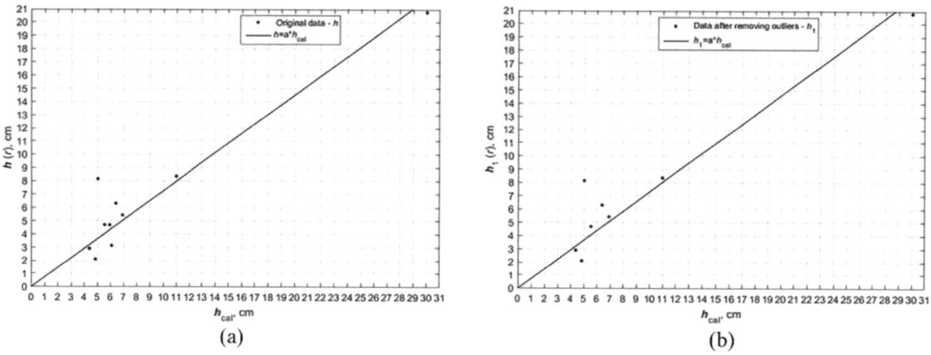

Figure 9.
Dependence of the maximum tsunami-wave amplitude on the calculated tsunami-wave amplitude at DART stations: (a) original data including all stations; (b) data after removing outliers, specifically excluding DART stations NZD and NZL.

Additionally, the $h(h_{cal})$ relationships follow a linear pattern, as illustrated in **Figure 9**, with their respective approximating equations:

$$h = 0.7251h_{cal}, \quad R^2 = 0.8902, \quad \sigma = 1.784\text{cm}, \tag{31}$$

$$h_1 = 0.7299h_{cal}, \quad R^2 = 0.8887, \quad \sigma = 1.958\text{cm}. \tag{32}$$

It is worth noting that Eqs. (29) and (30) and Eqs. (31) and (32) demonstrate satisfactory R^2 values (~9.0) and acceptable σ values (<2 cm), confirming the accuracy of height calculations. For instance, at r = 2.5 Mm, the calculated heights are around 4 cm, exceeding twice the σ value.

The time delay dependence for the tsunami's arrival at specific stations is illustrated in **Figure 10**. The fitting equations for this dependence, considering all stations as well as excluding outlier stations NZD and NZL, are given by:

$$\Delta t = 963.7 + 3466r, \quad R^2 = 0.6914, \sigma = 1566\text{s}, \tag{33}$$

$$\Delta t_1 = 600.2 + 3283r, \quad R^2 = 0.9971, \sigma = 138\text{s}. \tag{34}$$

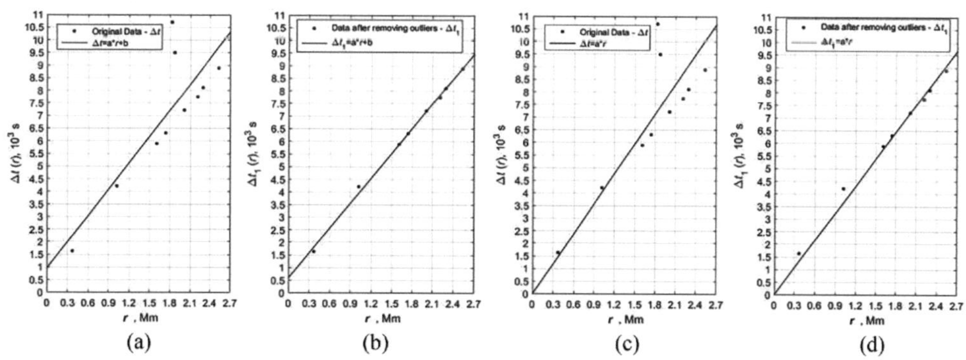

Figure 10.
Dependence of tsunami wave delay time on distance at DART stations: (a and c) original data including all stations; (b and d) data after removing outliers, specifically excluding DART stations NZD and NZL.

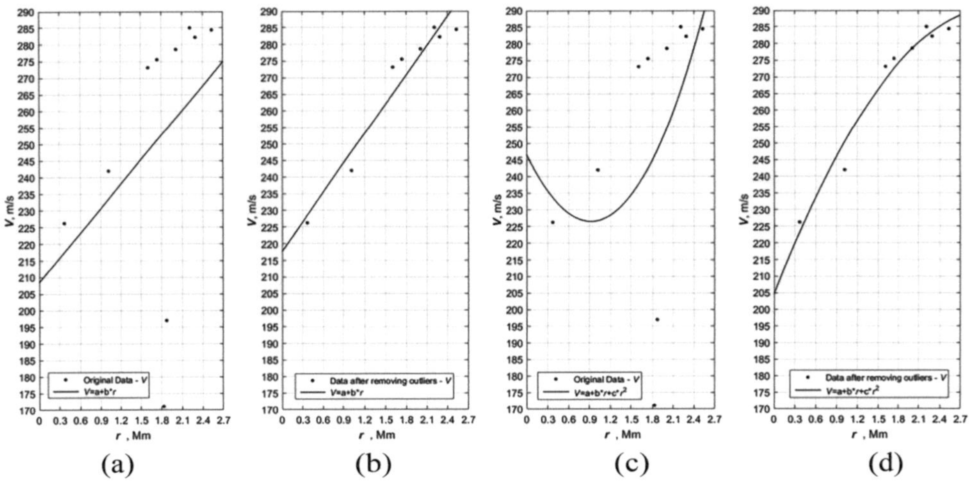

Figure 11.
Dependence of the tsunami-wave propagation speed on distance at DART stations: (a and c) original data including all stations; (b and d) data after removing outliers, specifically excluding DART stations NZD and NZL.

From Eqs. (33) and (34), the average tsunami speeds are approximately $\overline{V_{ts}} \approx 288.5$ m/s and $\overline{V_{ts1}} \approx 304.6$ m/s, respectively. It worth noting that $\Delta t_{10} \approx 600$ s (10 min), which exceeds σ, suggesting potential contributions from powerful infrasonic waves.

Alternatively, for all stations including and excluding NZD and NZL, the following approximating equations are applicable:

$$\Delta t = 3959r, \quad R^2 = 0.6758, \boldsymbol{\sigma} = 1513\text{s}, \tag{35}$$

$$\Delta t = 3585r, \quad R^2 = 0.9873, \boldsymbol{\sigma} = 266.3\text{s}. \tag{36}$$

From Eqs. (35) and (36), the average propagation speed values are $\overline{V_{ts}} \approx 252.6$ m/s and $\overline{V_{ts1}} \approx 278.9$ m/s, indicating that the tsunami waves may have been directly generated by the earthquake, with decreased wave propagation speeds.

The relationship between tsunami speed and distance is shown in **Figure 11**. The equations approximating these dependencies for all stations and excluding NZD and NZL stations are:

$$V_{ts} = 208.4 + 24.67r, \quad R^2 = 0.1474, \boldsymbol{\sigma} = 40.1\text{m/s}, \tag{37}$$

$$V_{ts1} = 217.7 + 29.45r, \quad R^2 = 0.9339, \boldsymbol{\sigma} = 6.1\text{m/s}. \tag{38}$$

When approximated by a second-degree polynomial, the relationships are:

$$V_{ts} = 246.6 - 43.55r + 23.68r^2, \quad R^2 = 0.2097, \boldsymbol{\sigma} = 41.3\text{m/s}, \tag{39}$$

$$V_{ts1} = 204.5 + 53.61r - 8.327r^2, \quad R^2 = 0.9640, \boldsymbol{\sigma} = 4.9\text{m/s}. \tag{40}$$

Figure 10 and Eqs. (37)–(40) reveal that the relationship between tsunami speed ($V_{(1)ts}$) and distance is variable. Eqs. (38) and (40) are considered acceptable, with Eq. (38) being preferred despite its slightly lower R^2 and higher σ compared to

Eq. (40). Variations in the corrected tsunami speed (V_{1ts}) are likely due to differences in ocean depths along the volcano-station paths. Assuming $V_{ts}(r)$ is constant, i.e., no distance dependence, we have V_{ts} = 228.1 ± 57.0 m/s and V_{ts1} = 255.7 ± 29.4 m/s, with fluctuations of ±25% and ±11%, respectively.

It is worth noting that the seismogram data shows erratic variations in the speeds of phases P1 and P2 with increasing distance, without a clear trend. Attempts to model $V_{P1}(r)$ and $V_{P2}(r)$ are shown in **Figures 6** and 7, and the corresponding relations expressed in Eqs. (26) and (27). There are no clear physical reasons to support Eqs. (26) and (27), leading to the conclusion that V and V_1 do not depend on distance. Therefore, we accept the estimates provided in this section.

7. Discussion

The discussion encompasses the propagation speeds and physical characteristics of various seismic waves and tsunami waves, including Rayleigh waves, Lamb waves, and underwater long waves, as well as their interaction with the Earth's lithosphere and oceanic systems.

7.1 Rayleigh wave and sheer wave speed

Rayleigh waves are surface seismic waves that propagate along the Earth's surface, with particle motion that decays exponentially with depth. These waves are sensitive to the elastic properties of the surface layers, and their speed is typically less than that of body waves such as P-waves and S-waves.

The speed of Rayleigh waves (V_R) can be expressed as a function of the shear wave speed (V_s) and compressional wave speed (V_p) [46]:

$$V_R \approx 0.87V_s + 0.13V_p. \tag{41}$$

Given the observed Rayleigh wave speed V_R = 3.76 km/s, and assuming a typical V_p = 5.5 km/s (common for crustal rocks), we can estimate the shear wave speed:

$$V_s \approx \frac{V_R - 0.13V_p}{0.87} = \frac{3.76 - 0.715}{0.87} \approx 3.50\text{km/s}. \tag{42}$$

This high shear wave speed suggests a stiff and elastic medium, typical of volcanic or igneous rocks near the eruption site [47]. The accurate estimation of V_p is critical for understanding the energy distribution of seismic waves in volcanic regions, as it directly influences the intensity and reach of the seismic impact.

7.2 Seismic oscillations and lamb waves

Our study identifies distinct seismic oscillations that were observed in the seismograms, specifically noting the presence of slow seismic waves with time delays an order of magnitude greater than the Rayleigh waves. These slow waves, observed with an average speed of approximately 306 m/s (excluding certain stations), were closely aligned with the propagation speed of atmospheric Lamb waves (V_L), which are a type of acoustic wave that propagates in the Earth's atmosphere and is often generated by explosive events, such as volcanic eruptions. They are influenced by the stratification of the atmosphere and can travel long distances with little attenuation [48, 49]. The speed of these Lamb waves was derived from the relation:

$$V_L \approx \sqrt{\frac{\gamma k_B T_a}{M_m}}. \tag{43}$$

where γ is the adiabatic exponent (approximately 1.4), k_B is Boltzmann's constant, T_a is the average air temperature in the surface waveguide, and M_m is the average mass of an air molecule. For $T_a \approx 230$ K and $M_m \approx 4.6 \times 10^{-26}$ kg, the Lamb wave speed, V_L, is approximately 313 m/s. This aligns well with the observed periods of seismic vibrations associated with the Lamb wave [2], confirming the Lamb wave's role in generating these slow seismic waves.

The energy carried by the Lamb wave was substantial enough to excite seismic waves within the Earth's lithosphere and oceans, contributing to the observed tsunamis. The seismic wave excitation within the lithosphere led to the activation of the Earth's normal mode ${}_0S_{29}$ [21, 50], with the Lamb wave capable of exciting a seismic wave with a similar period and speed.

The interaction between the Lamb waves and the Earth's surface was analyzed using the reflection coefficient at the interface between the air and the lithosphere. The reflection coefficient R_{12} is given by [11]:

$$R_{12} = \frac{v_2\rho_2 - v_1\rho_1}{v_2\rho_2 + v_1\rho_1}. \tag{44}$$

where ρ_1 and ρ_2 represent the densities of air and the lithosphere, respectively, while v_1 and v_2 denote the wave speeds in these media. It is typically assumed that $v_1.\rho_1 < < v_2.\rho_2$. When $v_1 = v_2$, the reflection coefficient R_{12} simplifies to $1 - 2\rho_1/\rho_2$, and the transmission coefficient T_{12} can be defined as $T_{12} = 1–R_{12} = 2\rho_1/\rho_2$. Using approximate values of $\rho_1 \approx 1.3$ kg/m^3 and $\rho_2 \approx 3 \times 10^3$ kg/m^3, the transmission coefficient D is estimated to be approximately 8.7×10^{-4}. This low value indicates that only a small fraction of the Lamb wave's energy is transmitted into the Earth's lithosphere, with most being reflected.

The pressure amplitude Δp_2 of the wave induced in the lithosphere by the Lamb wave is related to the amplitude of the air wave (Δp_1) and the velocity amplitude of oscillations of the Earth's surface (A_V) by the equation:

$$\Delta p_2 = T_{12}\Delta p_1 = \rho_2 v_2 A_V = \rho_2 v_1 A_V. \tag{45}$$

Thus, A_V can be expressed as:

$$A_V = \frac{\Delta p_2}{\rho_2 v_1} = T_{12}\frac{\Delta p_1}{\rho_2 v_1} = \frac{2\rho_1}{\rho_2^2 v_1}\Delta p_1. \tag{46}$$

Based on previous results [6], for the surface Lamb wave, we have,

$$\Delta p_1 = 4.863 r^{-0.5}, \quad R^2 = 0.9604, \sigma = 0.93\text{hPa}. \tag{47}$$

For $v_1 \approx 300$ m/s, in the SI system, Eqs. (46) and (47) yield

$$A_V \approx 10^{-9}\Delta p_1 = 4.863 \cdot 10^{-9}\sqrt{r}, \tag{48}$$

r, Mm	0.1	0.3	1	2	3	5	10	12	15	18
Δp_1, Pa	1537	887	486	344	281	217	154	141	125	115
A_V, μm/s	1.54	0.89	0.49	0.34	0.28	0.22	0.15	0.14	0.13	0.12

Table 6.
Dependence of pressure in the Lamb wave and the propagation speed of oscillations of the Earth's surface on distance.

where Δp_1 varied from tens to units of hectopascals as r increased from 0.1 Mm to 18 Mm. The estimation results, shown in **Table 6,** indicate that A_V changes from 1.54 μm/s to 0.12 μm/s over this distance range. Considering an oscillation period of $T \approx 268$ s [2], the range of displacement amplitude A_D of the Earth's surface, given by $A_D = A_V/\omega = A_V \cdot T$, ranges from approximately 4.1×10^{-4} meters to 3.2×10^{-5} meters, which aligns well with the observed slow seismic wave with P1 phase.

Slow seismic waves were generated simultaneously with the Lamb wave. Thus, at $r = 0$, there should be Δt_0 should be 0. The regressions for $\Delta t(r)$, as shown in **Figure 5**, corresponding to Eqs. (17) and (18). Preference is given to Eq. (17) due to its higher R^2 value and lower σ when excluding stations MSVF and TSN, compared to Eq. (18). **Figure 5** also displays the possible dependences of $V(r)$ and $V_1(r)$ for all stations and excluding MSVF and TSN. The R^2 values for both cases are very small, suggesting that the speed of slow seismic waves did not depend on distance, consistent with the non-dispersive nature of the Lamb wave, which propagates at an approximately constant speed of 310 m/s.

7.3 Underwater long waves and tsunami waves

Underwater long waves, or hydrodynamic seismic waves, are generated by the coupling of seismic energy into the water column, typically in the form of long-period waves that can travel across ocean basins. They play a critical role in the formation and propagation of tsunamis.

As estimated in Section 6, the average velocity of slow seismic wave with P1phase 307 m/s aligns with the Lamb wave speed (refer to Eq. (43)), while slow seismic wave with P2 phase 229 m/s is typical for gravitational long waves in water. For these waves, the speed is given by:

$$v = \sqrt{gH}, \tag{49}$$

where g is the acceleration due to gravity, and H is the water depth. At $v \approx 200$–259 m/s (as shown in **Table 4**, H is estimated to be approximately 4.08–6.85 km, which is consistent with the average depth of the Pacific Ocean where such gravitational waves propagate. This suggests that the seismic waves coupled effectively with the ocean's water column to generate long-period waves that contributed to tsunami formation [51, 52].

Furthermore, tsunamis are large ocean waves typically caused by seismic events such as underwater volcanic eruptions or earthquakes [52]. They are characterized by long wavelengths and periods and can cause significant damage upon reaching coastlines. The slight difference between the speeds of the hydrodynamic seismic waves and the tsunami waves (229 m/s vs. 228.1 m/s) may be due to varying ocean depths and nonlinear interactions within the water column, which affect wave propagation speeds [53].

8. Conclusion

The catastrophic eruption of the Hunga-Tonga-Hunga-Ha'apai (HTHH) volcano on January 15, 2022, provided crucial data on the interaction between volcanic activity and seismic waves. Our statistical analysis of seismic and tsunami waves from this event revealed key findings: the amplitude of surface mechanical vibrations decreases following a power-law with distance, and their speed follows a decreasing power-law dependence. Rayleigh waves, observed with an average speed of 3.77 km/s, consistently measured 3.76 $\pm$ 0.26 km/s, confirming theoretical predictions and validating the reliability of the seismic data.

Additionally, slow seismic waves associated with Lamb waves were detected with an average speed of approximately 306 m/s, closely matching the Lamb wave speed of 313 m/s. This alignment suggests that the Lamb wave significantly influenced seismic energy propagation. The observed velocities of the P1 and P2 phase slow seismic waves, averaging 307 m/s and 229 m/s, respectively, align with the linear distance dependence of recorded wave arrival delays. This supports the hypothesis that these velocities are independent of distance.

Moreover, the study also reveals that tsunami characteristics are influenced by seismic activity. The maximum tsunami height follows a decreasing power function, and its relationship with calculated height is well represented by linear functions. The average tsunami speed of 304.6 m/s suggests a possible contribution from Lamb waves, while the tsunami generated by hydrodynamic seismic waves would have an average speed of 278.9 m/s. The preferred model for tsunami speed with distance shows a linear increase of approximately 29 m/s per Mm, with a distance-independent estimate of 255.7 $\pm$ 29.4 m/s, primarily reflecting ocean depth effects. These findings enhance our understanding of tsunami formation and propagation, providing valuable insights for future geological and oceanographic studies.

Acknowledgements

Work by Leonid Chernogor was supported by the Ukraine state research project #0122 U001476 and #0124 U000478.

Conflict of interest

The authors declare no conflict of interest.

Thanks

The authors would like to thank IntechOpen for their support in publishing this article. We also extend our thanks to the academic editor, Professor Károly Németh, for his valuable suggestions, which greatly contributed to the improvement of the manuscript.

Author details

Leonid Chernogor and Yiyang Luo*
V.N. Karazin Kharkiv National University, Kharkiv, Ukraine

*Address all correspondence to: yiyangluo@163.com

References

[1] USGS. Available from: https://www.usgs.gov/

[2] Ovtchinnikov VM, Usoltseva OA. Seismic phenomena associated with the eruption of the volcano in the region of the Tonga Archipelago on January 15, 2022. Izvestiya, Physics of the Solid Earth. 2023;**59**:433-443. DOI: 10.1134/S10693513230396

[3] Shinbori A, Otsuka Y, Sori T, et al. New aspects of the upper atmospheric disturbances caused by the explosive eruption of the 2022 Hunga Tonga–Hunga Ha'apai volcano. Earth, Planets and Space. 2023;**75**:175. DOI: 10.1186/s40623-023-01930-4

[4] Pakoksung K, Suppasri A, Imamura F. The near-field tsunami generated by the 15 January 2022 eruption of the Hunga Tonga-Hunga Ha'apai volcano and its impact on Tongatapu, Tonga. Scientific Reports. 2022;**12**:15187. DOI: 10.1038/s41598-022-19486-w

[5] Poli P, Shapiro NM. Rapid characterization of large volcanic eruptions: Measuring the impulse of the Hunga Tonga Ha'apai explosion from teleseismic waves. Geophysical Research Letters. 2022;**49**(8):1-9. DOI: 10.1029/2022GL098123

[6] Chernogor LF, Shevelev MB. Statistical characteristics of explosive waves in the atmosphere generated by the Tonga supervolcano on January 15, 2022. Space: Science & Technology. 2024;**30**(1):66-79. DOI: 10.15407/knit2024.01.066

[7] Amores A, Monserrat S, Marcos M, Argüeso D, Villalonga J, Jordà G, et al. Numerical simulation of atmospheric Lamb waves generated by the 2022 Hunga-Tonga volcanic eruption. Geophysical Research Letters. 2022;**49**(6):1-8. DOI: 10.1029/2022GL098240

[8] Astafyeva E, Maletckii B, Mikesell TD, Munaibari E, Ravanelli M, Coisson P, et al. The 15 January 2022 Hunga Tonga eruption history as inferred from ionospheric observations. Geophysical Research Letters. 2022;**49**(10):1-10. DOI: 10.1029/2022GL098827

[9] Kulichkov SN, Chunchuzov IP, Popov OE, et al. Acoustic-gravity Lamb waves from the eruption of the Hunga-Tonga-Hunga-Hapai volcano, its energy release and impact on aerosol concentrations and tsunami. Pure and Applied Geophysics. 2022;**179**:1533-1548. DOI: 10.1007/s00024-022-03046-4

[10] Terry JP, Goff J, Winspear N, Bongolan VP, Fisher S. Tonga volcanic eruption and tsunami, January 2022: Globally the most significant opportunity to observe an explosive and tsunamigenic submarine eruption since AD 1883 Krakatau. Geoscience Letter. 2022;**9**:24. DOI: 10.1186/s40562-022-232-z

[11] Chernogor LF. Physical effects of the January 15, 2022, powerful Tonga volcano explosion in the Earth-atmosphere-ionosphere-magnetosphere system. Space: Science & Technology. 2023;**29**(2):54-77. DOI: 10.15407/knit2023.02.054

[12] Chernogor LF. Physics and Ecology of the Catastrophes. Kharkiv, Ukraine: V. N. Karazin Kharkiv National University Publ; 2012. [In Russian]

[13] Chernogor LF. Physical effects from the powerful Tonga volcanic eruption of January 15, 2022, in the earth–

atmosphere–ionosphere–magnetosphere system. Journal of Atmospheric and Solar-Terrestrial Physics. 2023;**253**: 106157. DOI: 10.1016/j.jastp.2023.106157

[14] Tarumi K, Yoshizawa K. Eruption sequence of the 2022 Hunga Tonga-Hunga Ha'apai explosion from back-projection of teleseismic P waves. Earth and Planetary Science Letters. 2023;**602**: 117966. DOI: 10.1016/j.epsl.2022.117966

[15] Ramírez-Herrera MT, Coca O, Vargas-Espinosa V. Tsunami effects on the coast of Mexico by the Hunga Tonga-Hunga Ha'apai volcano eruption, Tonga. Pure and Applied Geophysics. 2022;**179**: 1117-1137. DOI: 10.1007/s00024-022-03017-9

[16] Carvajal M, Sepúlveda I, Gubler A, Garreaud R. Worldwide signature of the 2022 Tonga volcanic tsunami. Geophysical Research Letters. 2022;**49** (6):1-4. DOI: 10.1029/2022GL098153

[17] Heidarzadeh M, Gusman AR, Ishibe T, Sabeti R, Šepić J. Estimating the eruption-induced water displacement source of the 15 January 2022 Tonga volcanic tsunami from tsunami spectra and numerical modelling. Ocean Engineering. 2022;**261**:112165. DOI: 10.1016/j.oceaneng.2022.112165

[18] Imamura F, Suppasri A, Arikawa T, Koshimura S, Satake K, Tanioka Y. Preliminary observations and impact in Japan of the tsunami caused by the Tonga volcanic eruption on January 15, 2022. Pure and Applied Geophysics. 2022;**179**:1549-1560. DOI: 10.17/s024-022-03058-0

[19] Lynett P. The tsunamis generated by the Hunga Tonga-Hunga Ha'apai volcano on January 15, 2022. Preprint available at Research Square. DOI: 10.21203/rs.3.rs-1377508/v1

[20] Lynett P, McCann M, Zhou Z, et al. Diverse tsunamigenesis triggered by the Hunga Tonga-Hunga Ha'apai eruption. Nature. 2022;**609**:728-733. DOI: 10.1038/s41586-022-05170-6

[21] Matoza RS, Fee D, Assink JD, Iezzi AM, Green DN, Kim K, et al. Atmospheric waves and global seismoacoustic observations of the January 2022 Hunga eruption, Tonga. Science. 2022;**377**(6601):95-100. DOI: 10.1126/science.abo7063

[22] Burt S. Multiple airwaves crossing Britain and Ireland following the eruption of Hunga Tonga–Hunga Ha'apai on 15 January 2022. Weather. 2022;**77**(3):76-81. DOI: 10.1002/wea.4182

[23] Chen CH, Zhang X, Sun YY, Wang F, Liu TC, Lin CY, et al. Individual wave propagations in ionosphere and troposphere triggered by the Hunga Tonga-Hunga Ha'apai underwater volcano eruption on 15 January 2022. Remote Sensing. 2022;**14**(9):2179. DOI: 10.3390/rs14092179

[24] Ern M, Hoffmann L, Rhode S, Preusse P. The mesoscale gravity wave response to the 2022 Tonga volcanic eruption: AIRS and MLS satellite observations and source backtracing. Geophysical Research Letters. 2022;**49** (10):1-13. DOI: 10.1029/2022GL098626

[25] Kubota T, Saito T, Nishida K. Global fast-traveling tsunamis driven by atmospheric Lamb waves on the 2022 Tonga eruption. Science. 2022; **377**(6601):91-94. DOI: 10.1126/science.abo4364

[26] Otsuka S. Visualizing Lamb waves from a volcanic eruption using meteorological satellite Himawari-8. Geophysical Research Letters. 2022;**49** (8):1-9. DOI: 10.1029/2022GL098324

[27] Vergoz J, Hupe P, Listowski C, Le Pichon A, Garcés MA, Marchetti E, et al. IMS observations of infrasound and acoustic-gravity waves produced by the January 2022 volcanic eruption of Hunga, Tonga: A global analysis. Earth and Planetary Science Letters. 2022;**591**: 117639. DOI: 10.1016/j.epsl.2022.117639

[28] Wright CJ, Hindley NP, Alexander MJ, Barlow M, Hoffmann L, Mitchell CN, et al. Surface-to-space atmospheric waves from Hunga Tonga-Hunga Ha'apai eruption. Nature. 2022; **609**:741-746. DOI: 10.1038/s41586-022-05012-5

[29] Yuen DA, Scruggs MA, Spera FJ, Zheng Y, Hu H, McNutt SR, et al. Under the surface: Pressure-induced planetary-scale waves, volcanic lightning, and gaseous clouds caused by the submarine eruption of Hunga Tonga-Hunga Ha'apai volcano. Earthquake Research Advances. 2022;**2**(3):1134. DOI: 10.1016/j.eqrea.2022.1134

[30] Aa E, Zhang SR, Erickson PJ, Vierinen J, Coster AJ, Goncharenko LP, et al. Significant ionospheric hole and equatorial plasma bubbles after the 2022 Tonga volcano eruption. Geophysical Research Letters. 2022;**20**(7):1-21. DOI: 10.1029/2022SW3101

[31] Aa E, Zhang SR, Wang W, Erickson PJ, Qian L, Eastes R, et al. Pronounced suppression and X-pattern merging of equatorial ionization anomalies after the 2022 Tonga volcano eruption. Journal of Geophysical Research, Space Physics. 2022;**127**(6):1-15. DOI: 10.1029/2022JA030527

[32] Ravanelli M, Astafyeva E, Munaibari E, Rolland L, Mikesell TD. Ocean-ionosphere disturbances due to the 15 January 2022 Hunga-Tonga Hunga-Ha'apai eruption. Geophysical Research Letters. 2023;**50**. DOI: 10.1029/2022GL101465

[33] Lin JT, Rajesh PK, Lin CCH, Chou MY, Liu JY, Yue J, et al. Rapid conjugate appearance of the giant ionospheric lamb wave signatures in the northern hemisphere after Hunga-Tonga volcano eruptions. Geophysical Research Letters. 2022;**49**(8):1-10. DOI: 10.1029/2022GL098222

[34] Saito S. Ionospheric disturbances observed over Japan following the eruption of Hunga Tonga-Hunga Ha'apai on 15 January 2022. Earth, Planets and Space. 2022;**74**:57. DOI: 10.1186/s40623-022-01619-0

[35] Shinbori A, Otsuka Y, Sori T, Nishioka M, Perwitasari S, Tsuda T, et al. Electromagnetic conjugacy of ionospheric disturbances after the 2022 Hunga Tonga-Hunga Ha'apai volcanic eruption as seen in GNSS-TEC and SuperDARN Hokkaido pair of radars observations. Earth, Planets and Space. 2022;**74**:106. DOI: 10.1186/s40623-022-01665-8

[36] Themens DR, Watson C, Žagar N, Vasylkevych S, Elvidge S, McCaffrey A, et al. Global propagation of ionospheric disturbances associated with the 2022 Tonga volcanic eruption. Geophysical Research Letters. 2022;**49**(7):1-11. DOI: 10.1029/2022GL098158

[37] Zhang SR, Vierinen J, Aa E, Goncharenko LP, Erickson PJ, Rideout W, et al. 2022 Tonga volcanic eruption induced global propagation of ionospheric disturbances via lamb waves. Frontiers in Astronomy and Space Sciences. 2022;**9**:871275. DOI: 10.3389/fspas.2022.871275

[38] Yamazaki Y, Soares G, Matzka J. Geomagnetic detection of the atmospheric acoustic resonance at 3.8

mHz during the Hunga Tonga eruption event on 15 January 2022. Journal of Geophysical Research, Space Physics. 2022;**127**(7):1-10. DOI: 10.1029/2022 JA030540

[39] Schnepf NR, Minami T, Toh H, Nair MC. Magnetic signatures of the 15 January 2022 Hunga Tonga-Hunga Ha'apai volcanic eruption. Geophysical Research Letters. 2022;**49**(10):1-9. DOI: 10.1029/2022GL098454

[40] Le G, Liu G, Yizengaw E, Englert CR. Intense equatorial electrojet and counter electrojet caused by the 15 January 2022 Tonga volcanic eruption: Space- and ground-based observations. Geophysical Research Letters. 2022;**49** (11):1-11. DOI: 10.1029/2022GL0992

[41] Chernogor LC, Luo Y. Statistical characteristics of seismic waves and tsunami generated by the 2022 Hunga Tonga volcanic eruption. In: International Conference of Astronomy and Space Physics. Kyiv, Ukraine: Kyiv University; 2024. pp. 84-86

[42] Global Seismographic Network (GSN), Incorporated Research Institutions for Seismology. Available from: https://www.usgs.gov/programs/earthquake-hazards/gsn-global-seismographic-network

[43] FDSN Network Data. Available from: https://www.fdsn.org/networks/detail/NZ/

[44] GeoNet Data Exploration. Available from: https://tilde.geonet.org.nz/ui/data-exploration/Dart

[45] Aki K, Richards PG. Quantitative Seismology. Mill Valley, California: University Science Books; 2002

[46] Lay T, Wallace TC. Modern Global Seismology. Cambridge, Massachusetts: Academic Press; 1995

[47] Gill AE. Atmosphere-Ocean Dynamics. New York, New York: Academic Press; 1982

[48] Kanamori H, Given JW. Analysis of long-period seismic waves excited by the May 18, 1980, eruption of Mount St. Helens—A terrestrial monopole? Journal of Geophysical Research. 1982;**87**(B7): 5422-5432. DOI: 10.1029/JB087iB07p05422

[49] Tiwari SH, Bagiya MS, Maurya S, Heki K, Dimri AP. On the role of volcanic plume heights in excitation of free oscillations of the solid Earth and atmosphere: Case study. Advances in Space Research. 2024;**73**(6):2988-2998. DOI: 10.1016/j.asr.2024.01.001

[50] Okal EA. T waves from the 1998 Papua New Guinea earthquake and its aftershocks: Timing the tsunamigenic slump. Pure and Applied Geophysics. 2003;**160**:2147-2181. DOI: 10.1007/s00024-003-2404-0

[51] Munk WH, Snodgrass FE, Carrier GF. Edge waves on the continental shelf. Science. 1956; **123**(3181):127-129. DOI: 10.1126/science.123.3181.127

[52] Satake K. Tsunamis: 2004 Sumatra-Andaman and the 2011 Tohoku-Oki earthquakes. Reviews of Geophysics. 2015;**53**(3):679-711. DOI: 10.1002/2015RG000482

[53] Titov VV, Synolakis CE. Modeling of breaking and nonbreaking long-wave evolution and runup using VTCS-2. Journal of Waterway Port Coastal Ocean Engineering. 1995;**121**(6):308-316. DOI: 10.1061/(ASCE)0733-950X(1995)121:6(308)

Section 3

Volcanic Hazard

Chapter 8

Historical and Paleolimnological Reconstruction of Degassing Maar Lakes: The Case of Lake Pavin (Auvergne, France), for the Last Two Thousand Years

Michel Meybeck, Emmanuel Chapron, Brice Molo and Pierre Lavina

Abstract

Degassing maar lakes were thought to be a new natural hazard until the Nyos Lake catastrophe (1986). Historical degassing events published for Lakes Pavin (France, 1785), Monticchio (Italy, 1770–1820), Albano Lake (Italy, 1829), combined with those of Monoun (1984) and Nyos (1986) in Cameroon, allow to propose a grid of degassing descriptors, on lake and waters changes, on atmospheric disorders and on impacts on populations. Pavin Lake degassing history has then been reconstructed from well-contextualized written source over five centuries, revealing boiling (1783), lake storm (1847), color change (1783, 1936), sudden limnic eruption (1551, 1785), emission of fog, vapors, and thick clouds (1550–1750), associated with reported sudden neurologic diseases, similar to those observed at Albano and Nyos. Their spontaneous healings have been reported as miracles by the church authorities. Pavin, described as a, « *marvelous lake with sudden storms with thunder and lightnings* », is feared and avoided by the local population. When eventually sounded (1770), Pavin is gradually normalized by local authorities who deny its misbehaviors until recently. Since 2005, Pavin geohistory is gradually unveiled by paleolimnology, geomorphology, and geochemistry, revealing sub-lacustrine slides, higher lake levels, overflows events, and its deep CO2 are beginning to be surveyed. A recent modelling of sudden lake degassing confirms that Pavin area is a potential hazard. Other world maar lakes could benefit from the multiple approaches implemented at Pavin to establish their past degassing history, necessary to risk management.

Keywords: degassing descriptors, maar lakes, Lake Pavin, limnic eruption, geohistory

1. Introduction

When the sudden eruption of Lake Nyos occurred on August 21, 1986, causing 1700 victims and thousands of injuries in the Grasslands of Cameroon [1], this phenomenon was new to the academic world of limnologists [2] and volcanologists [3–6]. Now termed limnic eruption [4], this phenomenon is interpreted as a violent and sudden degassing of CO_2, of magmatic origin, stored in stagnant deep waters of a very rare type of lake, the meromictic maar lakes, which otherwise behave like other lakes. It is quite different from the phreatic explosions produced by the volcanic magma when in contact with groundwater or surface water, resulting in an explosion of steam, water, ash, and rock debris. Maar is a volcanic term, originated from the Eifel volcanic province in Germany, applied to depressions that have been generated by violent phreato-magmatic eruptions followed by crater floor subsidence, then filled with water [7]. The meromixis is a very rare limnological property of some lakes, which are never completely mixed due to the presence of denser bottom waters originating from groundwater inputs or coastal intrusions [2]. The semantic associations "limnic eruption" and "meromictic maar lakes" reflect the sort of "natural monster" that is encountered throughout this paper, which mixes the permanent quietness that is attributed to still continental waters and the violence and unpredictability of volcanic processes.

The Nyos catastrophic event immediately attracted scientific teams from far away, mostly volcanologists, a few limnologists, and lake geochemists, who mainly worked separately on the field. In 1987, UNESCO organized a conference in Yaoundé to collect these past-event observations and tried to reach a consensus on the origins of the event [1]. Cameroon scientists' participation was at that time very limited. Also, very few social scientists were present, excepted Evguenia Shanklin, an US anthropologist, who made the first study of the perception of the catastrophe by the Grasslands population [8]. In the conference volume, her paper was the only one to address the question of possible past Nyos events and their perception by the local population through oral transmission and in a context of very limited historical records. She mentioned that actually the local population knew about the "misbehavior" of some lakes, a term that has been later used to qualify various lake deviations from their normal quiet state, such as color change, sudden eruptions, lake displacement, and their representations, often by various types of lake monsters and spirits [9–11].

For limnologists, maar lakes are rare at the global scale, probably one in 1000 or 10,000 lakes worldwide, but they may be common in recent volcanic areas such as Alaska, Japan, Ethiopia, Central America, the Andes, Cameroon, Indonesia, etc. (**Figure 1**). Before the Nyos catastrophe, they had been rarely studied, mostly in Europe: in Eifel (Germany) [7, 12], in Southern Italy [13], and in Auvergne-Velay (France), where they were mixed with other lakes generated by volcanic dams such as lava flows or eruption cones [14, 15].

For volcanologists, the term "volcanic lakes" covers a wide range, from crater lakes still receiving acidic hydrothermal fluids or young maar lakes exposed to phreato-magmatic explosions to quiescent maar lakes and caldera lakes as Crater Lake (Oregon). Their classification can be based on their genesis, their current activity, and their chemistry [16–18]. A classification of volcanic lakes into six classes according to their activity, from volcanic eruptions (class 1) to low activity (class 5) and no activity (class 6) is proposed. Most limnological studies before 1986 were concerning the class 6 lakes; other volcanic lakes were barely mentioned by limnologists. The specificity of Nyos-type lakes was only recognized recently by the community of limnologists [19].

The acidic volcanic lakes have been particularly studied for their singular chemical composition: many of them are extremely acidic, very mineralized, and are still constrained by internal inputs of volcanic fluids [16]. Their access is difficult, dangerous, and often restricted, and their limnological study through vertical profiles is rarely performed between volcanic crises. When the volcanic forces are decreasing,

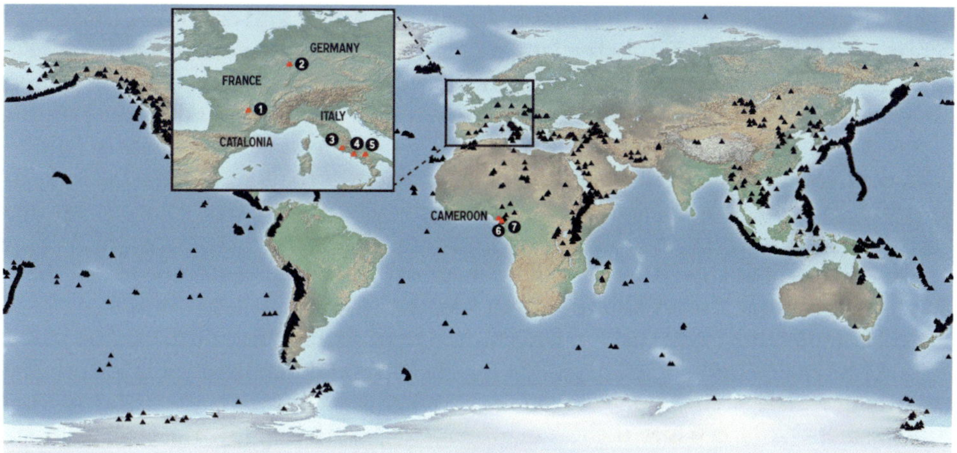

Figure 1.
World volcanoes and position of degassing maar lakes studied, or mentioned in this paper, in Western Europe and Cameroon. 1 Lake Pavin in Auvergne, France. 2 Laacher See, Ulmen, Weinfelder, Pulver Lakes in Eifel, Germany. 3 Lakes Albano and Nemi in Colli Albani, Italy. 4 Lake Averno in Phlegrean Fields, Italy. 5 Piccolo Lago di Monticchio and Grande Lago di Monticchio in Mount Vulture, Italy. 6 and 7 Lakes Monoun and Nyos in Oku volcanic field, Cameroon Volcanic Line. https://volcano.si.edu/. Data: Natural Earth, OpenStreetmap, Global Volcanism Program—Smithsonian Institution.

the aging maar lakes (classes 5 and 6) become gradually controlled by the external forces—solar energy, watershed inputs—as all other lakes. However, they are still marked by their volcanic origin, which controls their morphology, limits the inputs of water and material from their restricted annular drainage basins, and may still be responsible for sub-lacustrine hydrothermal inputs [19]. A recent inventory of African volcanic lakes by volcanologists gave a total of 220 lakes for Africa alone [18]. In this VOLADA data base, maar lakes are by far the dominant type (178 lakes).

Maar lakes sediment archives are characterized by their high resolution, in parts even annual, which gives detailed evidence of paleo climatic variations, regional seismic activity, volcanic activities, and human perturbations, as well as their biotic response within the lakes and their catchment areas, as studied in Western Europe maars [20].

After the Nyos catastrophe in August 1986, other potential degassing maar lakes have been rapidly looked for in Italy, Germany, and Japan and analyzed in hundreds of papers in the fields of volcanology, limnology, and geochemistry. However, their studies by human and social scientists (historians, human geographers, sociologists, anthropologists, literature specialists, and folklorists) are just beginning. Italian volcanologists [21, 22] have been the first ones to refer to pioneer studies of the Monticchio maar lakes in the late 1700s and to the Lake Albano *prodigium*, which occurred in 398 BC, according to Latin authors [23, 24]. More recently, a precious description of Lake Albano degassing in 1829 was found [24]. In Greek and Latin Antiquity, Lakes Nemi, Albano, and Averno have been considered sacred places, and the association between their past misbehavior and their well-known associated beliefs is beginning to be considered [25].

A few weeks after the Nyos catastrophe, Pavin Lake, in the Auvergne volcanic province (France), was rapidly identified as a potential degassing maar lake. A first national expertise of a degassing risk was conducted, including one of the authors (MM): we concluded to the absence of risk in present conditions, considering the relatively low CO_2 concentration in its deep and stagnant waters [26]. This lake has been the birthplace of French limnology since the 1890s [9] and continues to attract scientists from far away [27]. In the last decades, paleolimnological [28–30] and geological studies have been developed to reconstruct its past geohistory and the extend of its tephras, and CO_2 measurements have been made in the lake and its surroundings to assess the present rate of degassing [31–35]. The first historical analyses of Lake Pavin, over 450 years, revealed complex social responses to its past misbehaviors, lake normalization periods, and scientific controversies [9–11]. The social receptions of degassing risk at Nyos and Pavin reflect cultural and social differences [6]. We are referring throughout this paper to "Pavin" as most local people continue to call this place, which was not always identified as a lake in the past.

In this chapter, the Pavin past is reconstructed from written historical documents since 1551 and from paleolimnology. They show various types and intensities of lake misbehaviors attributed to degassing, none of which were taken into account in the first degassing risk assessment [26], on which the risk management is still based [6]. The local historical description covers the following European history periods: Renaissance, Enlightenments, modern scientific era (1890–1986), and contemporary post-Nyos era (since 1986). The paleolimnological records now extend over more than two thousand years.

Our study includes new historical sources such as the limnic eruption of 1785 and a description of a "Pavin event" in 1847. A companion paper [36] focuses on Pavin geoheritage values and on its geomythology, based on sources anterior to the 16th, generally in Latin, and on its numerous legends and tales.

We are addressing the following questions: (i) What terms have been used by scientists to report the past limnic eruptions and other major lake degassing events in four European maar lakes and their impacts on nearby populations between 1770 and 1829; are these descriptors comparable with those of Nyos and Monoun? (ii) Can these degassing indicators be used to re-analyze the past written accounts about Pavin since the sixteenth century? (iii) How have the perceptions and representations of Pavin have gradually changed from the eighteenth to twentieth centuries? (iv) How, gradually, the hypothesis of a degassing Pavin has been generated, debated, and accepted based on the paleolimnological, geological, and geomorphological elements? (vi) Which are the perspectives offered by Pavin analysis on the other maar lakes in Europe and other continents?

2. Maar lakes degassing reported in historical sources in Western Europe

Two hundred years before the Monoun and Nyos catastrophes, which have only been studied by scientists after the events, there have been few early descriptions of maar lakes degassing, by volcanologists, geologists, geochemists, and a medical doctor, and published as scientific communications in journals or reports between 1770s and 1829, with detailed contextualization, made in Italy (Monticchio Lakes and Albano Lake) and France (Pavin Lake) (**Figure 1**). The comparison of the different elements of these descriptions provides a coherent and converging set of descriptors of lake degassing and of their effects (**Table 1**). Another very special misbehavior of Lake Laach (Laacher See in Eifel, Germany) has also been reported in 1844 by a local witness, although not identified, therefore not included in **Table 1**. Also, the Albano *prodigium*, a slow lake level rise and overflow that threatened the city of Rome in 398 BC, as reported by many classical historians of Rome, four centuries after the event [23–25], is not part of this list. These historical scientific observations went largely unnoticed, and scientists at Monoun and Nyos thought they were the first to study this phenomenon [41].

	Pavin	Monticchio	Albano	Monoun	Nyos
Date	1785	1770–1820	1829	1984	1986
Ref.	[37]	[21, 22, 38, 39]	[40]	[41–43]	[1, 5, 43–46]
L1		X			X
L2					
L3					X
L4		X			X
L5	X	X			X
L6					X
L7					
L8		X			X
L9					
L10		X			
L11	X				
L12	X	X		X	X
A1		X	X		X
A2		X			X
A3					X
A4	X		X		X
A5	X		X		X
A6	X		X		
A7					
A8	X	X	X		X
A9	X		X		X
A10	X				
A11	X				
I1			X	X	
I2			X		X
I3			X		X
I4			X		X
I5			X		X
I6			X	X	X
I7			X		X
I8					X
I9		X			
I10					X
I11					X

Lake and water descriptors: L1 iron taste of water, L2 corrosive water, L3 absence of fish, L4 waves, L5 water spout, L6 red/orange/yellow colors, L7 glutinous aspect, L8 bubbling/boiling, L9 water spillover/inundation, L10 fish kill, L11 whirlpool, L12 waterspout.
Atmosphere descriptors: A1 roar/hiss, A2 sudden change, A3 explosion, A4 thunders, A5 lightning, A6 hail/ cold, A7 sulfur odor, A8 fog, A9 dense cloud, A10 wind storm, A11 serene weather outside crater.
Impacts on populations and living organisms: I1 dizziness, I2 paralysis, I3 loss of sight, I4 commotion, I5 skin damage, I6 death, I7 damage to fetus/babies, I8 crop damage, I9 fish kills, I10 massive livestock mortality, I11 bird absence or death.

Table 1.
Descriptors of degassing in Europe and Cameroon maar lakes: Pavin Lake (Auvergne, France), Monticchio Lakes and Albano Lake (Italy), Monoun and Nyos Lakes (Cameroon).

2.1 Pavin whirlpool, waterspout, and violent storm in 1785: The ignored description of a limnic eruption in Auvergne (France)

In 1803 (12 Vendémiaire, Year II), a "letter addressed by Citizen Cordier to Citizen Devilliers" is published in the Gazette nationale (also called Le Moniteur Universel), the high-ranking Parisian newspaper [37]. Cordier, a young geologist, reports to Devilliers, another geologist, the account of a geological tour carried out with two friends in Auvergne and Velay in which is found the following witness account about Pavin, one of the crater lakes he visited [all translations provided by the authors]:

> *"You will not read without interest the following description of a phenomenon that citizens Legrand and Enjelvin observed there [in Pavin] in August 1785; I got it from the latter.*
>
> *Attracted to these heights by the same motives as those who brought us there, tired by excessive heat, they sat on the steep banks which dominate the lake; their gaze, which at first wandered vaguely over the mountains, was soon drawn towards the lake by a sudden, dull noise. An extremely dense cloud had already formed near the middle of the bubbling water; it moved around, turning quickly on itself. In a short time, the wind became violent, the vapors increased excessively, and a waterspout 6 meters (18 feet) in diameter began to rise perpendicularly in the air. Barely had it exceeded the height of the enclosure when it ceased to rise; but then it began to pour down in abundance a thick fog, which quickly enveloped the mountain and its surroundings. Driven by a rapid movement, and supported by the waterspout, this stormy cloud swirled on the spot for a long time, and ended up spilling completely into the capacity of the crater. About this time, a terrible crash was heard, and continued for a quarter of an hour, accompanied by dazzling flashes of lightning which followed one another in rapid succession. However, the noise of the lightning gradually diminished, the cloud cleared, and eventually dissipated completely. Finally, the calm of the air was restored after two hours of this singular phenomenon. Descending to the bottom of the crater, the travelers found on the narrow banks of the lake a layer of very thin hail, which was 2 decimeters (7 inches) thick." Signed: Cordier Mines engineer.*

This description is fully contextualized, and both witnesses and the author are well educated and highly credible. Legrand is probably an abbot of St André de Clermont, the capital of Auvergne, and Enjelvin (1758–1815) is the director of the Pb-Ag mines of Pongibaud in Auvergne. Pierre-Louis Antoine Cordier (1777–1861) was 26 years old when he collected and wrote this story and already an experienced scientist, as he had been the assistant of the geologist Dolomieu during the scientific expedition in Egypt led by Bonaparte. He will be one of the prominent figures in the earth sciences of the nineteenth century (general inspector of mines, initiator of the mineralogy collection of the National Museum of Natural History in Paris, founder of the Société Géologique de France, and elected to the Academy of Sciences). What we can call today the limnic eruption of Pavin is not accompanied by any attempt at explanation or any reservation; it is presented in its simplest terms as an actual phenomenon. Cordier is not aware of other identical phenomena, as he would undoubtedly have pushed the comparison. The volcanic origin of Pavin was only known by scientists since the mid-1700s [9, 10] and mapped for the first time as a crater by Nicolas Desmaret (1725–1815), a pioneer of volcanology [47].

This description leads to the following comments: (i) This is the first detailed description by a scientist of a limnic eruption from start to finish. (ii) The eruption is triggered suddenly and lasts only 2 hours in total, with a high-intensity phase (thunder, lightning), which lasts a quarter of an hour. Such a phenomenon therefore had every chance of not occurring during very scarce visits, as Pavin was avoided at that time. (iii) Its fundamental elements are: bubbling, dense cloud, then violent winds, ascending and spiraling waterspout ("trombe lacustre"), succession of thunderclaps and incessant lightning, total invasion of the crater basin by swirling clouds, then overflowing of them, and finally formation of hail, collected on the edges of the lake. (iv) The witnesses were not affected by the cloud; it is likely that they were standing on the rim and that the cloud had dissipated when they came close to the water.

This letter was only recently found [48]. It appears to be the first—and so far, the only one—witness account of a limnic eruption made by scientists. It is also the most precise and detailed description to date as the Nyos eruption occurred in the middle of the night and was only witnessed by a few persons with limited scientific education. This description, published in Paris, had been so far totally ignored from the nineteenth century scientists, then by all Pavin analysts, including historians and volcanologists.

2.2 Monticchio Lakes (Italy) degassing, a matter of interest for Naple's scientists (1770s–1820s)

The twin lakes of Monticchio (Lago Picolo di Monticchio, LPM, and Lago Grande di Monticchio, LGM) are in the Mount Vulture Massif (Basilicate, Southern Italy) (**Figure 1**). They are just separated by a narrow piece of land. Both are meromictic, with a lack of oxygen below 15 m today and a marked positive temperature anomaly in deep layers [21, 49, 50]. The first scientific study of Monticchio Lakes has been realized by Domenico Tata, a prominent Naples scientist and pioneer of volcanology. He reports his observations in the noted *Lettera sul Monte Vulture* [38]. He comes to Mount Vulture in 1777, fully equipped (sounding lines, water sampling bottle) to study these lakes renowned for their marvelous nature. In LPM, he sees no water movements, but the water sampled at the lake bottom burst suddenly at 1 m high with a rotten egg odor, probably the first sampling of gaseous lake water ever realized. Such behavior has also been noted after Lake Monoun degassing by Sigurdson [41]. After Tata, several Naples scientists come to Monticchio [39, 49, 51]: Brocchi makes the first lake water analysis of CO_2 and H_2S. Tenore and Gussone describe a "permanent water fountain" that mixes the whole Lago Piccolo. According to local witnesses, as Father Tortonella, a maximum height of 5 m for the lake fountain was reported at LPM in November 1820. The lake activity gradually decreased and was not observed by Paci, the Naples geophysicist who made an in-depth report on the terrible 1851 earthquake in this region [51].

In their pioneer review of Monticchio historical degassing, Caracausi and his colleagues made the following summary [21]: "*we have come to the conclusion that triggering processes totally different from a landslide and /or a seismic shock should also be taken into due consideration. In fact, there have been gas bursts in both the Monticchio lakes up to no later than 200 years ago, which indicates that they have a recurring character, even if the temporal intervals have not been constant. A paroxistic activity occurred on 1 June 1810 when a 6-m-high water column formed in LGM. A new event was observed 2 months later on 31 July, when an intense roar was also heard; the column of water reached a height of about 3 m. Ten years later, a similar paroxistic event occurred in LPM; historical reports state that the water-spouts surged upwards out of the lake, reaching a height of 5 m. There were no fatalities, although a lot of dead fish were found on the land, a few meters from the shore, thus bearing witness to the explosive nature and gas toxicity of the event. It is worth noting that during the 10 years leading up to 1820 every so often LPM reached similarly dangerous conditions when all the fish living in LPM were killed."*

2.3 Lake Albano (Italy): A long degassing period in 1829 with impacts on population

Albano Lake is a major maar lake in the Colli Albani (Alban Hills) volcanic district, South of Rome (**Figure 1**). It is 3.5 km long, 2.3 km wide, 168 m deep, and is sometimes anoxic below 50 m. Its annular drainage area is very limited, and there are neither aerial inputs nor natural outlets. Major Latin historians such as Plutarch and Livius reported that, during early Roman times in 398 BC, a 1.45 km long tunnel had been dug out across the lake rim, following an elevation of the lake level by 70 m in 2 months and a subsequent overflow, threatening Rome, an event named the Albano *prodigium*, according to the tradition. The tunnel, a technical feat of Roman engineers, was still functioning in the early 1900s. The exact cause and effect of this event is still debated among geologists and volcanologists [23, 24, 52].

In 1829, a period of 8 months of degassing of Lake Albano is reported with great details by Luigi Bassanelli, a medical doctor in the city of Albano, located on the lake rim [40]. In his scientific letter to Pr Giacomo Folchi in Rome, he mentions the past local earthquakes (1762, 1777, 1778, 1779) and details the physical attributes of atmospheric changes in 1829. His report is also a unique medical testimony of the multiple effects of CO_2 intoxication on the population and the behaviors of animals, and of the social and religious responses. Despite a minimal description of Lake Albano itself, this detailed account includes: suddenness of event during serene weather, atmospheric storm with lightning ("electric atmosphere"), unusual cold in summer and hail, change of lake waters appearance, CO_2 presence, various diseases affecting the nervous system, deaths of babies, activation of traditional protection rituals (statue of the protector saint carried in procession), and population escape (developed in Appendix A and listed in **Table 1**).

2.4 The Laacher See (Germany) sudden and violent event in 1844

The Naples geophysicist Giacomo Maria Paci ([51], p. 79) also made an intercomparison between the Monticchio Lakes misbehavior and another misbehavior event which occurred in 1844, according to a Naples journal, in the Laacher See, the biggest Eifel maar lake in Germany (**Figure 1**). The original Laacher See event was actually published in Germany [53] by an unknown local witness, most probably a Benedictine monk from the Laach Abbey, established within the crater since the eleventh century:

> *"Today, early in the morning, around half past four, the following natural phenomenon occurred here: in calm and serene weather, suddenly the lake rose to a remarkable height and, at the same speed, it withdrew, while we perceived with a shudder rocks and abysses never seen. At the same time a dull sound, like thunder, was heard, while almost suffocating smoke swirled high in the air. On the other side of the lake, "there, where the well-known asphyxiating air perpetually rises from the earth", the earth opened up, and several trees were violently shattered as they fell towards the*

cavity. After this apparition, several dead fish were seen floating on the surface, and around the suffocating pit, many dead birds were dragged. It was all the work of a moment. It would therefore be desirable for specialists to take note of this interesting phenomenon and deem it worthy of further investigation."

However, this witness description is invalidated a few days later in the same journal, after a field investigation by a journalist: "most everything said here is false and is essentially limited to the very simple fact of the formation of crevasses and landslides... It is fairy tales." In addition, an unidentified expert interviewed by the journal writes that the phenomenon was linked to a sudden slide of the crater rim generated by on-going clay extraction: *"The sudden growth and fall of the lake, the rocks and abysses that became visible in the lake basin, the thunderous roar, the suffocating smoke swirling high into the air and everything else the"* eyewitness *"claims to have seen can only have been the product of an excited imagination. The naturalist should beware of this, even if it may well be fitting and appropriate for the poet."*

Such abrupt judgment, typical of the position of academic scientists, has been formulated many times for Pavin, until recently. Today, the suddenness of the lake level rise, then its drop, the thunder-like roar, the suffocating pit and the impact on fish and birds seem to us much more credible and suggest a violent lake degassing. The Laacher See is also the Eifel maar with the greatest number of legends, evoked in Section 6.3.

2.5 Monoun, Cameroon (1984), "a hitherto unknown natural hazard"

This catastrophic event occurred in Cameroon on August 15, 1984, at the Monoun Lake (0.53 km^2, 96 m deep) and officially claimed 37 victims. Haraldur Sigurdsson, a volcanologist from the University of Rhode Island, made the post-event investigation only a few months later and found no signs of a volcanic eruption. He detected "a hitherto unknown natural hazard" that could wipe out entire towns, and, in 1986, a few months before the Nyos disaster, he submitted his study to Science, the prominent scientific journal, which rejected the paper as far-fetched [44]. Sigurdsson eventually had it accepted in the Journal of Volcanic and Geothermal Research and therefore published a year after the Nyos event [42]. He was unaware of the Montichio, Pavin, and Albano historical degassings, unreported in volcanology or limnology textbooks at that time. No early warning was given to the scientific community or to local Cameroon authorities about such type of lakes. The reluctance of the scientific community to conceive this "new natural hazard" is exemplary.

Monoun rescuers, who came a few hours after the catastrophic event, observed "a misty cloud which did not disperse until three to four hours later" [43]. The Monoun eruption also occurred at night, and the gas cloud filled a low-lying area near the lake [42]. Travelers on a nearby road were killed as they entered the area, as were people who left higher ground to help those who had fallen in the valley. In the lake, the water fountains reached 5 m high.

Lake Monoun was then surveyed by scientists from several countries, and one of them, Minoru Kusakabe, said, "if the [surface] waters were stirred up by only three feet, the water could start bubbling and trigger an explosion." His colleague, Bill Evans, advised caution: "Let's not go splashing around too much out there" [44]. This corresponds to a near-saturation state (97%), very sensitive to any perturbation, as throwing a stone.

2.6 The Nyos (Cameroon) catastrophe, which shocked scientific communities (1986)

The catastrophe occurred on Lake Lwi, below the village of Nyos in the Cameroon Grasslands, on August 21, 1986. Today most scientific papers refer to Lake Nyos. The explosion occurred at night, and only a few surviving eyewitnesses gave a detailed account [4, 8, 45, 54]. The main features of the event can be summarized as such:

In the evening, an intermittent jet of water was seen, topped by a white plume. At 10 p.m. a major detonation—a series of rumbling or bubbling sounds lasting 15–20 seconds—occurred in the lake. One observer walked to the crater rim and saw a white cloud or mist rise from the lake, then a large water surge. Soon after, the carbon dioxide invaded the low-lying valleys and intoxicated the population and the animals over 20 km. Although people smelled the odor of rotten eggs or of gunpowder, they lost consciousness owing to the dense, ground-hugging, and toxic cloud of CO_2 mixed with water vapor and droplets, which smelled bitter and acidic. Livestock and other animals, birds, and insects also died in the event. Vegetation was "mechanically damaged," but no thermal or chemical damage could be identified with certainty. Just after the event, the surface of the lake turned reddish brown. This was eventually attributed to the oxidation of dissolved ferrous iron mixed to the surface from depth. A few months after the event, Rose-Marie Chevrier, a French geochemist, observed a sudden and intense flash of light in the crater and, again, a red spot in the water [46].

The medical symptoms of the Nyos victims and survivors have been analyzed by a specialized team [43]. Their study, focused on 845 hospitalized survivors, out of the 5000 total survivors, recorded: coughing (31%), headache (26%), skin lesions resembling burns (19%), fever (12%), weakness/malaise (11%), limb weakness (6%), severe eye symptoms, vomiting, diarrhea, and dyspnea (5% each). Many of them had a sudden loss of consciousness with a coma lasting a few hours, followed by rapid recovery that could be related to anesthetic concentrations of CO_2 mixed with air. No persistent respiratory disability was noted in survivors, implying low concentrations of SO_2 or H_2S. The most common abnormalities were skin damages: erythema, big skin bullae, more than 5 cm, skin changes resembling burns, often developed 2 or 3 days after the degassing. Only five patients—out of 161—attributed their skin lesions to burns. Long-term neurological damages could not be estimated. Surviving victims interviewed by the Cameroon Tribune also reported: "I felt dizzy," "I vomited blood," "I felt drunk" [5].

The most salient physical figure of Nyos limnic eruption is the erosion traces on a cliff up to 50–80 m above lake level, suggesting the occurrence of either landslides and/or violent waterspouts. An up to 4 m thick turbidite deposit was sampled at the lake bottom by gravity coring in 1988 and mapped with 3.5 kHz seismic profiles in the deep central basin of Lake Nyos. In 1991, further studies supported that Limnic Eruption Deposits (LEDs) can be recorded in meromictic maar lakes sedimentary infills [55].

There has not been a water overflow but an overflow of the CO_2 cloud, which caused victims over 20 km in downstream valleys. The official number of victims is 1700 and more than 5000 cattle heads [5]. The whole event lasted 4.5 hours.

2.7 Common descriptors of maar lake degassing and its impacts on populations and animals

When the descriptors of these degassing events and their impacts on populations are compared, many similarities are noted (**Table 1**). The differences are due to the type and/or intensity of degassing and to the position of observers. The actual state of the lakes during these events is not always detailed. The Nyos and the Pavin-1785 events represent what is now called a limnic eruption, a sudden degassing phenomenon, violent and short in both cases, while the Montichio and Albano-1829 descriptions represent more a prolonged state of degassing, suggesting different degassing causes and mechanisms [19, 24].

Toxic effects on humans are well detailed in Cameroon and Albano Lakes, ranging from dizziness, drunkenness, skin lesions, loss of sight, contusion, loss of consciousness, and coma to sudden death. Post-event effects on fetus and breast-fed babies are noted. At Albano Lake, the overall death toll on people is not given.

The early warning of animals is noted at Albano, although it cannot be said if they reacted to the CO_2 cloud or to micro tremors, radon emission, or else. At Nyos, the deaths of all, humans and animals, generated a post-event rumor: the secret experimentation of a neutron bomb supplied by a foreign country [56]. In the Roman Antiquity, the death or absence of birds was a well-known consequence of ground degassing in so-called mephitic places. As this was particularly observed in Lake Averno (a-ornos, i.e., without birds) in the Phlegrean Fields near Naples (**Figure 1**), *avernism* has been used to qualify such places [11, 25, 36].

The mist, fog, or clouds reported over these lakes are thought to be related to CO_2 and water vapor emissions, but only water analysis was realized at Monticchio to confirm this assumption. The gaseous micro particles emitted in the atmosphere can generate electrostatic storms with thunder and lightning. At Nyos, Monticchio, and Albano, a sulfur odor is noted in the atmosphere, although not dominant; for some it is gun powder smell (SO_2), for others rotten eggs (H_2S). In many meromictic lakes, H_2S is present in deep anoxic waters due to the decomposition of organic matter [2, 15].

The fish kills and the sudden change of water color and fluidity are the markers of the mixing of anoxic and iron-rich deep waters with surface water, the so-called rollover which results in the precipitation of orange-red iron oxide $Fe(OH)_3$, as the one that has been observed in Monticchio Piccolo Lago in January 2017 [48].

Limnic eruptions are characterized by an ascending column of water (Pavin-1785, Monticchio-1810 and 1820, Monoun-1984, Nyos-1986), reaching 5–80 m high. It generates a great emission of micro particles, which are associated with electrical storms. In the Pavin-1785 event, a great amount of hail is found on the shore, possibly due to the gas expansion. A colder atmosphere is also noted in Albano-1829.

3. Past accounts of tempestuous Pavin revealed by historical sources (1549 onwards)

Many historical accounts on Pavin have been found since 2009 [9–11], only the major ones are presented in this paper [57–81]. They represent several hundred pages, which are summarized in Appendix B and re-analyzed in **Table 2** using the grid of maar lakes degassing descriptors established in **Table 1**. The 1785 Pavin event has been fully described in Section 2.1.

3.1 The first pictorial representations of Lake Pavin in a royal land register (1579)

Pavin is figured on a landscape representation, the first ever made in France, hand painted in the royal land register of Besse [57]. In this magnificent illuminated manuscript, ordered by Catherine de Medicis, the Queen Mother and lord of Besse, Pavin cup shape and position "on top of a mountain" are very accurately represented (**Figure 2**).

The discharge of the lake outlet—the Couze Pavin River—is over-represented as if there were a lake overflow, spreading across the cascade 500 m below the lake.

Events	1551	1566	16th–17th	1750s	1783	1847	1936
Ref.	[59, 60]	[62]	[59, 60]	[66]	[68]	[71]	[72]
L1				X			
L2				X	X		
L3				X			
L4						X	
L5							
L6					X		X
L7					X		X
L8					X		
L9							
L10							
L11							
L12							
A1							
A2	X				(X)	(X)	
A3	X						
A4	X	X					
A5	X						
A6		X					
A7	X	X					X
A8				X			
A9				X		X	
A10	X					(X)	
A11	X						
I1							
I2			X				
I3			X				
I4			X				
I5			X				
I6							
I7							
I8		X					
I9							
I10			X				
I11				X			

Table 2.
Analysis of major witness accounts of Pavin Lake through the degassing descriptors listed in **Table 1** *(details in Appendix B and Section 2.1 for the 1785 event).*

Figure 2.
Pictorial representation of Pavin Lake area in 1579 in the royal land register of Besse, featuring the lake and the city of Besse [57] (actual size of this partial picture: 20 × 30 cm).

The mountain featured behind the lake is the Montchal, today identified as a volcano with its crater. The Vassivière Mountain is the last one on the upper right side. The area between Vassivière and Besse—the valley of the Couze Pavin River—is the location of many miraculous healings recorded between 1547 and 1750s. This is the first landscape and first lake ever depicted in France. In 1663 the lake is localized for the first time in Joan Blaeu's world atlas published in Amsterdam, an indication of its famous reputation [10].

3.2 Geographic setting of Pavin and its surroundings

Lake Pavin is in the center of the Auvergne lake district, located at mid-altitudes between 800 m and 1200 m, on top of the hydrological divide between the Allier-Loire and the Dordogne-Garonne river basins (**Figure 3**) [9, 14]. The lake altitude is 1196 m above sea level today. The Pavin area is located at the limits between the historical regions of Haute and Basse Auvergne near the city of Besse (4 km east of Pavin) known across Auvergne for its famous Catholic pilgrimage dedicated to Notre Dame located on the Vassivière Mountain at 1.5 km west of the lake at 1300 m asl (**Figure 3**). Before the twenty-first century, all these lakes were frozen during the whole winter, up to 6 months of ice cover for Pavin in the 1700s [58]. The mountainous region of Besse was isolated, and the local road was avoiding the valley below Pavin until the 1820s. The Couze Pavin River, originating from the Southern Mont Dore massif and Lake Pavin, joins the Allier River, some 30 km downstream of Pavin, at the main city of Issoire. Until the 1800s, the population in this mountainous part of Auvergne was largely uneducated and isolated from the main communication routes, particularly during the 6-month snow season.

3.3 Geological setting of Pavin

For volcanologists, the Pavin volcanic zone is located between two ancient volcanic districts of the French Massif Central (**Figure 4**), the Mont-Dore stratovolcano, aged 0.2–1 Ma, and the Cézallier stratovolcano, aged 3–8 Ma, at only 20 km of the northern

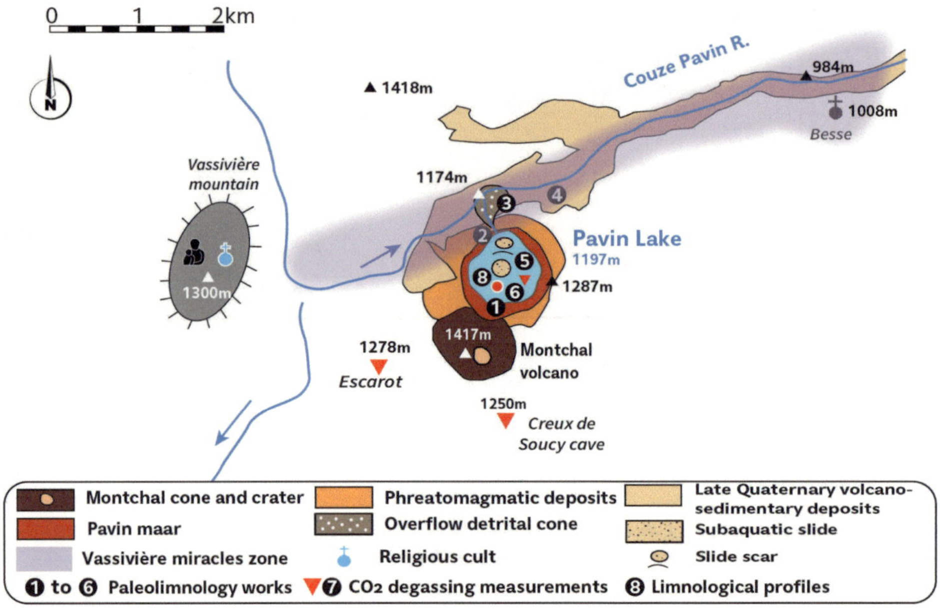

Figure 3.
Schematic map of Pavin geosite and its current scientific approaches (see also **Figure 6***). 1. Paleo lake levels revealed by LIDAR and morphological analysis of lake rim. 2. Natural outlet incision analysis. 3. Pavin detrital cone stratigraphy below outlet. 4. Scattered overflow deposits in Couze Pavin River thalweg. 5. Scars and deposits of subaquatic sediment sliding revealed by fine-scale bathymetry, sub-bottom profiling, and radiocarbon dating of long piston sediment cores. 6. Sedimentary archives (7000 BP-Present) from lake core stratigraphy and geochemistry. 7. Present CO_2 measurements in lake and surroundings. 8. Limnological profiles in the lake waters and measurement of CO_2 emission from the lake surface. Besse city and the sacred site of Vassivière are culturally linked to Pavin by the Vassivière cult and its attributed miracles.*

side of the Cantal stratovolcano, aged 2–13 Ma. In contrast, Pavin maar results from a very recent phreato-volcanic eruption 6730 years BP, which occurred a few hundred years after three other nearby eruptions, the Montcineyre, the Estivadoux, and the Montchal; this recent volcanic area is almost aligned with the N-S Chaîne des Puys. Pavin maar depression, cut into the Mont Dore basalts and into the northern part of the Montchal volcanic cone, generated the latest cataclysmic phreato-magmatic eruption of Plinian type in the Massif Central; its tephras found in numerous peat bogs are considered to be the last ones emitted in continental France [35].

3.4 Written sources and attempts to solve the mysteries of Pavin

The type of historical descriptions of Pavin and its surroundings have evolved a lot since their first publication in the sixteenth century. They also depend on the geographic position of the witness with regards to the lake and on their social position. As the lake was avoided until the mid-1800s [6, 10] and its violent events can be quite short—two hours for the 1785 waterspout—full descriptions of events by in situ witnesses are rather rare. Therefore, the post-event descriptions are dominant.

Heterogeneous sources on Pavin history were found one by one: an official complaint to the King of France (1566), a pictorial representation in a royal manuscript (1575), a visit by a medical hydrologist (1605), the religious registers of local miracles (1615, etc.), a very detailed manuscript description of the lake written by a local Besse scholar (1750s), publications in local and national newspapers (1783, 1803, 1986),

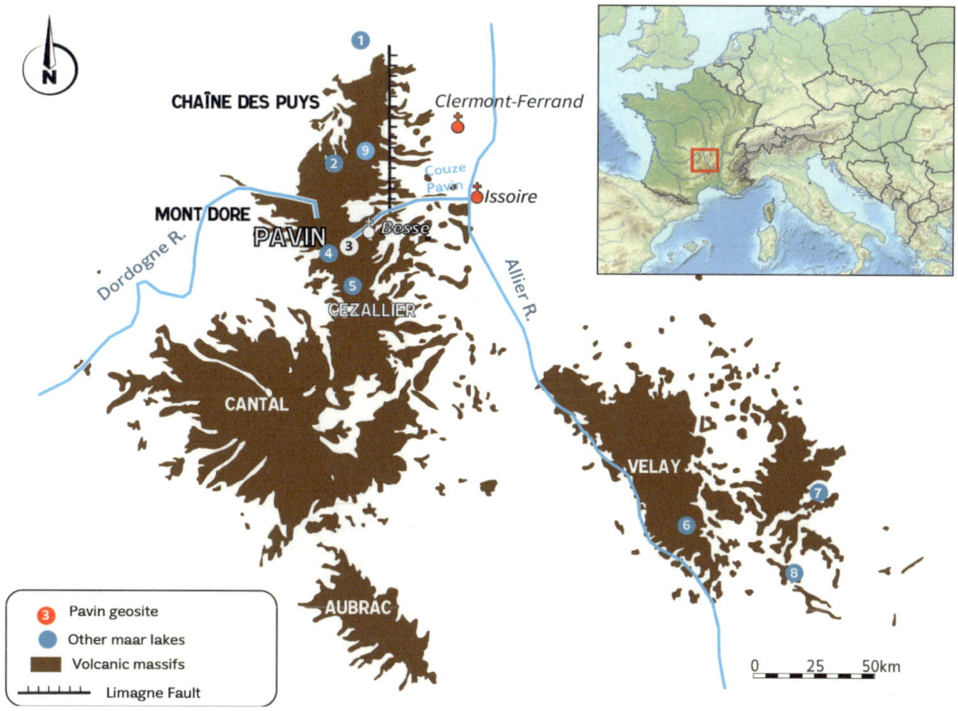

Figure 4.
Pavin Lake in the Central Massif volcanic provinces (France): Cantal, Mont Dore, Cézallier, Chaine des Puys, Aubrac and Velay. Maar lakes: 1 Tazenat, 2 Servières, 3 Pavin, 4 Chauvet, 5 La Godivelle d'en haut, 6 Bouchet, 7 St Front, 8 Issarlès. 9 Aydat Lake (lava flow dam). Only Lake Pavin is presently meromictic and believed to have degassed in the recent past.

maps (since the mid-1600s), multiple travelers' accounts (eighteenth century and after), guide-book (nineteenth century), recent interviews, and contemporary legends collected by the authors. An unexpected and precious source of information on the Pavin area is the register of miracles, attributed to the power of an early medieval wooden statue, the Vassivière Madonna, revered since ages. There, a famous local pilgrimage was re-initiated in 1547. Since then, all miraculous healings occurring in the Vassivière-Pavin-Besse area have been carefully registered by the Catholic Church authorities, including a striking tempestuous event reported in 1551. Numerous Pavin legends have orally been transmitted across centuries and then transcribed between 1820 and 1925; they are discussed in a companion paper [36].

3.5 Analysis of historical accounts and descriptions about Pavin

We have selected here the most detailed and pertinent descriptions, generally made by identified eyewitnesses (**Figure 5**). They are well contextualized, providing dates and positions, and are often reported by local elites with higher educational backgrounds. These sources have been largely or totally ignored by Pavin analysts until now [6].

Pavin misbehavior is reported here through seven events of various intensities. Several violent storm events (1551, 1750s, 1785) combining lake and atmospheric disorders have been described through a triple motif: atmospheric and lake storm + thunders + lightning's, generally associated with the suddenness of the event occurring within a "serene

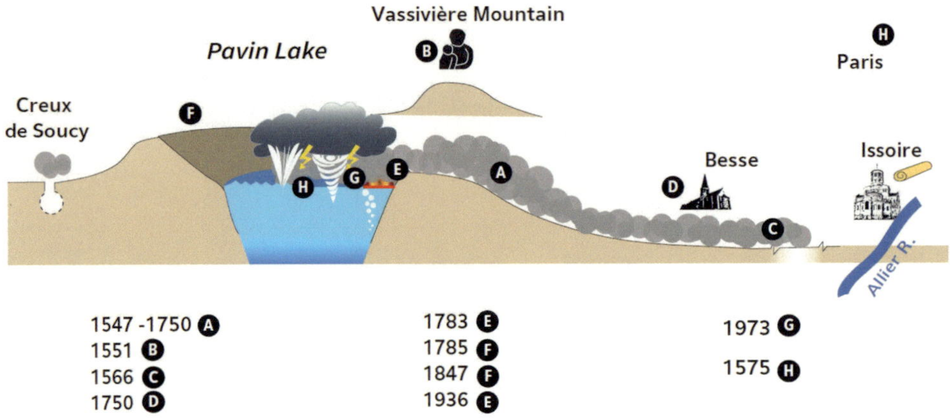

Figure 5.
Principal historical observations of Pavin degassing and positions of witnesses (see Annex B for details). A, 1547–1750s: neurological damages in Vassivière-Pavin-Besse area, reported in Vassivière chronicles. B, 1551: violent explosion witnessed from Vassivière. C, 1566: storms with thunder and hail causing damage to crops in the valley. D, 1750s: detailed description of Pavin behavior by Besse observer. E1, 1783: sudden color change (within 4 hours) and whole lake boiling witnessed at outlet. E2, 1936: color change (in less than 12 hours) witnessed at outlet. G, 1785: limnic eruption with whirlpool and waterspout (6 m diameter, 50 m high). F2, 1847: lake storm, observed from rim (?). G, 1972: boiling circle in lake (10 m wide with 20 cm eddies) crossed on boat. H, 1575: Pavin storm generated by a tempestary stone, published by Parisian cosmographer Belleforest. Many historical manuscripts were kept in Issoire monastery, 30 km downstream of the lake, and exposed to its misbehavior.

atmosphere" (clear sky), often with smoke emission, whirlpool, and/or waterspout. They can be understood as limnic eruptions. Medium-intensity degassing events are accompanied by fog and devastating clouds (1566, 1750s). Low-intensity events include rollovers with boiling waters and sudden color and texture changes (1783, 1847, and 1936). The impacts on the local population are inferred from the Vassivière miracle register. The detail of the descriptions depends on the position of the observers (**Figure 5**).

In 1575, the description of Pavin—not listed here—published by Parisian cosmographer Belleforest [57] attributes the sudden storms of tempestuous Pavin, in association with thunders and lightning, to a stone thrown into the lake (see Appendix B, 1575). A thorough analysis reveals that Belleforest had actually reproduced an earlier description (thirteenth century) of an unlocated and unnamed tempestuous "mountain pit" where demons generated storms when a stone was thrown into it [36].

4. The perception and response of the local population to Pavin disorders (1547–2024)

Historical sources also allow us to reconstruct the perception, the representation, and, in general, the social response to Pavin disorders, which have been officially considered for several centuries and then gradually called into question. While in the nineteenth and twentieth centuries official science erased the past behaviors of Pavin, whose original name, *pavens*, means "the terrifying" (Appendix B, 1605), the memory of the fear he generated and the need for protection remain very present today in an exceptional religious practice, which has been taking place since the mid-1500s, without the connection with the stormy lake ever having been recognized by the church. The chronological summary of Pavin descriptions, perceptions, representations, and analyses is proposed in **Table 3**.

4.1 The Vassivière register of miracles, an account of degassing impacts on local population (1547–1700s)

The past effects of Pavin degassing on local populations are derived from our re-interpretation of well-known local miracles collected since 1547 by the religious authorities from Besse, then validated by a complex process set up by the Catholic

1551	Sudden violent storm, thunder, and lightning observed from Vassivière [60, 61, 66, 73].
1549–1700s	Healings (contusion, loss of sight, sudden paralysis...) observed between Vassivière and Pavin [60, 61, 66, 73].
1566	Complain to Charles IX about a *mountain pit exhaling clouds, thunder, and hail that damage crops in valleys* [62].
1575	Belleforest's description of a tempestuous lake on top of a mountain whose storm is triggered by a thrown stone [61].
1579	Pavin is pictured for the first time in the royal Besse manuscript, overlooking the valley of the Couze Pavin [57].
1672*	A stone thrown into Pavin does not trigger any storm [67].
1750s	Tempestuous Pavin detailed description; every day in summer the lake produces misty fog and occasionally few violent storms which threaten the whole region, hence its name: *lacus pavens,* the terrifying lake [66].
1770*	Pavin is sounded for the first time (96 m) by a royal engineer [58, 77].
1783	Boiling lake (rollover): the watercolor and aspect of the lake outlet changed within a few hours. The whole lake is boiling. Corrosive water generates skin buboes. A sub-lacustrine slide is inferred [68].
1785	Limnic eruption by fair weather: a spiraling ascending column of water, 6 m wide, is seen. When reaching the edge of the crater, a terrible storm breaks out, with thunder and incessant lightning. The storm lasts 2 hours, then the sky is clear again. A hail deposit, 20 cm thick, is observed on the shore [37].
1788*	*Pavin is just a simple lake; I would like to dive in its deep waters to unveil its secrets* [77].
1829	*Often thick clouds and hot exhalations would spread on its shore...; Its name, Pavens, imposed by the Romans, tells well that it was a threat in this area* [75].
1835*	*Pavin stories are absurd tales and legends invented by fear and transmitted by credulity* [70].
1847	Storm developed at the lake surface with shocking waves and whirlpools [71].
1859*	First salmonidae fishes successfully introduced in Lake Pavin by Rico and Lecoq [6, 10].
1892	CO_2 emission from Creux de Soucy cave is assessed for the first time by a speleologist [82].
1896	CO_2 presence and impact on visitors near the lake are evoked for the first time [80].
1890–1909*	Development of touristic access to the lake; first limnological studies [9, 10, 14].
1929	Degassing hypothesis: the CO_2, stored in Pavin deep waters and generated by volcanism could suddenly burst out and generate lake overflows as those traditionally feared by Issoire people [30 km downstream] [81].
1936	Color change (rollover) overnight, from very transparent waters to yellowish-orange glutinous waters that clogg the lake outlet grid; colloidal sulfur smell is perceived; fish kill [72].
1950s	Limnologists discover Pavin meromicticity by O_2 profiles, then by water chemistry (Fe, Mn, etc.) [9, 10].
1971*	Pavin stories are *tales of maid women;* the Pavin name originates from *pallenc*, palissade in Occitan [78].

1972	Column of bubbles emitted from the southern part of the lake, 10–15 m wide, 20 cm high (August 1972); no specific smell was observed. Witness never dared to report the event. [Interview in Besse, May 2024].
1993*	First degassing risk assessment at the request of national authorities [26].
2005–2009	Detrital deposits were found in the Couze Pavin thalweg and attributed to violent lake overflow [83]. A confidential commission is set up by regional authorities.
2008*	Limnologists report: *"No spontaneous gaseous eruption can take place; The present situation is stable and only important landslides, disturbing bottom sediments could generate a partial degassing."* [84].
2009*	The Pavin and Meromictic Lakes conference held in Besse rejects the hypothesis of Pavin misbehavior.
2010–2024	Paleolimnological, geomorphological, and geological evidence of Pavin Lake events during the last two millennia, coherent with perception and description of past Pavin "misbehavior" [6, 9–11, 29, 30, 36, 48, 85–91].
2016	First synthesis on Lake Pavin [27]; geohistory and paleolimnology cover several chapters [9–11, 35, 91].
2021*	Volcanologists report: *The so-called Lake Pavin misbehavior (Meybeck, 2016a, 2016b) has not been well documented or understood* [32].
2023	*Only three lakes in Europe (i.e., lakes Pavin in France, Albano, and Monticchio in Italy) and one in Africa (i.e., Lake Kivu in the D.R. Congo) evidenced markers of limnic eruptions in their sedimentary archives* [87].
2024	First modeling of a sudden release of CO_2 during a limnic eruption with risk assessment for Pavin visitors and Couze River valley population [34].

Table 3.
Chronological summary of past Pavin misbehaviors, normalizations (*), and analyses (detailed accounts in Appendix B).

bishop of Clermont, including names of witnesses, their locations, and dates [59, 60, 73, 74].

The Vassivière Mountain, located 1.5 km West of Pavin at 1300 m asl (**Figures 2** and **3**), is the location of continuous and major Christian worship for the last 500 years. According to some local historians, it was probably active in the Late Antiquity [73]. There, a wooden statue of a Black Madonna with Child was worshipped since the Early Middle Age in a first church dismantled in 1320 for "profane behavior" of the population. From 1320 to mid-1500s, the statue, put in a depository in an ancient wall, was still worshipped. The devotion to the statue was re-vitalized after 1547, when the miraculous healing of a merchant, suddenly caught blind between Vassivière and Besse, was officially declared by church authorities [59, 60]. A chapel was rebuilt in the 1550s, and a new pilgrimage started at a time when Pavin misbehavior was particularly noted and feared. The original statue was destroyed at the Revolution and quickly replaced by a copy, continuously revered in the 1800s and 1900s [11, 73, 74]. One of the authors, MM, assisted several times as a kid to the ritual masses of August 15 in honor of N.D. de Vassivière in the mountain sanctuary. The most significant ceremonies (1638, 1881) attracted up to 20,000 pilgrims from all over Auvergne to this tiny hamlet, abandoned in winter, where snow was sometimes lasting in August [59]. The Vassivière Guild, composed of all parishes, 20–30 km around, oversaw the rituals, the collection of offerings, and the acquisition of the many relics.

The close relation between the ancestral Vassivière worship and Pavin misbehaviors has been suggested in some local religious sources, although not officially recognized by the church and the miracles attributed to the Madonna were never contested by the local population. In 1688, Cladière reports in his chronicle: "*The inhabitants of Besse carry the Image [statue] in every danger... the corrupted air was purified in an instant*" [60]. In the early 1800s the "tradition" collected at Besse by Nodier, Taylor and Cailleux [75] clearly connects Pavin disorders to the Vassivière pilgrimage: "*Some would also tell you that if She [the Vassivière Virgin] was not brought back [from Besse to Vassivière], one could see at once a luminous vapor surging from the earth and wrap the church, with gliding cherubs, musicians archangels and all Christian mythology spirits, who bring her back with pomp in her uninhabited land*". In 1857, a Vassivière poem still evoked the mighty storms generated by demons: "*Satan, on the stormy peak, led his infernal rounds... For eight days, the storm roared, and during this time the devils, unleashed in the neighborhood, damned the inhabitants. ... Lucifer, without being exorcised, flies away...the demons come out of their hideouts, the air regains its serenity*" [69].

The detailed Vassivière "miracle register," maintained from 1547 until the 1800s, contains 109 miracles, many of them occurring locally in the area between Vassivière and Besse [73, 74]. One of the most remarkable miracles relates a sudden and violent storm in 1551 (Appendix B, 1551).

In 1907, Dr. Noir, a Parisian medical doctor, noted [76]: "*the group of the most numerous and most interesting miraculous cures consists of 33 cases where the illness clearly has the characteristics of hysteria. Most of these patients are unfortunate people who are at once paralyzed, blind, deaf, and mute and who suddenly recover movement and the use of their senses after prayer to the Virgin.*" The doctor could not offer any explanation for these neurological damages. The similarities with the impacts on populations generated by the degassing of lakes Albano and Nyos are striking (**Table 1**). We believe that these diseases were most likely generated by CO_2 emissions from Pavin, invisible, sudden, and sneaky, and, for centuries, their spontaneous healings have been considered miracles by the local population. The Vassivière records rarely mention human life losses since there was no miraculous healing but occasionally report a massive mortality of cattle.

The multi-secular belief in miracles attributed to N.D. de Vassivière is so deeply grounded in traditions and Auvergne culture [74] that no historian has attempted a re-assessment of this unique and long-lasting tradition from a non-religious point of view [6].

4.2 The successful sounding of Pavin in 1770

In the 1700s, the reputation of the unfathomable nature of the marvelous lake seemed unbearable to the authorities. After a failed attempt in 1726 by locals, a successful sounding is carried out in 1770 by Chevalier, a royal engineer of bridges and roads [9]. He is equipped with the latest scientific instruments (navy-sounding instrument, Réaumur thermometer, and leveling instruments) to achieve precise measurements and examine the possible relationships between the Pavin and the Creux de Soucy cavity (**Figures 3** and **5**). Chevalier finds the bottom at 288 feet (96 m), discovers the hidden springs that feed Lake Pavin, identifies the lava cliffs (Mont Dore lava flow), and establishes that the tiny and secret lake found at the bottom of the Creux de Soucy is at a higher elevation than Pavin level, making communication possible, a hypothesis denied today. The normalization of Pavin Lake can begin. Ironically, a few years later, two major descriptions of Pavin misbehaviors are published; both will be ignored by authorities and Pavin analysts [6, 48].

4.3 The normalization of Pavin's tempestuous character by scientists and authorities (1770–1993)

Many attempts to solve Pavin mysteries have been made since 1788, by literary voyagers, a botanist, a naturalist and geologist, authors of scientific guidebooks, local historians, a school professor, folklorists, a renown Auvergne historian, writers, and many others [6, 9, 10, 36]. Generally, they are based on one or two sources only.

In 1770, the sounding by Chevalier puts an end to several Pavin mysteries listed by Belleforest (Appendix B, 1575). Legrand d'Aussy in his *Voyage en Auvergne* (1788) [77] and Abbé Delarbre [58] went further and invalidated the famous devastating clouds and the sudden storms mentioned by the cosmographer, although Legrand raises many questions on Pavin deep waters [6, 10]. Meanwhile, the local and regional authorities pushed for a complete normalization of Pavin. All were ignoring the 1785 limnic eruption, only published in 1803 [6], but reported orally by local guides to a growing number of visitors. Naturalist Henri Lecoq [70] will name them "Pavin stories" (Appendix B, 1835). In the 1820s, a new local road is constructed in the Couze Pavin valley; the previous one avoided passing below the lake. A first boat is built and brought to the lake, allowing the successful introduction of fish (trouts, salmon, chars)—absent until then—by Lecoq and Rico in 1859, an ultimate reason to invalidate all Pavin stories [6].

The first access pathway to the lake and the walkway around the lake (1859–1909) finally allowed the visitor access and the development of tourism. Meanwhile, the city of Besse pushed for the scientific analysis of Pavin by the first limnologists in 1890–1914 [6, 9].

In the twentieth century, all volcanologists and limnologists avoided evoking the tumultuous Lake Pavin history [9], including one author of this paper [15, 26]. It must be recognized that, until recently, the very few known historical descriptions of past Pavin events were often entangled with legendary accounts told by local guides, such as those of Delarbre and Lecoq, particularly the story of the tempestary stone (Appendix B, 1575), still told orally in the Pavin area in the 1960s (MM personal observation). Historians and social scientists had also not much addressed Pavin: most were following the statements of Legrand d'Aussy in 1788, "Pavin is just a simple lake" [77], and the official Pavin descriptions, as the one of Jouan (Appendix B, 1566), were regarded as "legends invented by fear and transmitted by credulity" [71] or as "tales of maid women" [58], an expression endorsed later by renown Auvergne historian Pierre-François Fournier [78], for whom the lake name comes from the Occitan pallenc, which means palisade, an allusion to its steep basaltic cliffs (and not from the Latin *pavens*). This judgment made by a well-respected scholar 15 years before the Nyos eruption, has probably limited further investigations by other historians [6].

4.4 Deep-water CO_2 degassing, a hypothesis made by writers ignored by scientists

At the end of the nineteenth century, the hypothesis of lake degassing is made for the first time by Ajalbert, a noted Parisian writer visiting Auvergne [80]: "*suddenly one can no longer breathe, the throat is strangled as at the bottom of caves where, for lack of air, one would suffocate. It is the upper part of a crater which jumps under the force of the underground gases, like the cork of a bottle of champagne.*"

In 1929, this hypothesis was further developed by Jean Larat, a literature teacher at Issoire, 30 km downstream from the lake (**Figures 3** and **5**): "*the volcanic activity has not completely disappeared and it still results in continual releases of carbonic acid. At the bottom of Lake Pavin, these gases remain trapped in the crevices of the volcano under the effect of the* *heavy mass of water which covers them. But if the fall of a stone, reaching the bottom of the lake, releases the gaseous masses it contains, the rise of these gases which explode on the surface of the lake then produces a sort of bubbling of the water and the formation of real waves. It's a storm in miniature, a small storm on the lake. Here, the legend is therefore not entirely devoid of any reality... Nor will we try to dissipate the fear which consists of seeing a geological commotion project, one day, the enormous mass of the lake onto the slope of the ravine of the Couze* [Couze Pavin River]. *According to legend, it must, in its furious descent from the mountain, after having carried away numerous villages, submerge the town of Issoire*" [81]. This striking analysis, published in a local journal at a time when neither the meromictic character of the lake nor the presence of CO_2 in deep waters were yet established, has been ignored by all Pavin analysts and scientists.

4.5 The degassing controversy among scientists (1986–2024)

When the Lake Nyos catastrophe occurred in 1986, Pavin was quickly identified as a maar lake with potential degassing risk [6] for the nearby population and for the 200,000 annual visitors coming to see the beautiful site, most of them standing a few meters above the lake level, at the outlet, and on the circular hiking path.

The first risk assessment of Pavin is carried out in 1987–1992 at the request of national authorities by volcanologists and geochemists, including one of the authors (MM), and led by Guy Camus, a Clermont-Ferrand volcanologist. CO_2 was measured for the first time in its deep waters, and we concluded in 1993 that there was no risk of degassing under current conditions [26]: "*it is considered that poison gas outburst of limnic origin (upset and ebullition of CO2-rich water from the monimolimnion) is not to be held today as a realistic hazard in the French Massif Central.*" Pavin history was totally ignored. A second assessment (2005–2008) was carried out confidentially at the request of the provincial authorities by a committee of experts, including one of us (PL), following his discovery of a patchy detrital recent sedimentary formation deposited downstream the lake in the Couze Pavin thalweg over 10 km when updating the local geological map [48] (see **Figure 3**). For its discoverer, it would have been generated by a recent sudden overflow of the lake [83], hardly dated, a catastrophic hypothesis as soon questioned by local volcanologists and rejected when it was submitted to the Journal of Auvergne natural history. A similar rejection consensus was reached once more at the International meeting on Pavin and other meromictic lakes held in Besse in 2009 and attended by French specialists of Nyos, local volcanologists, and scientists from Pavin [6], including three of the authors (PL, EC, and MM). There, the hypothesis of past violent degassing/overflow presented during a field visit (led by PL) was again ruled out. At the meeting, Pavin history analyses made by Delarbre [58] and Lecoq [70] were still considered references by Pavin scientists.

These three official scientific risk assessments (1993, 2005–2009, 2009) ignored the comparisons with other degassing maar lakes (Monticchio, Albano) and the exceptional geohistory of Pavin. The Besse meeting in 2009 triggered an interest in Pavin history and paleolimnology, while Pavin limnological studies were also developing.

5. Pavin degassing approached by scientists

Since the 1970s, Pavin has been studied using very diverse scientific approaches in limnology, hydrobiology, geochemistry, microbiology, sedimentology, paleolimnology, geology, and geomorphology [27], as schematized in **Figure 6**.

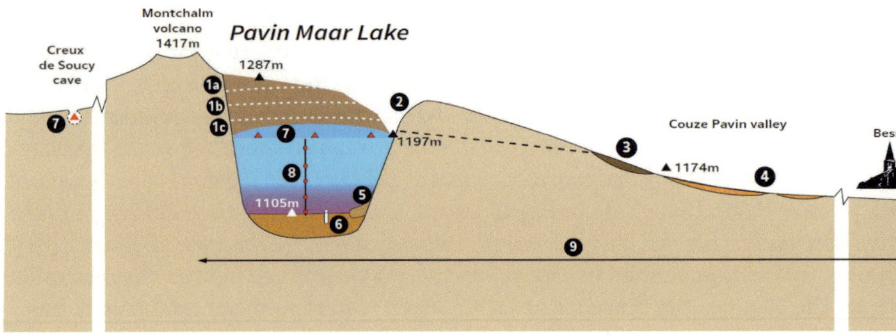

Figure 6.
Scientific approaches used to analyze Pavin functioning, its current degassing risk, and its past behavior, as overflows and rollovers. 1a, 1b, 1c: Paleolevels (+18 m, +8 m, +4 m, respectively) revealed by LIDAR and morphological analysis of lake rim. 2: Outlet incision geomorphological analysis. 3: Pavin detrital cone stratigraphy below outlet. 4: Scattered overflow deposits in Couze Pavin River thalweg. 5: Scars and deposits of subaquatic sediment sliding revealed by fine-scale bathymetry and sub-bottom profiling. 6: radiocarbon dating of long piston sediment cores and sedimentary archives (7000 BP-Present) from stratigraphy and geochemistry. 7: Present CO_2 flux measurements at lake surface and in Pavin surroundings. 8: Profiles of water geochemistry, physics, biodiversity, and microbiology in monimolimnion, chemocline, and mixolimnion. 9. Degassing risk model of the Pavin area covering the crater zone and Couze Pavin valley.

5.1 Pavin for limnologists: A "lake monster," a window on primitive Earth

Lake Pavin is one of the very few lakes that have been studied continuously by limnologists for more than 100 years [9]. They have gradually discovered its limnological peculiarities, which make the lake attractive for dozens of scientists from many disciplines [27] and so different from all other maar lakes of Auvergne-Velay. Pavin rare peculiarities (origin, morphology, hydrology, chemistry, ecology, algal biodiversity, sedimentology) already found in the late nineteenth century by Delebecque [14] and by local naturalists from Clermont-Ferrand [9, 10], have been confirmed by recent investigations [27, 92].

It took limnologists a hundred years to realize its hybrid nature, due to sub-lacustrine hydrothermal inputs linked to the recent volcanic origin of the lake [27, 93–97]. The well-oxygenated surface layer, the mixolimnion (0–60 m 40 years ago, today 0–50 m) is controlled by photosynthesis, water, and material inputs by atmosphere and lake basin, as in any other lake. The lower part is regulated from the hydrothermal control (heat, water, salts, and CO_2 inputs) and by the constant rain of siliceous diatom algae detritus produced in the euphotic zone (0–15 m). This bottom layer, the monimolimnion (65–92 m formerly, today 55–92 m), is much richer in dissolved salts and totally anoxic. Between these two layers, there is a transition layer of a few meters, the chemocline, also termed the redoxcline, in which gradients of ionic concentrations (e.g., Fe and Mn species) and dissolved gases (CO_2, CH_4) can reach two or three orders of magnitude. Today, the in situ and continuous profiles of O_2, temperature, and electrical conductivity profiles reveal the great complexity and dynamics of the chemocline.

Pavin chemocline and monimolimnion are now a natural laboratory, intensively studied over 50 years [93–102], offering insights on the chemical and microbiological functioning of the Archean Earth, 2.5 billion years ago. Also, its varved diatomite sediments provide precious archives at various time scales, including for the analysis of the Anthropocene, as in a few other lakes [103–105]. Until recently, many limnologists were rather skeptical about the historicity of its past misbehavior: "*the current pressure of CO_2 and CH_4 is much lower than the hydrostatic pressure: spontaneous degassing is improbable... however, other events must be taken into consideration such as a large landslide*" [84].

5.2 Pavin, a model for paleolimnology of degassing maar lakes

The first obvious morphological feature of Pavin in relation to its past history is its specific outlet. Natural outlets are very rare in maar lakes, whether in Auvergne, Eifel, or Italy, and none of them are as incised as Pavin. Actually, the rim of several European maar lakes has been excavated by tunnels in Antiquity or Middle Age (Albano, Nemi, Laacher See) to generate artificial outlets, probably to limit overflows [11, 23, 24]. These enormous works around 400 BC (Albano, Nemi) or in the eleventh century (Laacher See) can be considered as technical feats. In 1783, the Pavin outlet was described as a terrifying gorge [68], and the original steep outlet notch incised in the lake rim, suggesting intense erosion, is well pictured in Lecoq's watercolors in 1835 [9, 48, 106]. Today the outlet has been completely modified by the construction of an access road and a visitor platform [9, 10].

Detailed geomorphic analysis of the Pavin area also revealed a lake outlet fan, multiple scars of movements on lake slopes, and rockslides [31], which result from the instability of Pavin Maar and Lake. Integration of terrestrial LIDAR data with multibeam bathymetric data, sub-bottom profiling, and multiproxy study of radiocarbon-dated long sediment cores in the deep central basin of Lake Pavin suggests at least three former periods characterized by higher lake levels (**Figure 6**) [85]. Two abrupt lake level drops around 600 AD and 1300 AD were probably synchronous to crater outburst floods in the Couze Pavin valley and subaquatic sediment slides in Lake Pavin [86].

A team of volcanologists [87], well trained in the study of degassing maar lakes, recently reviewed how "Lake Nyos-type" behavior can be archived in the geological record. They are presenting Pavin as a model for a multidisciplinary paleolimnology approach and list what has been found there by a team led by one of us (EC) for a decade [29, 30, 55, 86–90]. (From [87] with a few additional elements and citation numbers from the authors of this paper):

> "**Local CO2 degassing** *currently occurring in Pavin area: "It was not until 2005-2006 that CO_2 degassing points were discovered in many places in the region [83].*
> **Lake overflow deposits**: *A long mudflow deposit downstream of the lake outlet has been suggested and attributed to a recent (and undated) violent overflow of the lake [107].*
> **Multidisciplinary paleolimnological study**: *acoustic surveys, sediment cores stratigraphy and geochemistry (spectrophotometry Total organic carbon, Hydrogen index, Si/Ti and Fe/Mn ratios).*
> **Detrital sediments indicator**: *the titanium concentration makes it possible to monitor the evolution of the supply of detrital sediments [30].*
> **Past lake levels**: *A significant drop in lake level since 600 AD is suggested by changes in littoral subaquatic environment and sedimentary organic matter [29, 91].*
> **Major subaquatic sediment slides**: *around 600 AD and 1300 AD reflecting regional seismicity probably favored an intensification of Lake Pavin meromicticity and the development of limnic eruption [88, 90, 91].*
> **Outlet erosion**: *These landslides caused the outlet to erode and the crater rim to rupture, leading to a sudden drop in water level and the release of a debris flow downstream [88].*
> **Five historical earthquakes**: *recorded in the region between 1348 and 1921 AD, they triggered limited sedimentary events characterized by changes in magnetic susceptibility and/or spectro-colorimetry values [107].*

Gas-rich sediments reworking: *it could cause [55, 88, 91]: (i) the release of an enormous amount of gas into the water column and consequently a limnic eruption [28, 108, 109], (ii) the triggering of violent waves and the overflow of the lake which could lead to the formation of a debris flow as well as a sudden drop in the lake level, thus affecting the stability of sediments and gases in the water column and leading to a limnic eruption [28].*

Thunder sounds and orange water color: *documented by Meybeck [10, 48] they are linked to the rollover of 1783 and to the waterspout of 1785 AD , similar to a limnic eruption [48].*

Conclusion: *Signs of limnic eruptions in Pavin include reworked sediments of reversed ages, deposit colors reflecting a mixture of highly gas-laden water in an anoxic environment and water in an oxic environment and/or at the surface, sediments gas-rich, sediments rich in iron hydroxide, strong peaks of Ti and Fe oxidized during events."*

Pavin paleolimnological (sedimentary) records of major events (large subaquatic landslides around 600 AD and 1300 AD) are associated with clusters of regional earthquakes, but also records of minor events (small slides or thin turbidites contemporary to historical earthquakes in 1783 AD, 1844 AD, 1863 AD, and eventually in 1921 AD), overflows (after 1765 +/− 55 cal BP), and eventually of rollovers (in 1783 AD).

A regional lake study [88] highlighted that numerous erosive events are characterizing sedimentary basin fills of the Auvergne volcanic lakes and are thus limiting paleo environmental reconstructions during the Roman period. Only one subaquatic slide has been identified and dated to 170 +/− 40 Cal AD (1770 +/− 60 Cal BP) in Lake Aydat (**Figure 4**), located 30 km north to Lake Pavin near the Limagne active fault (UNESCO world heritage), suggesting that the oldest reports of potential Lake Pavin misbehaviors may as well be a consequence of the active geological setting of this volcanic province rich in maar lakes.

If the mantellic or bacterial origin of CO_2 stored in the deep waters of Pavin has been discussed [26–28], regional investigations of volcanic lakes from Mont Dore and Chaine des Puys volcanic provinces (**Figure 4**) highlighted the impact of two main active faults in this territory: a first one below the Mont Dore stratovolcano and a second one at the edge of the Chaine des Puys and Limagne basin [89]. Based on multidisciplinary studies of lacustrine deposits from Tazenat, Chauvet, Pavin, Montcineyre, Aydat, Chambon Lakes, and Limagne palustrine sequence, clusters of paleo-earthquakes are suggested since 2700 Cal BP [88].

5.3 Current CO_2 measurements in Pavin and its area and modeling of sudden lake degassing

In the last two decades, CO_2 measurements have been realized in the Pavin area (**Figure 3**), confirming the CO_2 emission around the lake, particularly at the Escarot Mofette, 1 km southwest of Pavin where deep-seated CO_2 levels may be highly variable) [33, 83]. The nearby Creux de Soucy cavity, located on Montchal lava bed 1.5 km south of Pavin (**Figure 3**), is also a site where CO_2 can be very variable. In 1575, the cosmograph Belleforest mentioned that it was emitting "vapor," as observed in Pavin [61]. In 1860, the mayor of Besse made a first attempt to explore the cavity; it failed because of "lack of air." In June 1892, the famous speleology pioneer, Alfred Martel, came fully equipped to explore the Creux de

Soucy but failed to reach the bottom: a 6 m layer of gas was resting in the cave. Six months later, the CO_2 had vanished, allowing the cavity exploration for the first time by Clermont-Ferrand biologists [82]. During the second half of the twentieth century, other occurrences of CO_2 have been noted in this cavity, a sign of the great variability of CO_2 emission in the Pavin area. Today, speleologists and scientists check its occurrence before working there, and they believe that the cave water is not discharging underground to Pavin.

Rafflin and Boudoire from the Clermont-Ferrand Volcanology Laboratory (LMV) and several Italian teams from the Istituto Nazionale de Geofisica e Vulcanologia (INVG) have carried out the first comprehensive assessment of the degassing risks of Lake Pavin [34]. They acknowledge that "*past degassing events can be inferred from historical records describing activity in 1551, 1783 and 1936 according to Meybeck*" [9, 10], a statement in contrast to other LMV volcanologists who are denying the historicity of Pavin misbehavior and doubting its violent overflow events (**Table 3**, 2021) [32]. Their risk analysis starts with seasonal CO_2 profiles inferred from water chemistry that lead to the quantification of CO_2 stocks in deep water, then by measurement of CO_2 flux escaping from the lake surface. The potential degassing is simulated according to three scenarios, including a sudden limnic eruption emitting 450 t of CO_2, contained in the lower layer from 80 m to 92 m, out of a total of 1750 t, triggered by an external event (slide, earthquake, etc.). The dispersion of the CO_2 cloud in the entire Pavin-Couze Pavin area is then simulated by the INVG model, taking into account the topography established by LIDAR and the local distribution of winds. For the first time, the effects of degassing are quantified. Concentrations encountered at 1.5 m above ground are simulated, particularly those at the outlet platform. Impact on the 2000 daily summer visitors is finally assessed using the reference toxicity scale developed by the American Department of Industry. The risk is greatest for visitors standing on the lake outlet platform and for those hiking around the lake. In the sudden degassing scenario, the toxic cloud would reach the platform in 15 minutes and would be fatal during 1 hour (CO_2 > 15%). People in the Couze River valley at Besse, 4 km downstream (**Figure 3**) would be affected by the CO_2 cloud after 30 minutes but exposed to much lower concentrations, generating breathing difficulties (0.5% < CO_2 < 1.9%). Impacts would also be more severe for people with respiratory problems. Several risk management measures are proposed [34]: "*Areas around the lake would experience lethal CO_2 contents in the air less than 15 min after the beginning of a limnic eruption... The spatio-temporal evolution of the potential CO_2 cloud raises issues regarding the impacts of such a hypothetical event in the close vicinity of the lake and, given the area is populated and highly visited, needs to be considered in future risk mitigation strategies by (i) continuous monitoring of CO_2 in the Pavin area, (ii) rapid emergency response in the event of a limnic eruption, (iii) installation of gas detectors, sirens and information panels, (iv) an evacuation route and signage to identified safety points.*"

5.4 A new look at Pavin is now needed

The debate on Pavin misbehavior, initiated after Pavin sounding in 1770, closed by Lecoq in the nineteenth century, forgotten by the twentieth century limnologists, has been reactivated after 2005, *i.e.*, two decades after the Nyos catastrophe. In the last

20 years the parallel analyses of Pavin geohistory and paleolimnology have proven to be consistent. Furthermore, they are similar to historical descriptions of past degassing in other lakes. At Pavin several levels of degassing intensity are reported; most of them do not leave geological traces, such as bubbling, little waterspouts, fog, emission of vapors and clouds, effects on animals, and even limnic eruptions (**Tables 1** and **2**; Appendix B). The impacts on populations related to numerous Vassivière miracles include many neurologic damages and skin diseases [76], also mentioned at Nyos and at Albano.

Our on-going geomythological analysis of Pavin before the 1500s is suggesting that in Late Antiquity and Early Middle Age, Pavin was probably also very active [36]. The impacts on populations, evoked in some ancient lives of Auvergne saints (*Vita Sanctorum*, seventh-tenth century) and in some legends transcribed during the 1800s [11, 36], may thus be rooted in cascading natural hazards documented in and around Lake Pavin (earthquakes, landslides, tsunamis, crater outburst floods, lake level drops, limnic eruptions). Archeological and further historical investigations are now required to precise their possible origin since the Roman period.

We also believe that Pavin threats and damages on local populations are expressed in the unusual and rich iconography found in the local church of Besse: the medieval capitals, the 16th misericords (wooden sculptures), and the 17th chapel dedicated to Sta Martha, a famous dragon slayer [11]. Other social and religious responses concern the Vassivière cult and the protective ritual, initiated in the mid-1500s and continuously practiced until today: the statue of the miraculous Black Madonna with Child is brought each summer from Besse church to the mountain sanctuary across Pavin zone by a devout assistance convinced of its power of protection and healing (**Figure 3**). End of September, the statue is brought back to Besse, waited by thousands with great joy and gunshots, a ritual famous across Auvergne (La Dévalade). One striking result of our study is this intimate connection between lake degassing and religion, a point which could be investigated in many similar lakes.

The famous French geographer Elysée Reclus knew about Pavin and continued to refer to its "enchantments" in 1885 without being able to understand their origin [110]. A new, multidisciplinary, and open-minded look at the lake must now be taken by scientists (Limnologists, geochemists, volcanologists, etc.) but also by social scientists (Auvergne historians, historians of religions, human geographers, anthropologists, etc.).

Until now, Pavin visitors are still largely unaware of its past history, but this could change now that a scientific consensus on its past degassing is being reached. An appropriate reaction of local Auvergne authorities (city of Besse, Puy de Dôme départment, Parc Naturel Regional des Volcans d'Auvergne) to the re-assessment of degassing risk [34] is now expected. According to the local press (La Montagne, September 29, 2024), the lake CO_2 and the Escarot Mofette (**Figure 3**) are now surveyed since spring 2024, nearly two decades after the early warning by one of the authors (PL), and historical degassing is now mentioned, replacing "Pavin stories." The social response at the local scale will probably diffuse gradually, although slowed down for several reasons [6]: (i) authorities have been so far denying any risk on the basis of scientific reports and very reluctant to admit the historicity of Pavin misbehavior, e.g., on information panels presented to the visitors; (ii) admitting any potential risk may probably lead to a revision of tourism management at Pavin and may concern properties in the down lake area potentially exposed to the modeled CO_2 cloud. However, these changes would probably be balanced by the increased attraction of Pavin if it is developed as an exceptional geosite with great geoheritage value.

The diversity of historical sources and the variety of scientific approaches that have been used here for the reconstruction of Pavin past were unsuspected when this

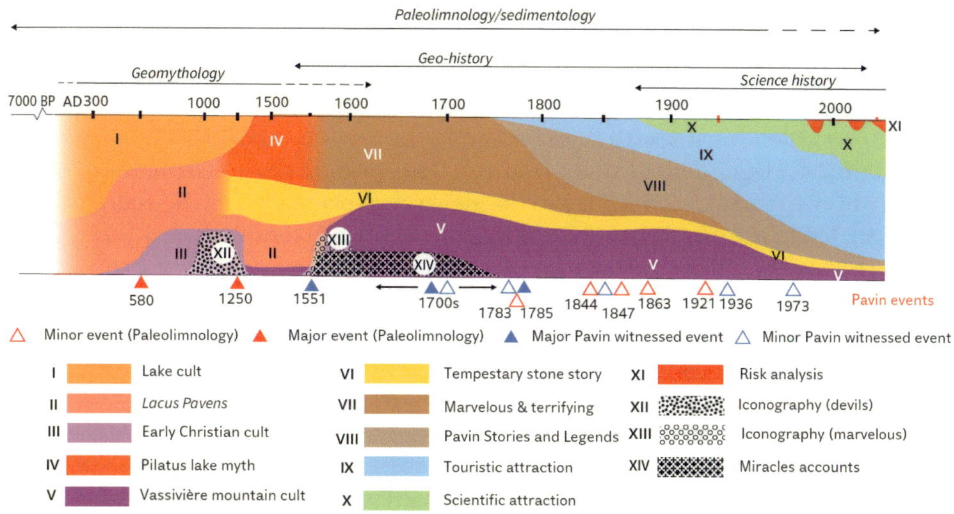

Figure 7.
Reconstruction of Pavin past misbehaviors and graphical summary of the evolution of Pavin perception, representation, and memorialization from antiquity to the present. The analysis of Pavin geomythology—before 1550—and legends is not detailed here [36]. (Key dates of major Pavin events, normalization attempts, and degassing hypothesis: see text, **Tables 1** *and* **2***; Appendix B).*

analysis started in 2009. This was made possible because we shifted the traditional points of view previously used by scientists (including one of us, MM) and by historians of Auvergne [78] or of Pilatus lakes at the Middle Age [111], and we did not rule out the hypothesis of its degassing as in Nyos.

Our current understanding of the complex and evolving social, religious, and scientific relationships between Pavin and society is presented through a temporal fresco (**Figure 7**).

6. Degassing maar lakes, from "new hazard" to global geoheritage

6.1 Past misbehavior in other European maar lakes, a growing evidence

Descriptors of maar lakes degassing (**Table 1**) can now be tested at Pavin before 1550, and in other maar lakes, where archives, archeological artifacts, literary and legendary sources, or traditional rituals are abundant. Preliminary assessments show that perception and representation of the degassing risk in antiquity, the Middle Age and the Renaissance can be quite different. The re-analyses of ancient sources at Lake Albano [23–25, 112], Lake Montichio [21, 22, 113, 114] and Lake Averno [25] are promising and reveal that degassing perception can be quite old, related to many Greek and Roman myths [25], to Christian myths and associated with religious cults of these periods.

The Eifel maars in Germany have a great potential for such reconstruction. A first analysis of local legends has revealed that at least four of them could be linked to lake misbehavior events [11, 115]: (i) Ulmener Maar where "giant fishes" (probably waterspouts) were described by Rhenan cosmographer Sebastian Münster (1550s), who reports their association with the death of new-born babies [116], and where legends of witches living by the lake and possibly intoxicated by it are still locally celebrated, (ii) in Weinfelder Maar (or Toten Maar) where the population of a whole village suddenly disappeared between two censuses in the 1500s and in which a "Pilatus rock" is still shown on the lake cliff, (iii) Pulver Maar, famous for its legend of a shepherd making circumambulation around the lake to prevent its eruption [11], (iv) Laacher See which has the richest set of legendary accounts: legends of Early Middle Age sudden storms with thunder and lightning—as reported in the 1844 event (see Section 2.4), legend of a violent lake overflow, prior to the construction on the lakeshore of the Maria Laach Abbey in the eleventh century, where the sudden death of livestock led to the execution of local witches in the sixteenth century [117], finally where an historical sudden rise of lake waters led to the construction of a tunnel in the twelfth century to stabilize the lake level (as for Albano and Nemi Lakes).

Other legends of the Rhine Valley, analyzed by Scottish author Lewis Spence, mention two stories, according to Christian tradition [118]: (i) a pit-dragon with pestilential breath terrifying the whole country, then defeated during the evangelization of this area, and (ii) a sudden atmospheric storm with big waves and thick clouds encountered in the 800s33 on the Rhine River by Emperor Charlemagne. Unlike for Italian lakes, the re-analysis of these sources has not yet been attempted, and the possibility of past violent degassing events has not been explored in paleolimnology works.

The preliminary analysis of the legends and beliefs of maar lakes (or other lakes which can be assimilated to them as Guatavita Lake) in other continents revealed many similarities with Pavin: tempestary stone stories, pit dragons, lake cults, lake fear, sudden lake storms associated with lake spirits, etc., in Ethiopia (Shetan Lake, *i.e.,* Satan Lake), Columbia (Guatavita Lake), and Northern Japan (Ainu people lake legends) [36]. The presence of Mamy Wata, a famous water myth in West Africa, has also been invoked after the Nyos catastrophe [5, 56]. Italy undoubtedly has the most important set of written archives, from the pre-Roman to Christian periods. Potential sources of historical archives exist on all continents, such as the missionary archives that were regularly sent to the Vatican and the General Archive of the Indies in Sevilla—for Latin America. The re-analysis of the Motif-Index of Folk Literature of Stith-Thompson can also be considered.

6.2 Lake degassing, a new risk, still underrated

We recognize here how much the human community owes to the people of Nyos and Monoun, who have unfortunately attracted attention on "degassing maar lakes" (DML). Before these catastrophes, some DML were already known, feared, or revered, but modern sciences and humanities failed to recognize them.

Even if the degassing of DML, observed by some scientists at Monticchio (1770–1820), Pavin (1785), and Albano (1829), had been recognized by scientists of the last century and integrated into academic courses, it is unlikely that the attention of scientists and authorities could have avoided the disasters of Monoun (1984) and Nyos (1986). The Cameroonian authorities made little attempt to attract the attention of scientists after Monoun. Also, Sigurdsson's paper on Monoun, first rejected by Science, was still in press when the Nyos disaster occurred [6]. A similar reluctance of academic science to admit the reality of degassing occurred for the Laacher See in 1844, for Pavin at the end of the eighteenth century, and it still does today.

In addition, each of DML constitutes a potential risk toward local populations and their visitors, which remains to be addressed in all its dimensions. In Cameroon, the issues of people resettlements, of their information and warnings, of their grief and mourning have been barely addressed, four decades after the catastrophe, but an efficient lake degassing device has been installed at Nyos.

Recent calls to better take this risk into account [21, 34, 87] must be supported. Our current knowledge on DMLs covers only a very small part of active volcanoes in the world (**Figure 2**). The inventory of maars initiated in Africa [18] should be extended to other continents. Meromictic lakes should be particularly targeted: Cameroonian and European maar lakes convinced or suspected of degassing are presently meromictic. These inventories should combine paleolimnological, geohistorical, and geomythological approaches.

6.3 Degassing maar lakes: A world heritage

Degassing maar lakes are extremely rare natural objects that have multiple patrimonial values: scientific, cultural, literary, folkloric, anthropologic, sociological, etc., as demonstrated for Lake Pavin and outlined here for a few others. The interest in DML seems to have no limits. The famous Messel Pit in Germany, 47Ma old formation, and now classified as a world heritage by UNESCO, holds one of the richest fossil deposits ever found. Recently, geologists understood that it was an ancient maar lake, the Lake Messel, which became permanently meromictic with anoxic bottom waters. Sudden or sneaking degassing could have killed instantly animals on the shore or in the lake itself. Their remains are now extremely well preserved in anoxic sediments with their soft parts, feathers, colors, copulation figures, etc.

The global inventory of the DML could lead to a global network of selected maar lakes geosites of world heritage value [36]. The first on the list are, of course, Nyos and Monoun, Albano, Nemi, Averno and Monticchio, Pavin, Laacher See, Ulmener Maar and a few other Eifel maars, Ainu lakes, and Lake Guatavita, from which the El Dorado legend is derived, although its maar origin is still debated. So far only the Eifel maars are recognized as a geopark by UNESCO, but their legends remain to be studied and valorized. This first set could be followed by a larger set of DML from all continents.

Appendix

A.Lake Abano degassing in 1829

Summary of Bassanelli's description [40] *(made by the authors.)*

> "**Start of event**: *on May 22nd, 1829 around 4 a.m., suddenly, in a serene sky, at 5:45 a.m. a furious wind rose accompanied by heavy rain. Shortly after 1:30 p.m., downpours of water coming from the south-southwest side of the lake fell accompanied by a smell of sulfur. A few days after these tremors, we began to hear loud and terrible detonations, in the atmosphere and in the ground, like distant cannon fire. [Bassanelli counted the detonations each month].*
>
> **Population response**: *At 6 p.m. on May 23 a big shock which put everyone in anguish, fearing for the destruction of the city. It is said that this volcano is waking up., the families were afraid and left for Rome; 300 leave. On the night of the 23rd,*

a memorable procession was held. The relics of St. Pancrace (local protector)) are carried out and brought to safety. The following days people sleep outside. On May 31, another shock and on June 1, an even bigger one: it cracked houses, caused stones to fall, broke windows, accompanied by a terrible atmospheric rumbling.

Vapors and electric atmosphere*: Most of the time the air was dusty, especially in June and July, filled with nocturnal electric clouds. At the end of July, just before the tremors, we could observe small white clouds, similar to the Milky Way, and strong winds. They were stretched in bands, often colored, dark blue or white. From July 8 to 12, electric clouds produced lightning on the south / southwest side. Sometimes these phenomena developed in a line, like an aurora borealis, with columns of fire and flying dragons, almost always from the south side, without connection with tremors... The lightning, whether generated by ambient electricity or not, predicted nothing. On some days, most people have smelled sulfur in the atmosphere... The wind almost always came from the south/southwest and on certain days was very strong, impetuous, overflowing. It was instantaneous changes from serenity to rain with thick, dense fog, even on the most beautiful summer day. This happened all day on June 28 and on the evening of the 29th there was a terrible storm with hailstones (chabili), a confirmation of a complete atmosp-heric imbalance. Very dirty rain (poiggie) occurred on October 21, 22, 27 and on November4, 16, 23 ...Some days the continuous shaking was accompanied by specific gases and an unusual cold in summer.*

Animal behavior*: The inhabitants were well warned by prior animal warning signs of the tremors: braying of donkeys, barking of dogs, meowing of cats, and crowing of roosters.*

Lake waters*: The edges of the lake, at the port, from the beginning of May (before the tremors) were always turbid, flaky. During these and even after, they remained opaque, and never presented their magnificent blue color of the previous year. Until December, the waters remained cloudy and turbid.*

CO_2 presence*: In many cavities under the city there was a spontaneous formation of carbon dioxide, making breathing impossible, killing animals and putting out the lights; in others, this non-respirable gas developed during the tremors and persisted for a certain time.*

Effects on people*: People suffered from profuse sweating and nervous system disorders. Some had a feeling of warming in their legs, accompanied by intense nervousness, tightness in the chest. I felt it, particularly on the first morning of the earthquake, May 22. Others, at the first emission of gas, had an intense inner emotion, felt a few minutes before the shock. Others, during and after the shocks, began to cry or experienced severe tremors. The most common diseases affected the nervous system, the joints and catharsis affecting the throat and lungs, sometimes cured, and sometimes leading to death. Also mild to severe ophthalmia. There were many intermittent fevers especially in summer and fall. Hive was very pronounced, with unbearable skin burning. There were a large number of abortions at 6, 7, and 8 months, even among robust women, and fetal growth delays of 2 to 3 months, attributed to delayed fetal nutrition. Most deliveries were difficult and most babies came with skin diseases (herpes and tumors). I observed that many of the babies breastfed during the tremors died of stomach problems, vermin* *and feces, others died of convulsions. The other children were tormented by diarrhea, by dental conditions with cough, by ophthalmia... Nervous conditions affected all individuals, even more so women, producing strong convulsive movements, vomiting, heart pain and more, and even a loss of the senses."*

B.Main historical sources concerning Pavin and its area

B.1 1551: The terrible storm was witnessed at 1.5 km from Pavin

This first recorded Pavin event, witnessed by six persons of high social positions from Besse, is fully contextualized, with an exact date (28 August 1551). It has been carefully recorded by the church authorities, kept in their church, and published for the first time by a Jesuit born in Besse, Father Michel Coyssard (1615) [59] then by a Benedictine also originating from Besse (1688) [60], reproduced in many others Vassivière pilgrimage histories [9, 10]. Witnesses were standing on the Vassivière Mountain (**Figure 3**) when a mentally sick child was brought for healing by the miraculous statue, a new chapel being under construction:

"The air, which was at this moment very calm and tranquil, changed so suddenly that all attendants were astonished: an impetuous wind rose all of a sudden with such violence that they were almost all knocked on the ground, scattering their hats and coats. What put them most in fright and fear is that at the same time the whole air was like ignited by the length of terrible lightning, that occurred without stopping, and by terrible thunder that could be heard around them...this was just the effects of the furiousness of the demon who was being exorcized, as one could smell for a long time after an awful sulfur smell. The weather went back to clear state as it was before, lightning and thunder stopped and the atmosphere went tranquil again".

The very sudden, brief and violent change occurring in a serene atmosphere is much similar to the 1785 limnic eruption (Section 2.1, **Table 1**), although it is not reported that it originates from the nearby Pavin, masked by the crater rim. The sulfur smell is noted, as in the 1936 Pavin event; see further.

B.2 1566: A hazardous abyss, generating storm, thunder, and hail presented to Charles IX

On March 31, 1566, King Charles IX, touring his kingdom of France, and his mother Catherine de Medicis are informed in Clermont, the capital of Auvergne province (**Figure 4**) about "*a mountain, called Puy de Dome, on which there is a big abyss from which escapes ordinarily a big storm of thunder and hail, which damages grains in the valleys.*" [62].

Although this description made by the royal chronicler Abel Jouan wrongly places the terrible abyss (a lake is not mentioned) on top of the Puy de Dôme, the famous mountain near Clermont, the attribution is most likely at the Mont Dore mountain and concerns Lake Pavin. According to this official complaint, Pavin misbehavior is permanent; damaging grains in the downstream valleys, which could correspond to at least 20–30 km downstream, may be as far as the city of Issoire (**Figure 4**). The triplet wind storm, thunder, lightning, and the occurrence of hail are very close to the 1785 (Section 2.1).

B.3 1575: The sudden lake storm in "a lake on top of a mountain," triggered by a tempestary stone

In the Cosmographia Universalis, an impressive work realized by the royal cosmographer François de Belleforest [61], there is no direct mention of Pavin, nor of Paven, its common spelling, until 1800:

> *"...Near this Mont D'or [Now Mont Dore] stands the city of Besse, at a half quarter league from which one can see a great lake located nearly on top of a mountain, without a measurable depth and without any water inlet. Actually, it is both admirable and terrible to see as, if a stone is thrown into it, it is ascertained that soon thunder, lightning, rainstorm and hail would occur. Not far from this lake there is an abyss named Creux des Soulcis...from which gas is emitted".*

This universal geography is very well organized with a final index of topics and locations featuring about 10,000 entries, where Pavin is found as a "marvelous lake located on top of an Auvergne mountain." This topographic position is a recurrent marker of Pavin, which probably dates from the Late Antiquity [36]. It is striking to find small natural objects as Pavin (44 ha) and Creux de Soucy (a few feet wide cavity) described with details that could only be known by a local scholar. In this work, there are only fifteen lakes named and described; all of them are very peculiar, such as the Asphalt Lake (Dead Sea). The Auvergne description in this work is the most complete one at the time. At that period, no one knows that the Auvergne Mountains are volcanoes and that Pavin is a volcanic crater; their origins will be discovered two centuries later.

Pavin Lake storm is triggered, according to this description, by a "tempestary stone." This story is one of the best-known motifs of Pavin, constantly replicated over the following centuries by many geographers [9–11]. It was still reported locally in oral tradition in 1969 (MM, personal observation). We first put forward two hypotheses [10]: (i) it could describe a lake close to saturation state and ready to degas at any time, as was described at Lake Monoun after its explosion in 1986 [42], or (ii) it is the metaphor of a landslide or a rockfall which would have triggered the eruption as suggested at Nyos. Additional investigations show that in fact this motif was already existing, word per word, in the thirteenth century and diffused by religious scholars in Vienne, 160 km east of Pavin, in the Rhône River valley. They were mentioning "a tempestuous lake in mountains where demons were producing major disorders in air and in water" [36]. For this reason, these scholars were also linking this demonic place to the final resting place of Pontius Pilate, the Roman magistrate who chaired the trial of Jesus of Nazareth: according to a religious tradition, his body was attached to a big stone, thrown into the Tiber River, where it provoked a violent rejection! This story, made famous in the Christian world near 1300 AD by Jacques de Voragine in his Golden Legend [63], has been very popular in Christianity for centuries. Religious scholars and profane authors have attributed many locations to this maleficent and tempestuous lake: in the Pilatus Mountain (Central Switzerland), in Mont Canigou (Catalonia) and in Monte Sibillini (Abruzzo, Italy). Also, a Pilatus rock is found in the Weinfelder Maar (Eifel), famous for its legend. Medieval Pavin behavior could have generated this myth of Pilatus Lake, famous across Western Europe, a hypothesis difficult to support due to the scarcity of medieval sources [36].

B.4 1605: *Lacus pavens*, the terrifying lake, is mentioned by a hydrologist

In his book on thermal resources of Auvergne Jean Banc, a medical hydrologist visiting Besse, he mentions the term *lacus pavens*, in Latin the terrifying lake, and reports its storms generated by the tempestary stone, a story told by a local peasant [65]. It is the earliest citation of the lake with its original name. "Paven" will be replaced by "Pavin" in the early 1800s in maps and written accounts. Until the twentieth century, the proponents of a normal lake have always doubted this origin, preferring the Occitan "pallenc," meaning palisade, a hypothesis evoked since the 1700s, then abandoned in the following manuscript source.

B.5 1750s: The Godivel Manuscript details permanent degassing and occasional storms of "Pavent"

This manuscript, written in the 1700s by a member of the notorious Godivel family from Besse, found and transcribed in 1884 [66], contains Pavin's three-page-long detailed description, which is summarized as such:

> *"The lake is one of the most beautiful to be found in Europe. Its depth is thought to be unlimited: some courageous and curious [men] went with a craft in the middle of the lake and let run 313 fathoms of ropes [without reaching the bottom]. The outlet discharge is important, even during summer drought. The water is clear, a little bitter and smells like ferruginous springs in the Pyrenees. Fish has never been seen [in the lake] nor any water grass. Birds are rare in the surrounding woods: the reason is that the frequent fog, which is seen escaping from the lake at any time, keeps them out and makes this place uninhabitable...This place is feared because in summer several exhalations and vapours are emitted and eventually result in clouds and storms that threaten the whole area. The lake has always been called the Pavent Lake. This paving stone that can be seen all around it could have given rise to this name, which originally comes from the verb paveo which was used in the past and means to pave in French. ... It seems more likely that it is called the Pavent Lake because of the fear and fright that it causes in all those who look at it, even if they had seen it every day of their lives. Pavent would therefore come from these two patois words Par-ven or from the verb paveo which means I am afraid, either because this fear is excited by the mere sight and consideration of the lake, or because several exhalations and vapors rise from this lake during the summer which shortly after degenerate into clouds and finally into storms which hold the whole country in fear."*

Several lake descriptors are mentioned here: the taste of water, the absence of fish and the rarity of birds in the woods, the permanent fog. The occasional exhalations and vapors, which shortly after degenerate into clouds and storms, also reported in 1566, terrorize the whole country. Several possibilities to explain the origins of the "Pavent" toponym are proposed here, but the terrifying lake—*lacus pavens*—is clearly privileged.

It must be noted that no reference is made here to a tempestary stone. Actually, this motif, reported until the 19th and still present today, has never been mentioned by an identified witness of Pavin misbehavior! For instance, in 1672 de Fretat [67], a Jesuit cartographer of Auvergne who qualifies Pavin as the 11th wonder of Auvergne clearly asserts that "a stone thrown into the lake does not trigger any storm, as reported by Belleforest," then by its many followers in France and in Western Europe [10]. He also reports that he has sounded the lake at 60 fathoms, about 110 m, although without mentioning how he proceeded. His work remained ignored so far [9, 10].

B.6 1783: Paven is suddenly boiling, orange, and glutinous in summer 1783

This detailed account of this sudden Pavin event is published only a week after the event in a two-page "letter to the editor" sent to the Feuille Hebdomadaire de la Haute et Basse Auvergne, a pre-revolutionary weekly magazine [68]. The author is Chatelain Godivel, another member of the notorious Besse family and magistrate of this city. It is abridged as such:

> *"Paven Lake, at half league from Besse, is one of Auvergne marvels. It represents a terrible gorge in a mountain...Its water has always been very clear, never clouded by rains nor by storms or inlets, which are invisible. The outlet discharge is very stable and could work a watermill. On August 21 [1783], on a very clear day, I have crossed this stream at eight in the morning to go to Vassivière [Location on figure 3]. When going back at four in the evening..., I found the outlet water very turbid and yellowish...My son climbed to the lake, some 150 fathoms [180 m] from there, and found the lake surface completely troubled, carrying a sort of yellow-orange and glutinous silt. He took the water in his hand and when this substance dried it produced on his hand some pimples with a slight pain that lasted some time. The same day the land keeper came to tell me: "Sir, there is something extraordinary in Paven which I have never seen, the water is boiling everywhere but there is no wind, it is yellowish and even rusty in some places". The day after, I called my farmer Gelac who stays close to the lake. He told me that the water had been dirty two or three times this summer, in the evening...I suppose that there has been inside the lake some slide, so important that such volume of water was clouded and infected by some venomous material."*

This journal, the first news magazine in Auvergne, has a section devoted to "Arts and science," and aims to be read by nobles, bourgeois, and ecclesiastics. When the lake change occurred, the witnesses were only 1.5 km away and did not remark any thunder or lightning; it was probably a moderate lake mixing (boiling waters) and non-violent degassing. For the first time, a scientific hypothesis, a sediment slide, is put forward. This Pavin event is probably a limited mixing of deep anoxic and iron-rich waters with surface waters, leading to oxidation of ferrous iron and ferric hydroxide precipitation, as observed in Nyos Lake, in Monticchio Piccolo Lake and in Albano Lake, in 1829 (Section 2).

In 1877, one finds a veiled reference to this event in a manuscript, where the 1783 event is evoked in relation to the 1783 Calabria earthquakes [68].

B.7 1835–1847: Contrasted Pavin descriptions made by Henri Lecoq

In 1835, this prominent Auvergne scientist—he is also a botanist, a hydrologist, and a noted businessman—makes his first visit to the lake and collects "Pavin stories" told by his local guides [70]. They can be understood as a mixture of secular traditional beliefs dating from Belleforest, as the tempestary stone, with allusions to the Pavin event of 1785, featuring a whirlpool:

> *"No boat crosses its deep waters. No inhabitant of the surroundings would dare to linger on this liquid plain. A whirlpool exists in the middle, and would swallow up the imprudent person who wanted to drive his boat there. A stone thrown far enough to reach this chasm, boils the waters and produces a storm. The lead of the probes melts in the middle of the lake and its depth reaches no limits. Such are the absurdities we heard told."*

This description will be popularized throughout the nineteenth century and picked up by most authors of guidebooks [10]; it is still found today in some of them [6]. During his many visits to Pavin, Lecoq has not attempted to use the latest scientific instruments, unlike lake study pioneers such as Tata at Monticchio (Section 2.2). In 1871, he sums up 35 years of visits to Pavin in a book on Massif Central hydrology, much descriptive [71]. He mentions once more the Pavin stories as "legends invented by fear and transmitted by credulity." However, he also describes, without further analysis, a strange lake storm as he was sounding the lake with the first boat ever brought to Pavin, thus giving credit to some Pavin stories:

> *[September 13, 1847, great storm] "We noticed from time to time a phenomenon that could have led one to believe in whirlpools and smoke from the lake that rises during stormy weather. We could indeed see the wind driving the water in various directions, pushing the waves in different directions and from their shock real whirlpools and white dust were born that rose while turning and gave rise to large smoke." Lecoq invokes the wind effect, actually due to its high rims, Lake Pavin crater is actually very well protected from external wind.*

B.8 1936: The sudden color change of Pavin

This summer 1936 event has been only revealed by its witness in 1987 [72], while press journalists were investigating Pavin past (Section 4.5). The witness, Paul Joanny, a pharmacist at Clermont, had not dared to report the event for 50 years [6]! In 1936 he was a young fisherman from Besse; he has also been interviewed in 2012 by one of the authors (MM):

> *"My father had a boat on the lake. On Sundays, I go and get the crayfish by putting down a basket. I had left the lake spotless, smell and consistency. The next day, I went back with my father: the whole lake was coloured, I saw a big yellowish foam floating in certain parts of the lake, from yellow to orange, with streaks of blue. A fisherman named Boccard, arrived in turn; he told me he had already seen that once or twice. The cap of the basket was caught in this gelatinous mass that smelled a lot of sulphur. (In pharmacy, when I was studying, we prepared colloidal sulphur for the preparation of the arsenic poisoning test: the same smell floated in Pavin). I took a bottle of water: the teacher thought it was algae. The foam floated on half of the lake. It was forming before my eyes: the wind was gathering the deposit towards the spillway of the lake. The river was no longer flowing. We unblocked the spillway; the fishermen reported dead trout. It was not the water that had changed colour, it was the yellow-orange deposit, as in the pharmacy studies program."*

In 1987, this witness account has not been considered by scientists despite the striking comparison with the 1783 event, rediscovered in 1986 and well noted by the journalist.

A similar reluctance to report past Pavin misbehaviors is still found today: during our investigations in Besse (PL and MM, May 2024), an eyewitness from Besse (J-L V.) also reported another event that occurred in summer 1972 while he was asked to take back to the lake outlet a rented rowing boat left near the south shore of the lake:

> *"I choose to go across the lake straight to the outlet. At one moment, I suddenly crossed a boiling area 10 to 15 m wide with eddies about twenty cm high. I expected to see a monster emerge and I rowed harder to get past this dangerous area."*

Fifty years later, this witness still does not want to disclose publicly his name, fearing ridicule. These two delayed testimonies illustrate the difficulties of assembling contemporary accounts of Pavin misbehaviors. The major reason is the lack of official recognition of the turbulent past of Pavin, which has been developed over more than two centuries by scientists, historians, and local authorities [6].

Acknowledgements

The authors wish to thank Aurelien Baro for the graphics and all those who helped us find precious historical sources, particularly Christiane De Vriendt and Martine Olivier (Besse), Max Reynaud (Toulouse), Horst Kämpf (Potstdam), Karl-Heinz Schumacher (Mendig) and Rita Voltmer (Trier) and Loredana Lancini (Le Mans), Laurence et Geneviève Fraisse, Marcelle Planeix –Fraisse (Egliseneuve).

Author details

Michel Meybeck[1]*, Emmanuel Chapron[2], Brice Molo[3] and Pierre Lavina[4]

1 METIS, UMR 7619, CNRS, Sorbonne Université, Paris, France

2 (GEODE) UMR 5602 CNRS, Université Toulouse, Toulouse, France

3 CERMES 3, CNRS-EHESS-Paris, Paris, France

4 Terre Géopatrimoine and Volcans, Barbetoux, Saint-Dier-d'Auvergne, France

*Address all correspondence to: michel.meybeck@upmc.fr

References

[1] Sigvaldason GE. International conference on Lake Nyos disaster, Yaoundé, Cameroon 16-20 March, 1987: Conclusions and recommendations. Journal of Volcanology and Geothermal Research. 1989;**39**(2):97-107

[2] Wetzel RG. Limnology. 1st ed. Philadelphia: Saunders College Pub.; 1975. 743 p

[3] Blong RJ. Volcanic Hazards. A Sourcebook of the Effects of Eruptions. Academic Press; 2013

[4] Kling GW, Clark MA, Wagner GN, Compton HR, Humphrey AM, Devine JD, et al. The 1986 lake Nyos gas disaster in Cameroon, West Africa. Science. 1987;**236**(4798):169-175

[5] Molo B. Cameroun : de l'assistance internationale en situation de crise à la politique étrangère de gestion des risques et catastrophes. 1986-2012. In: Esseba C, Bidias R, editors. Comprendre les relations internationales contemporaines. Paris: L'Harmattan; 2020. pp. 59-91

[6] Molo B, Meybeck M. Faire face à l'inconnu. Des Hommes face au dégazage des lacs tueurs : la naissance d'un risque. In: Alary V et al., editors. Percevoir, représenter, communiquer les risques naturels. PUBP: Clermont; 2024. pp. 149-184

[7] Thienemann A. Physikalische und chemische Untersuchungen in der Maaren der Eifel. Verhandlungen des Naturhistorisch Vereines der preussischen Rheinlande und Westphalens. 1914-1915;**70**:249-302 and **71**:273-389

[8] Shanklin E. Exploding Lakes and maleficent water in grassfields legends and myth. Journal of Volcanology and Geothermal Research. 1989;**39**:233-246

[9] Meybeck M. Pavin, the birthplace of French limnology (1770-2012), and its degassing controversy (1986-2016). In: Sime-Ngando T, Boivin P, Chapron E, Jézéquel D, Meybeck M, editors. Lake Pavin: History, Geology, Biogeochemistry, and Sedimentology of a Deep Meromictic Maar Lake. Springer; 2016. pp. 3-27

[10] Meybeck M. A rich but fragmented history (200 AD-2016). In: Sime-Ngando T, Boivin P, Chapron E, Jézéquel D, Meybeck M, editors. Lake Pavin History, Geology, Biogeochemistry and Sedimentology of a Deep Meromictic Maar Lake. Springer; 2016. pp. 29-52

[11] Meybeck M. Dragons, fairies, miracles and worship at Pavin and other European Maar-Lakes. In: Sime-Ngando T, Boivin P, Chapron E, Jézéquel D, Meybeck M, editors. Lake Pavin History, Geology, Biogeochemistry and Sedimentology of a Deep Meromictic Maar Lake. Springer; 2016. pp. 53-79

[12] Scharf BW, Björk S, editors. Limnology of Eifel Maar Lakes. Advances in Limnology (Archiv. für Hydrobiol.). Vol. 38. 1992. 348 p

[13] Margaritora FG. Limnology in Latium: The volcan lakes. Memorie dell'Istituto Italiano di Idrobiologia. 1992;**50**:319-336

[14] Delbecque A. Les Lacs français. Paris: Chamerot et Renonaire; 1898. 436 pp

[15] Pourriot R, Meybeck M. Limnologie générale. Paris: Masson; 1995. 976 pp

[16] Rouwet D, Christenson B, Tassi F. In: Vandemeulebrouck J, editor. Volcanic Lakes. Heidelberg: Springer; 2015

[17] Christenson B, Németh K, Rouwet D, Tassi F, Vandemeulebrouck J, Varekamp JC. Volcanic lakes. In: Rouwet D, Christenson B, Tassi F, Vandemeulebrouck J, editors. Volcanic Lakes. Springer; 2015. pp. 1-20

[18] Rouwet D, Németh K, Tamburello G, Calabrese S, Issa. Volcanic lakes in Africa: The VOLADA_Africa 2.0 database, and implications for volcanic hazard. Frontiers in Earth Science. 2021;**9**:717798

[19] Tassi F, Rouwet D. An overview of the structure, hazards, and methods of investigation of Nyos-type lakes from the geochemical perspective. Journal of Limnology. 2014;**73**(1):55-70

[20] Negendank JFW, Zolitschka B. Paleolimnology of European Maar Lakes. Berlin: Springer Verlag; 1993

[21] Caracausi A, Nuccio PM, Favara R, Nicolosi M, Paternoster M. Gas hazard assessment at the Monticchio crater lakes of Mount Vulture, a volcano in southern Italy. Terra Nova. 2009;**21**(2):83-87

[22] Cabassi J, Tassi F, Vaselli O, Fiebig J, Nocentini M, Capecchiacci F, et al. Biogeochemical processes involving dissolved CO_2 and CH_4 at Albano, Averno, Monticchio meromictic volcanic lakes (Central-Southern Italy). Bulletin of Volcanology. 2013;**75**:1-19

[23] De Benedetti AA, Funiciello R, Giordano G, Diano G, Caprilli E, Paterne M. Volcanology, history and myths of the lake Albano maar (Colli Albani volcano, Italy). Journal of Volcanology and Geothermal Research. 2008;**176**:387-406

[24] Rouwet D, Chiodini G, Ciuccarelli C, Comastri A. Costa A Lago Albano, the "anti-Nyos-type" lake: The past as a key for the future. Journal of African Earth Sciences. 2019;**150**:425-440

[25] Lancini L. Un terrible et affreux combat. L'apport de l'approche géomythologique à la compréhension ancienne du volcanisme. Travaux du Comité français d'Histoire de la Géologie. 2020;**3**:7-23

[26] Camus G, Michard G, Olive P, Boivin P, Desgranges P, Jézéquel D, et al. Risques d'éruption gazeuse en Auvergne. Bulletin de la Societe Geologique de France. 1993;**164**:767-778

[27] Sime-Ngando T, Boivin P, Chapron E, Jézéquel D, Meybeck M, editors. Lake Pavin, History, Geology, Biogeochemistry, and Sedimentology of a Deep Meromictic Maar Lake. Berlin: Springer; 2016. 421 p

[28] Aeschbach-Hertig W, Hofer M, Kipfer R, Imboden DM, Wieler R. Accumulation of mantle gases in a permanently stratified volcanic lake (Lac Pavin, France). Geochimica et Cosmochimica Acta. 1999;**63**(19-20):3357-3337

[29] Chassiot L, Chapron E, Di Giovanni C, Albéric P, Lajeunesse P, Lehours AC, et al. Extreme events in the sedimentary record of maar Lake Pavin: Implications for natural hazards assessment in the French Massif Central. Quaternary Science Reviews. 2016;**141**:9-25

[30] Chassiot L, Miras Y, Chapron E, Develle AL, Arnaud F, Motelica-Heino M, et al. A 7,000-year environmental history and soil erosion record inferred from the deep sediments of Lake Pavin (Massif Central, France). Palaeogeography, Palaeoclimatology, Palaeoecology. 2018;**497**:218-233

[31] Thouret JC, Boivin P, Labazuy P, Leclerc A. Geology, geomorphology and slope instability of the maar lake Pavin (Auvergne, French Massif

Central). In: Sime-Ngando T, Boivin P, Chapron E, Jézéquel D, Meybeck M, editors. Lake Pavin: History, Geology, Biogeochemistry, and Sedimentology of a Deep Meromictic Maar Lake. Springer; 2016. pp. 155-174

[32] Thouret JC, Boivin P, Miallier D, Donnadieu F, Dumoulin JP, Labazuy. Post-eruption evolution of Maar Lakes and potential instability: The Lake Pavin case study, French Massif Central. Geomorphology. 2021;**382**:107663

[33] Gal F, Leconte S, Gadalia A. The "Escarot" gas seep, French Massif Central: CO_2 discharge from a quiescent volcanic system-characterization and quantification of gas emissions. Journal of Volcanology and Geothermal Research. 2018;**2018**(353):68-82

[34] Rafflin V, Boudoire G, Massaro S, Stocchi M, Costa A, Grassa F, et al. Modelling CO_2 dispersion in the air during potential limnic eruption at the lake Pavin (France). Journal of Volcanology and Geothermal Research. 2024;**447**:108024

[35] Juvigné E, Miallier D. Distribution, tephrostratigraphy and chrono-stratigraphy of the widespread eruptive products of Pavin volcano. In: Sime-Ngando T, Boivin P, Chapron E, Jézéquel D, Meybeck M, editors. Lake Pavin: History, Geology, Biogeochemistry, and Sedimentology of a Deep Meromictic Maar Lake. Springer; 2016. pp. 143-154

[36] Meybeck, Chapron E. Geocultural and geomythological values of degassing maar lakes. The Pavin example (France). Geoheritage 2025. Submitted

[37] du Citoyen Cordier L. Ingénieur des mines, au Citoyen Devilliers. Gazette Nationale (Le Moniteur Universel). 1803;**102**:410-411. (15 Vendémiaire An II)

[38] Tata D. Lettera sul Monte Volture a sua eccellenza il signor D. Gugliemo Hamilton. Napoli: Stamperia Simoniania; 1778. 62 p

[39] Tenore, Gussone. Peregrinazioni effetuate nelle state del 1838 dai Signore Tenore e Gussone (1838). In: Ciarallo A, Capaldo L, editors. Principitato Citeriore e di Basilica. Viaggo al Vulture, Edition Osanna; Venosa; 1995. 157 p

[40] Bassanelli L. Sopra il tremuoto che ha sofferto la città di Albano con le sue vicinanze dal giorno 21 di maggio a tutto il dì 6 di dicembre 1829. Lettera del dottor fisico Luigi Bassanelli all'ecc. mo sig. dottor Giacomo Folchi pubblico professore di medicina nell'archiginnasio romano. Giornale Arcadico di Scienze, Lettere, ed Arti. 1829;**44**:37-74

[41] Sigurdsson H. Gas bursts from Cameroon crater lakes: A new natural hazard. Disasters. 1988;**12**(2):131-146

[42] Sigurdsson H, Devine JD, Tchua FM, Presser FM, Pringle MKW, Evans WC. Origin of the lethal gas burst from Lake Monoun, Cameroon. Journal of Volcanology and Geothermal Research. 1987;**31**(1-2):1-16

[43] Baxter P, Kapila M, Mfonfu D. Lake Nyos disaster, Cameroon, 1986: The medical effects of large scale emission of carbon dioxide? British Medical Journal. 1989;**298**:1437-1441

[44] Krajick K. Efforts to tame second African "Killer Lake" begin. Science. 2093;**299**:805

[45] Westerman F. La vallée tueuse. Paris: Christian Bourgeois Éditeur; 2015. 390 p

[46] Tazieff MH, Faivre-Pierret RX, Le Guern F, La. Catastrophe de Nyos, République du Cameroun. Vol. 811.

Gif sur Yvette: Centre des Faibles Radioactivités, CNRS Contrib; 1987. 76 p

[47] Desmarest N. Lacs volcaniques. In: Dictionnaire de géographie physique, Lacs Volcaniques. Vol. 5. Paeis: Agasse; 1798-1828. pp. 687-688

[48] Meybeck M. Les dégazages ignorés du Pavin, le terrifiant lac-maar d'Auvergne. Géochroniques. 2019;**150**:18-36

[49] Ciarollo A, Capaldo D. Viaggio al vulture. Venose: Edizioni Osanna; 1995

[50] Chiodini G, Cioni R, Marini L, Principe C, Braco B. Water and gas chemistry of the Lake Piccolo of Monticchio (Mt Vulture, Italy). Current Research on Volcanic Lakes, IAVEI Newsletter. 1997;**10**:3-9

[51] Paci GM. Relazione dei tremuoti di Basilicata del 1851, Stabilimento Tip. Vol. 2, 1853. Naples: del Real ministero dell'interno. pp. 113-299

[52] Funiciello R, Heiken G, De Benedetti AA, Giordano G. Volcanic activity of the Lake Albano maar in Roman history and mythology. In: Funiciello R, Giordano G, editors. The Colli Albano Volcano. Special Publications, Geological Society: London; 2010. pp. 331-338

[53] Koblenzer Anzeiger (The Journal of Coblence). Koblenz, July 3 and July 10, 1844

[54] Le Guern F, Shanklin E, Tebo S. Witness accounts of the catastrophic event of August 1986 at Lake Nyos (Cameroon). Journal of volcanology and Geothermal Research. 1992;**51**(1-2):171-184

[55] Chapron E, Albéric P, Jézéquel D, Versteeg W, Bourdier JL, Sitbon J. Multidisciplinary characterization of sedimentary processes in a recent maar Lake (Lake Pavin, French Massif Central) and implication for natural hazards. Natural Hazards and Earth System Sciences. 2010;**10**:1815-1827

[56] Molo B. Une géohistoire des catastrophes rumorogènes au Cameroun : les éruptions limniques de Njindoun et Nyos, 1984-1986. Géocarrefour. 2019;**93**(1). Available from: https://journals.openedition.org/geocarrefour/12993

[57] Godivel A. Terrier pour la Royne mère du Roy, de sa terre, ville et mandement de Besse, faict par Antoine Godivel, notaire royal en la-dite ville. Chantilly: Manuscrit Bibliothèque du Musée Condé; 1579

[58] Delarbre A, (abbé). Notice sur l'Auvergne et en particulier sur la ville de Clermont. Landrier: Clermont-Ferrand; 1805. 256 p

[59] Coyssard M. Abrégé de l'histoire et miracles très-bien averez de Notre-Dame de Vassivière. Près du Mont d'Ore en Auvergne, à une lieue de Besse; Lyon; 1615

[60] Cladière J-J. L'histoire de la Sainte-Chapelle de Vassivière près du Mont-d'Or en Auvergne, célèbre par les miracles que Dieu y fait continuellement par l'intercession de sa très pure et glorieuse Mère. Par un religieux Bénédictin de la congrégation de St Maur. Damien Boujon, Clermont (1688 or 1687). Re-impressed in 1837, Clermont-Ferrand

[61] Belleforest (de) F. Cosmographie Universelle, 3 volumes, Nicolas Chesneau, imprimeur-libraire, Paris. Vol. 2. 1575. 1127 p. (Bibliothèque Sainte-Geneviève-Sorbonne, Paris, réserve, cote RXVIB 3=321-1)

[62] Jouan A. Recueil et discours du voyage du Roy Charles IX, Charles Etienne, imprimeur du Roy. Paris; 1566

[63] Voragine J. The Golden Legend. Medieval Sourcebook: The Golden Legend (Aurea Legenda). Compiled by Jacobus de Voragine, 1275. Englished by William Caxton, 1483. Available from: http://www.fordham.edu/halsall/basis/goldenlegend/

[64] Meybeck M, André MF. Histoire du Creux de Soucy. Revue des Sciences Naturelles d'Auvergne. 2015;**79**:1575

[65] Banc J. La Mémoire retrouvée des merveilles des eaux naturelles en faveur de nos Nymphes Françaises et des maladies qui ont recours à leurs emplois salutaires. Paris: L. Pierre Sevestre; 1605

[66] Jaloustre E. Le Beffroi de Besse. Clermont-Ferrand: Mémoires de l'Académie Royale des Sciences, Belles-Lettres et Arts de Lyon; 1884

[67] Frêtat (de) Aimable. Carte de la Haute et Basse Auvergne (1672) (map)

[68] Godivel (chatelain). Lettre ouverte au Rédacteur de la feuille. Feuille Hebdomadaire de la Haute et Basse Auvergne, contenant ce qui peut intéresser cette province. In: Chabrit M, editor. Antoine Delcros imprimeur. Clermont-Ferrand; 1783. p. 152

[69] Bouillet J-B. Description historique de Besse. Clermont-Ferrand: University Library (BCIU), Manuscrit, 40 f, Ms 977-fo; 1875. pp. 39-40

[70] Lecoq H. Description de l'Auvergne. Paris: Le Mont d'Or et ses environs. Paris: J-B Baillère; 1835. 437 p

[71] Lecoq H. L'Eau sur le Plateau Central de la France. Paris: J-B Baillère; 1871. 394 p

[72] Joanny P. Un Pharmacien clermontois raconte un souvenir de jeunesse. Interview by J Gandebeuf. Le Républicain Lorrain. 1987

[73] Jaloustre E. Histoire du pèlerinage de Notre-Dame de Vassivière. Impr. Dumont, Clermont-Ferrand; 1910. 150 p

[74] Mosnier SM(A). Les Saints d'Auvergne. Paris: Le Thielleux; 1899

[75] Nodier C, Taylor JD, De Cailleux J. Voyages pittoresques et romantiques dans l'ancienne France; 1829. (Re-edited 1987, Presse du Languedoc, Montpellier Tome 1, 110 p)

[76] Noir J. Besse en Chandesse et ses environs. Folklore médical : La Foi qui guérit au sanctuaire de Vassivière. In: Histoire et géographie médicale d'une vieille commune d'Auvergne. Paris: Le Progrès Médical; 1907. pp. 33-44

[77] Legrand d'Aussy JB. Voyage en Auvergne. Paris: E. Onfroy; 1788. 555 pp

[78] Fournier PF. Le Nom du lac Pavin, Essai d'étymologie. Bulletin Historique et Scientifique de l'Auvergne. 1971;**85**(631):159-164

[79] Sebillot P. Les Eaux douces, Folklore de France, (1904-1906), (Re-edited in 1983). Paris: Imago. 304 p

[80] Ajalbert J. L'Auvergne. Paris: Ancienne maison Quentin; 1896. 362 pp

[81] Larat J. Les peurs du lac Pavin. Le Soleil du Centre. 1929 (Bibliothèque Henri Pourrat, Clermont-Ferrand)

[82] Martel E-A. Les Abymes. Chap. XXII Le Creux de Souci (Puy-de-Dôme); 1894. pp. 387-393

[83] Lavina P, Del Rosso T. New geological findings in Pavin area. In:

Oral Communication, Symposium Pavin and Other Meromictic Lakes, Besse-en-Chandesse. 2009

[84] Jezequel D, Sarazin G, Fonty G, Tassin B. Le lac Pavin, le volcan, l'eau et la vie. Pour la Science, L'eau Attention Fragile. 2008;**Feb**:52-59

[85] Serieyssol K, Beauger A, Miras Y, Chassiot L, Arricau V, Chapron E. Diatoms-based paleolimnology of Lake Pavin over the past 7000 year. Journal of Paleolimnology. **72**(3):99-315. DOI: 10.1007/s10933-024-00330-2

[86] Chapron E, Ledoux G, Simonneau A, Albéric P, St-Onge G, Lajeunesse P, et al. New evidence of holocene mass wasting events in recent volcanic lakes from the French Massif Central (Lakes Pavin, Montcineyre and Chauvet) and Implications for Natural Hazards. In: Yasuhiro Y, Kiichiro K, Ken I, Yujiro O, Urgeles R, Mosher D, Chaytor J, Strasser M, editors. Submarine Mass Movements and Their Consequences. Dordrecht: Springer Netherlands; 2012. p. 25564. DOI: 10.1007/978-94-007-2162-3_23

[87] Mouhamed AN, Rouwet D, Tassi F, Fantong WY, Mouncherou OF, Ekomane E, et al. Evidence of "Lake Nyos-type" behavior in the geological record: A review. Earth-Science Reviews. 2023;**247**:104603

[88] Chapron E, Foucher A, Chassiot L, Fleurdeus W, Arricau V, Perdereaux L, et al. Evaluating Holocene natural hazards in the French Massif Central from a regional lake sediment approach. Quaternary International. 2022;**636**:134-153

[89] Chassiot L, Chapron E, Di Giovanni C, Lajeunesse P, Tachikawa K, Garcia M, et al. Historical seismicity of the Mont-Dore volcanic province (Auvergne, France) unraveled by a regional lacustrine investigation: New insights about lake sensitivity to earthquakes. Sedimentary Geology. 2016;**339**:134-150

[90] Chassiot L, Chapron E, Miras Y, Schwab MJ, Albéric P, Beauger A, et al. In: Sime-Ngando T, Boivin P, Chapron E, Jézéquel D, Meybeck M, editors. Lake Pavin Paleolimnology and Event Stratigraphy. Lake Pavin: History, Geology, Biogeochemistry, and Sedimentology of a Deep Meromictic Maar Lake. 2016. pp. 381-406

[91] Chapron E, Chassiot L, Lajeunesse P, Ledoux G, Albéric P. Lake Pavin sedimentary environments. In: Sime-Ngando T, Boivin P, Chapron E, Jézéquel D, Meybeck M, editors. Lake Pavin: History, Geology, Biogeochemistry, and Sedimentology of a Deep Meromictic Maar Lake. Springer; 2016. pp. 365-379

[92] Rioual P. Limnological characteristics of 25 lakes of the French Massif Central. Annales de Limnologie-International Journal of Limnology. 2002;**38**(4):311-327

[93] Michard G, Viollier E, Jézéquel D, Sarazin G. Geochemical study of a crater Lake: Lake Pavin, France. Identification, location and quantification of the chemical reactions in the lake. Chemical Geology. 1994;**115**:103-115

[94] Viollier E, Jézéquel D, Michard G, Pèpe M, Sarazin G, Albéric P. Geochemical study of a crater Lake (Lake Pavin, France): Trace elements behavior in the monimolimnion. Chemical Geology. 1995;**125**:61-72

[95] Viollier E, Michard G, Jézéquel D, Pèpe M, Sarazin G. Geochemical study of a crater Lake: Lake Pavin, Puy-de-Dôme, France. Constrains afforded by

the particulate matter distribution in the element cycling within the lake. Chemical Geology. 1997;**142**:225-241

[96] Busigny V, Planavsky NJ, Jézéquel D, Crowe S, Louvat P, Moureau J, et al. Iron isotopes in an Archean Ocean analogue. Geochimica et Cosmochimica Acta. 2014;**133**:443-462

[97] Lehours AC, Bardot C, Thenot A, Debroas D, Fonty G. Anaerobic microbial communities in Lake Pavin, a unique meromictic lake in France. Applied and Environmental Microbiology. 2005;**71**:7389-7400

[98] Lehours AC, Evans P, Bardot C, Joblin K, Fonty G. Phylogenetic diversity of Archaea and Bacteria in the anoxic layer of a meromictic Lake (lake Pavin, France). Applied and Environmental Microbiology. 2007;**73**:2016-2019

[99] Cosmidis J, Benzerara K, Morin G, Busigny V, Lebeau O, Jézéquel D, et al. Biomineralization of iron-phosphates in the water column of Lake Pavin (Massif Central, France). Geochimica et Cosmochimica Acta. 2014;**126**(2):78-96

[100] Busigny V, Mathon FP, Jézéquel D, Bidaud CC, Viollier E, Bardoux G, et al. Mass collection of magnetotactic bacteria from the permanently stratified ferruginous Lake Pavin, France. Environmental Microbiology. 2022;**24**(2):721-736

[101] Bidaud CC, Monteil CL, Menguy N, Busigny V, Jézéquel D, Viollier E, et al. Biogeochemical niche of magnetotactic cocci capable of sequestering large polyphosphate inclusions in the anoxic layer of the Lake Pavin water column. Frontiers in Microbiology. 2022;**12**:789134

[102] Stebich M, Brüchmann C, Kulbe T, Negendank JF. Vegetation history, human impact and climate change during the last 700 years recorded in annually laminated sediments of Lac Pavin, France. Review of Palaeobotany and Palynology. 2005;**133**(1-2):115-133

[103] Krishnaswami S, Lal D, Martin JM, Meybeck M. Geochronology of lake sediments. Earth and Planetary Science Letters. 1971;**11**:407-414

[104] Martin JM. The Pavin Crater Lake. In: Stumm W, editor. Chemical Processes in Lakes. New-York: John Wiley; 1985. pp. 169-188

[105] McCarthy FM, Patterson RT, Head MJ, Riddick NL, Cumming BF, Hamilton PB, et al. The varved succession of Crawford Lake, Milton, Ontario, Canada as a candidate global boundary Stratotype section and point for the Anthropocene series. The Anthropocene Review. 2023;**10**(1):146-176

[106] Lecoq H. L'Auvergne pittoresque. Atlas colorié. Le Mont-Dore et ses environs. Seize vues coloriées. Paris: Chaillon; 1835

[107] Del Rosso-d'Hers T. Montchal-Pavin, mise au point sur les travaux récents (2004-2010). Eruption. 2010;**23**:41-43

[108] Ewans WC, White LD, Tuttle ML, Kling GW, Tanyileke G, Michel RL. Six years of change in lake Nyos, Cameroon, yield clues to the past and cautions for the future. Geochemical Journal. 1994;**28**:139-162

[109] Mott RW, Woods AW. A model of overturn of CO_2 laden lakes triggered by bottom mixing. Journal of Volcanology and Geothermal Research. 2010;**192**(3-4):151-158

[110] Recluz E. Géographie Universelle. Vol. II. Paeis, France: Hachette; 1885

[111] Berlioz J. Crochet de fer et puits à tempête. La légende de Ponce Pilate à Vienne (Isère) et au mont Pilât au XIIIe siècle. Le Monde alpin et rhodanien. Revue Régionale d'ethnologie. 1990;**18**(1):85-104

[112] Grandazzi A. Une nouvelle hypothèse sur le miracle du lac Albain. Bulletin de l'Association Guillaume Budé. 2003;**1**(1):96-106

[113] Pietrafesa F. *La badia di Monticchio*. Laurenziana: Napoli; 1980

[114] Spicciarelli R, Marchetto A. The contrasting evolution of twin volcanic lakes (Monticchio, Mt. Vulture, Italy) inferred from literature records: Literature record of Monticchio lakes. Advances in Oceanography and Limnology. 2019;**10**(1):44-56

[115] Meybeck M. La longue et terrible histoire des lacs maars. Relief. 2019;**10**(3):58-63

[116] Mûnster S. Cosmographie Universelle. Paris: Michel Sonnius; 1565-1566. 1430 pp

[117] Voltmer R. Jagd auf "böse Leute". Hexenverfolgungen in der Region um den Laacher See (16.–17.Jahrh.). Plaidter Blätter. Jahrbuch des Plaidter Geschichtsvereins. Available from: historicum.net

[118] Spence L. Hero Tales and Legends of the Rhine. North Chalmsfort: Courier Corporation; 1995

Chapter 9

Eruption Scenarios of a Monogenetic Volcanic Field Formed within a Structurally Controlled Basement Terrain: Harrat Lunayyir, Saudi Arabia

Károly Németh, Mostafa Toni, Vladimir Sokolov, Abdulrahman Sowaigh, Mahmoud Ashor and Fawaz Moqeem

Abstract

Lunayyir Volcanic Field is one of the youngest (~600 ka to Historic) Cenozoic mafic intracontinental volcanic fields in the western margin of the Arabian Microplate. It consists of about 150 volcanic edifices that worked around 700 vents indicating strong vent alignments and potential longevity of activity to build each of its volcanoes. The volcanoes are strongly aligned following the Neoproterozoic basement structure determining the surface valley network that commonly captured lava flows. The volcanoes are typical magmatic effusive and explosive types of volcanoes ranging from minor fissure vents of spatter ramparts to complex multi-crater scoria cones. Effusive hazards are dominated by transitional lava flows with restricted pāhoehoe and ʻaʻā lava emplacement while explosive eruptions are spatter cone formation with lava fountaining, scoria cone growth and destruction, paroxysmal sub-Plinian and/or violent Strombolian style activity, directed ballistic curtain formation and a possibility of development of passive volcanic fog (vog). According to these eruption types, effusive eruption scenarios are the most likely and long-term impactful events, while explosive eruptions are dominated by low intensity low impact scenarios. The area however shows evidence of extensive tephra blanket that explosive phases capable to reach volcanic explosivity index (VEI) 4 capable to carry ash away over 250 km.

Keywords: monogenetic, basaltic, scoria, spatter, pāhoehoe, ʻaʻā lava, fissure, seismic swarm

1. Introduction

Research on monogenetic volcanism became an increasingly important part of volcanology. Monogenetic volcano is a volcanic edifice with a small cumulative volume (typically ≤1 km^3) that has been built up by one continuous, or many discontinuous, small eruptions fed from one or few (probably in the dozen range) distinct magma batches erupted in a short time interval (hours to month most) [1]. Small-scale basaltic magmatic systems with magma fluxes of 10^{-1} to 10^{-3} km^3/ky are common on Earth and tend to form volcanic fields over long time (ka to Ma) despite their individual volcanoes are typically short-lived. Individual volcanoes in such fields are referred to as monogenetic, emphasizing the short duration over which the individual volcano was active, the small volume of eruptive magma involved, and the relative simplicity of the volcanic edifice as a reflection of the simple and "one-off" nature of the eruption itself [2]. This type of mostly basaltic volcanism within a volcanic field setting creates a variety of volcanoes, such as spatter cones, scoria cones, maar–diatremes, tuff rings and cones and associated lava fields [3]. Monogenetic volcanic fields occur only where the magma supply to the volcano is low or where vents are not close enough or large enough to develop magmatic plumbing systems for continuous feeding of magma. Monogenetic volcanic fields can provide snapshots of the underlying region beneath the surface and may be useful in studying the generation of magma and the composition of the mantle since the single eruption produced would match that of the commonly deep source from which it erupted [1]. The magma supplying such fields is thought to have rapidly ascended from its source region, with only short resident times (decades or less) in shallow crustal magma storage systems such as dyke and sill networks. Monogenetic volcanic fields show great diversity of spatio-temporal vent distribution patters that are rather indicative to the lithospheric structure [4–6] especially in old continental lithosphere such as in the case in the western margin of the Arabian Peninsula, that is part of the Neoproterozoic Arabian Shield. Western Saudi Arabia hosts one of the world's largest alkali basalt provinces that covers a total area of about 180,000 km^2 [7]. Volcanic rocks and volcanic edifices normally directly overlying the Precambrian basement complex of the Arabian Shield, part of the western Arabian plate. The province has traditionally subdivided into 19 regions, each characteristically dominated by eruptive products of coherent lava and less voluminous pyroclastic rocks mostly mafic in geochemical composition [8–10]. The subdivision however is commonly ambiguous both geologically and by usage of local names to specific geographical region. The subdivision of volcanic fields is also complicated in areas where the number of post-Miocene volcanic edifices numbers are reaching several hundred such as the in the region between Madinah and Makkah (**Figure 1**) Harrat Rahat (**Figure 2**) as older volcanic edifices are difficult to reconstruct correctly especially to locate their individual vents [11–13]. Out of the 19 volcanic fields, 13 are those that host dozens of vents and cover significant surface areas. Distinguishing volcanic fields can also pose problem as large volcanic fields nearby show transitional appearance in the field and often putting the boundary between them is not straight forward such as the case in between Harrat Rahat and Khaybar, Harrat Khaybar and Ithnayn, and Harrat Khaybar and Kura (**Figure 2**) for instance [10]. Extensive geological and geochemical work was carried out in the western Arabian monogenetic volcanic fields in the 1970s to early 1990s [9, 12, 14, 15], leading to several volcanic source models and hypotheses about the relationship of this type of volcanism and the Red Sea Rift [9]. While this question is still under debate it is more and more evident that volcanism generated the harrat are results of deep sourced melt propagation to the surface along basement structures reacted to tectonic stress allowing small batches of melt reaching the surface periodically [16, 17]. This style of volcanism,

Figure 1.
Tectono-stratigraphic units of the Arabian Peninsula with the marked location of its Cenozoic, mostly younger than 10 Ma volcanic fields in western Arabia that erupted on the basement rocks of the Neoproterozoic Arabian Shield.

commonly called as monogenetic [1] or dispersed volcanism of deep sourced magmas (e.g., basalt), is one of the most common types of volcanism on Earth especially in intraplate settings, and one of the most challenging to develop a workable and useful volcanic hazard estimate. The surface manifestation of vent distribution and associated lava inundation in such old, structurally complex terrains can generate complex volcanic terrain morphology where monogenetic volcanoes can erupt within narrow fault-bounded gorges, their edifice grow over steep rugged terrains inhibiting gradual cone collapse and spreading, while their lava effusions can feed extensive lava fields captured and channelized into narrow gorges. Such regions can offer a deep insight of the volcano-morphological evolution of a terrain [18] and offer good scaling to see the type of volcanic hazard and eruption impact future volcanic events may pose to the terrain. In this chapter such volcanic field is documented in the NW part of the Arabian Shield (**Figures 1** and **2**) called as Lunayyir (Shaqa) Volcanic Field (LVF) (**Figure 2**). Within these extensive young (younger than 600 ka) volcanic field [19], particularly in its northern sector distinct basins are inundated by lava flows, while cones grown into narrow gorges or grown over steep structural network.

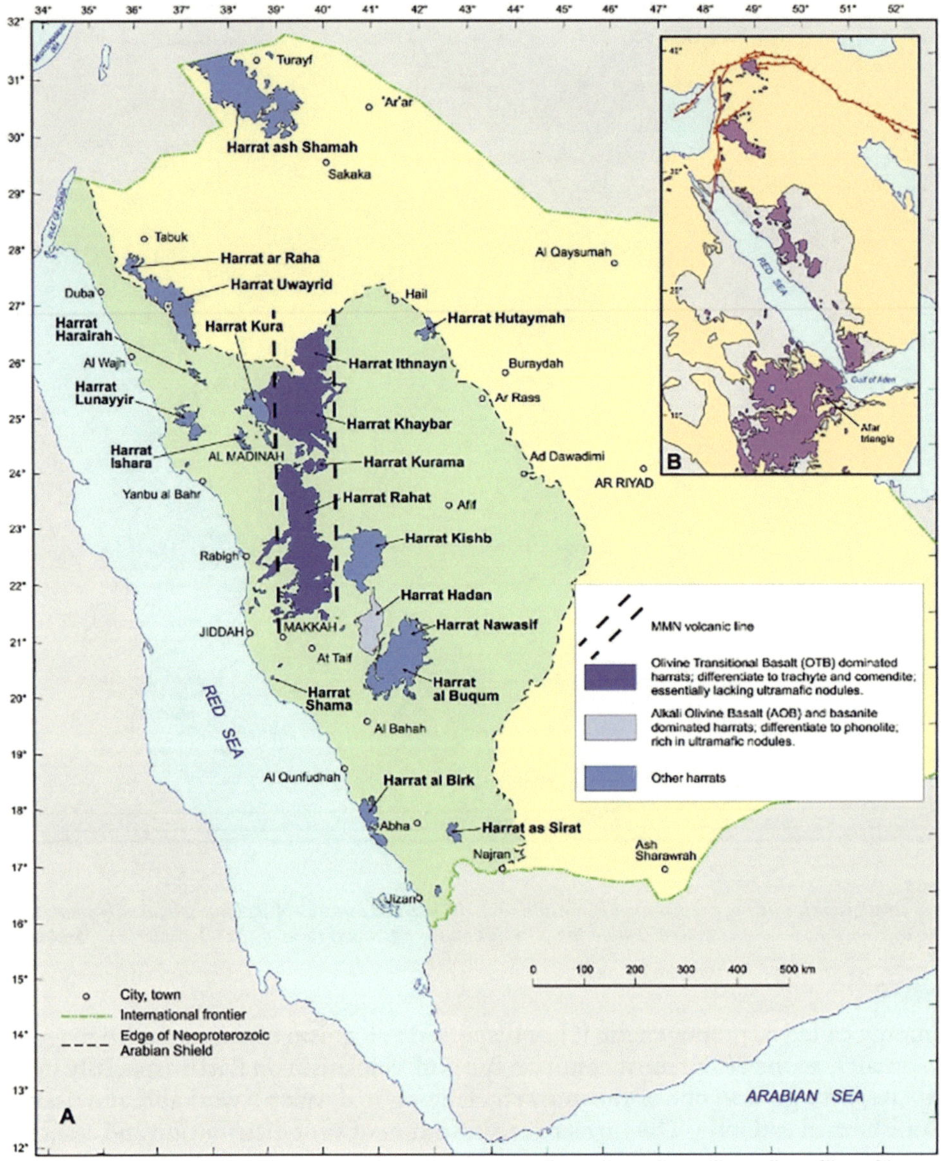

Figure 2.
Volcanic harrat (volcanic fields) of the western Arabian Peninsula cover extensive regions with lava flows and typical monogenetic volcanic fields. Small inset map shows the broader geotectonic situation as orogenic belts in the northern margin and left-lateral strike-slip plate margin in its western boundary. The location of the Nubian – Arabian Shield marked by gray while extensive Pleistocene to Holocene volcanic regions is shown by pink fields.

The study area was defined as a distinct basin that is filled with lava flows exhibit young surface textures and mapped as Holocene eruptive products (**Figure 3**).

Vents are defined as map view projection of the outlet of a volcanic conduit that transport magma to the surface. Magma then exits through a crater that is the surface manifestation of a vent erupted pyroclasts through explosive magma fragmentation or as a liquid that was fed low viscosity basaltic lava flows. Volcanic edifice, or volcano in this sense is the 3D landform that may have numerous vents that may or may not

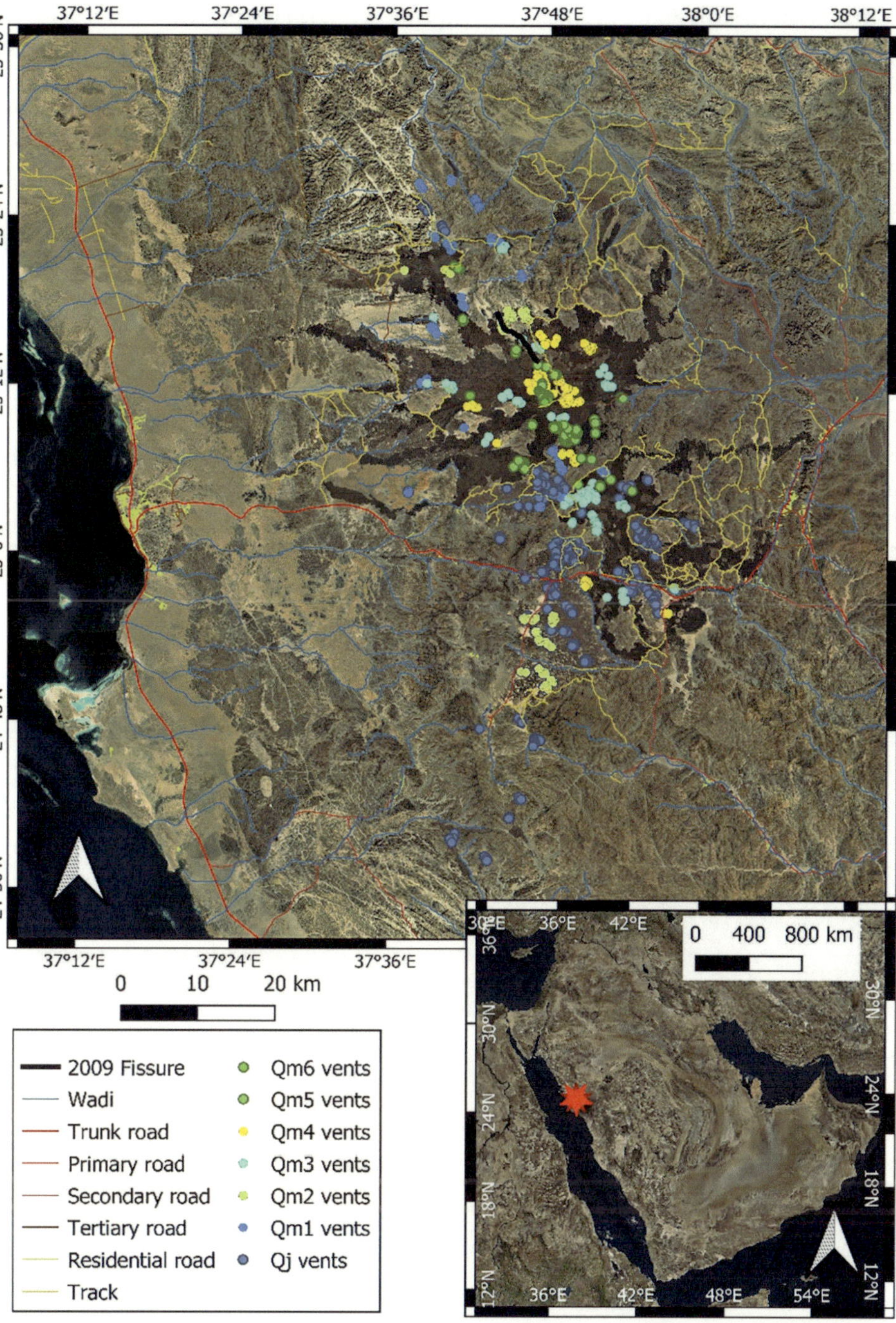

Figure 3.
Lunayyir Volcanic Field is in the northern Red Sea coast in Saudi Arabia. On a satellite image the lava flows clearly visible filling the narrow valley network between elevated Neoproterozoic hosts. Red star marks the location in an inset map showing the Arabian Peninsula.

erupted at the same time. It is important to understand this distinction as vent numbers can be significantly higher than actual volcanic landform numbers (e.g., volcano). Along fissures, compound and complex volcanoes numerous exit points (vents) can form that is the case at Harrat Lunayyir. Eruption scenario in the other hand refer to a specific type of volcanism that can be inferred from the preserved volcanic facies and rocks. Identification of eruption scenarios are the key to volcanic hazard studies as with correct identification a graphic timeline of events can be drafted what to expect from future eruptions, what the most and less likely scenario, or what is the most and less impactful scenario a region must be ready for future eruptions. A specific eruption scenario naturally can give hints about the sudden or gradual built up of a volcanic event or its manageable or catastrophic impacts. This research eventually culminates into a simplified but evidence-based scenario builder. Complex vent distribution can generate various vent distribution maps.

2. Geological setting

The Arabian Plate lies between the Gulf of Aden, the Red Sea, the Gulf of Aqaba and the Dead Sea transform zone, the Taurus and the Zagros Mountains, and the Arabian Gulf (**Figure 4**). The tectonic structure of the plate has been widely discussed in the literature [20–29]. Saudi Arabia is situated in the center of the Arabian plate (**Figure 4**).

In this area, the plate comprises the following formations: (I) a crystalline basement of Precambrian continental crust about 40–45 km thick and mostly 870–550 million years (Ma) old; (II) an overlying sequence of younger Phanerozoic

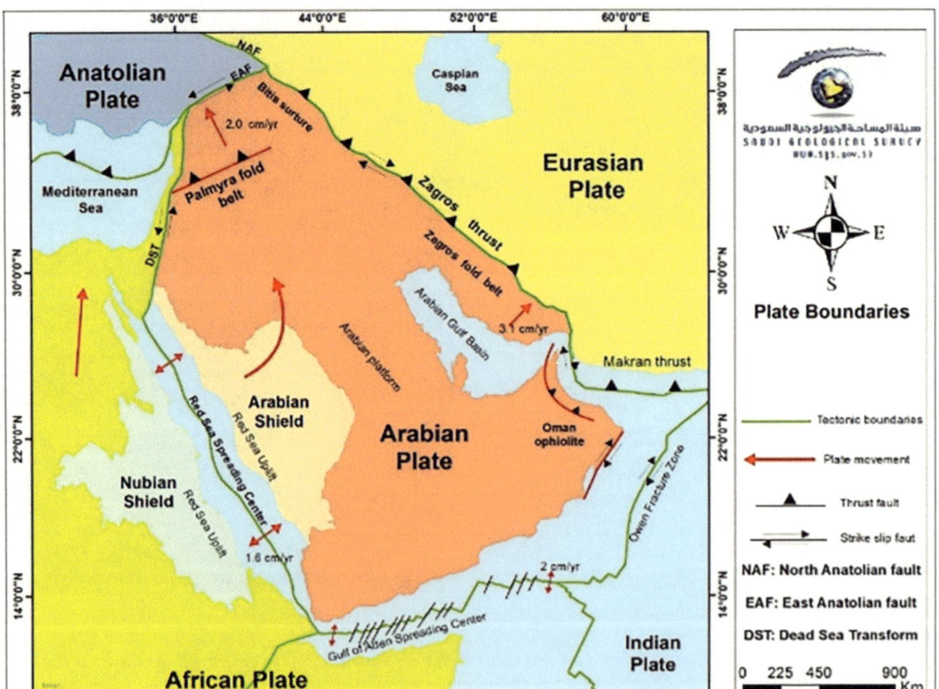

Figure 4.
Geotectonic situation of the Arabian Microplate. Tectonic framework is based on "Johnson (1998) Tectonic map of Saudi Arabia and adjacent areas. Ministry of Petroleum and Mineral Resources, Deputy Ministry for Mineral Resources Technical Report USGS TR-98- 3, (IR 948)".

sedimentary rocks that range in age from Cambrian (540 Ma) to the Pleistocene and in thickness from zero to 10 km; (III) Cenozoic flood basalts, (IV) Paleogene-Holocene intracontinental deposits, (V) oceanic basins along the Red Sea and Gulf of Aden (**Figure 1**). In contrast to the generally flat Arabian Platform, the Arabian Shield is characterized by a ragged and elevated relief reaching an elevation of 2000 m. Initially, the Arabian Shield was united with the Nubian Shield on the African Plate, but later they were separated by the process of rifting, which began approximately 30 Ma [30] and caused the opening of the Red Sea (**Figure 4**). During the period of separation between Arabia and Africa, new oceanic crust and sedimentation in the Red Sea and Gulf of Aden basins is created; the western and southern margins of the Arabian Plate were uplifted and partly covered by subaerial flood basalt, resulting in the creation of the Red Sea Escarpment and extensive lava fields locally named as harrat (**Figure 2**). These harrat however are well-defined distributed volcanic fields commonly termed as monogenetic volcanic fields from their vents, lavas inundated much larger areas then their vent location defines. The northern and northeastern margins were sutured to rocks in Iran and Turkey, causing crustal shortening and the formation of the Zagros fold-and-thrust belt. The crustal structure of the Arabian Shield is affected by the Neogene and Quaternary volcanism. In the Cenozoic time at around 31 Ma a mantle plume formed beneath what is now the Afar Triangle [31]. Crustal extension commenced along the site of the future Red Sea with the eruption of the older lava fields (harrat) of Saudi Arabia between 30 and 20 Ma (tholeiitic-transitional phase of volcanism parallel to the margin of the Red Sea) including Harrat Ishara, Hadan, Nawasir-Al Buqum and As Sirat [9, 10, 32–35] (**Figure 2**). Around 24 to 22 Ma in response to crustal stretching, major gabbro dikes were emplaced in a fault zone corresponding to the foothills of the Red Sea [31]. During the past 10 million years (transitional-alkaline phase of volcanism) the youngest lava fields were formed (**Figure 2**). The lavas mainly lie along a 900 km line within the Arabian shield that extends south from the Great Nafud Desert, through the cities of Al Madinah and Makkah, and then as far south as Al Qunfudah on the coastal plain (**Figures 1**, **2** and **4**). The northernmost 600 km of this trend takes the form of a north-south graben structure through which the main Cenozoic basaltic lava fields have been erupted. This zone has been named the Makkah-Madinah-Nafud (MMN) (**Figure 2**) volcanic line, and includes Harrat Rahat, Khaybar and Ithnayn [10]. The MMN volcanic line may be a weakly propagating rift zone where crustal extension has averaged about 0.054 mm per year over the past 10 million years and is distinct from the main Red Sea rift zone. It forms the axis of uplift in western Saudi Arabia, and geothermal phenomena are observed along this trend. The fact of asymmetric distribution of harrat with respect to the Red Sea is a strong argument showing that the recent volcanism in Saudi Arabia is probably not directly associated with the rifting processes within the Red Sea Rift and rather originates from mantle sources, such as plume or asthenosphere upwelling [36, 37]. Approximately twenty eruptions in Western Saudi Arabia and Yemen were recorded in historical times during the last 1500 years [35]. Among the recent eruptions it is necessary to mention two historic eruptions that were recorded (i) on northern Harrat Rahat near Al Madinah city [38–40], and (ii) on Harrat Lunayyir [41, 42]. Late prehistoric scoria cones and lava flows are known on northern Harrat Rahat, and harrat Lunayyir, Khaybar and Kishb [34, 35]. The MMN line and areas further south along the Red Sea coast show geothermal phenomena that take the form of shallow water wells with elevated temperatures, hot springs and occasional steaming [43, 44].

3. From terrain analysis to volcanic geology

The focus area is in the northern section of the Lunayyir Volcanic Field (**Figure 5**). This area is seemingly a region that evolved through vents well-clustered along NW-SE trending alignments distinguished from other part of the region. Narrow fault-bounded gorges in addition captured the extensive lava flows and formed zones of lava inundated regions separated by crystalline basement rocks of tonalite, trondhjemite, monzogranite and syenogranite ridges commonly crosscut by slightly more mafic dykes.

On various satellite imagery, the structural architecture of the region is clearly visualized having steep sided rugged ranges of basement rocks and the narrow gorges filled with dark basaltic lava (**Figure 6**).

Through a representative cross section, it is very evident that a major morphological step separates the western region of the area from the deep interior of the volcanic field that is also very uniform in morphology (**Figure 7**).

The highest point reaches over 1,000 meters and form characteristic steep sided mountain ranges separated by box valleys commonly filled with lava flows. The same morpho-structural features are also evident on a slope angle map generated from the ALOS PALSAR 12.5 m resolution digital elevation models showing slopes reaching over the general angle of repose of loose sediments posing potential high-risk regions for rock falls, landslides, or other geophysical mass flow initiation points (**Figure 8**).

The terrain ruggedness is also demonstrated by the landform characterization by applying the geomorphon theories [45] that divide the landforms to 10 basic categories based on the elevation data in every single datapoint (**Figure 9**).

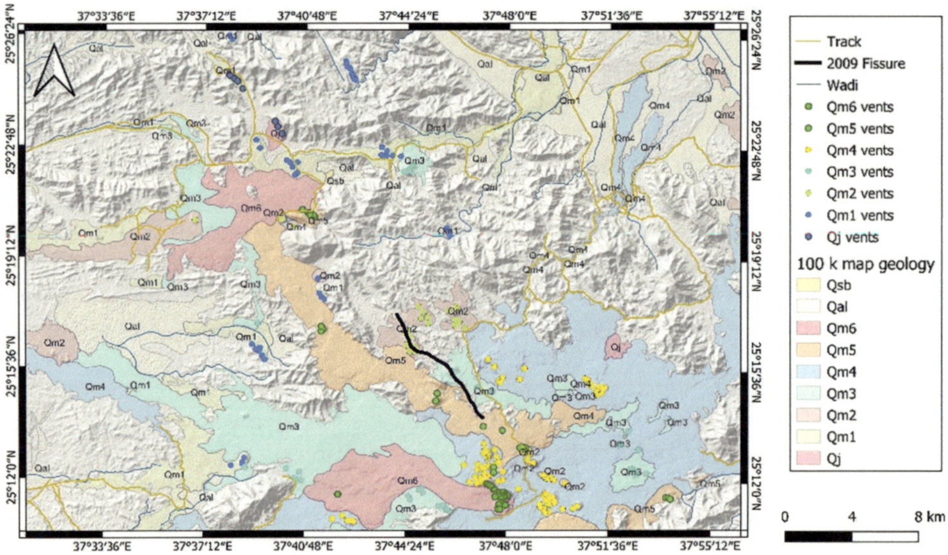

Figure 5.
Geological map of the northern Harrat Lunayyir based on a 1 to 100,000 scale mapping. Qj represents the oldest volcanism about 600 ka old, while the youngest Qm6 volcanoes probably historic and can be as young as about 1 ka. Areas where the hillshade map (based on ALOS-PALSAR 12.5 m resolution digital elevation model) clearly visible are Neoproterozoic crystalline basement of dominantly tonalite, syenogranite or monzogranite penetrated by extensive andesitic dyke swarms. The map shows the fissure opened in 2009 accompanied with a seismic swarm interpreted to be a result of dyke intrusion that stalled few hundreds of meters below the surface, hence defined as a failed eruption.

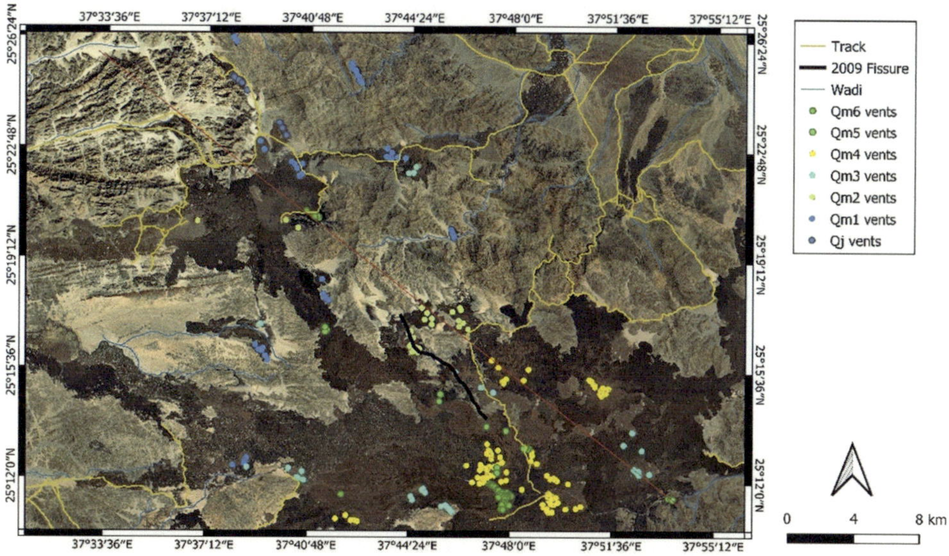

Figure 6.
Northern Harrat Lunayyir region on a Bing Satellite image. Red line represents a NW-SE topographic cross section shown on **Figure 7**.

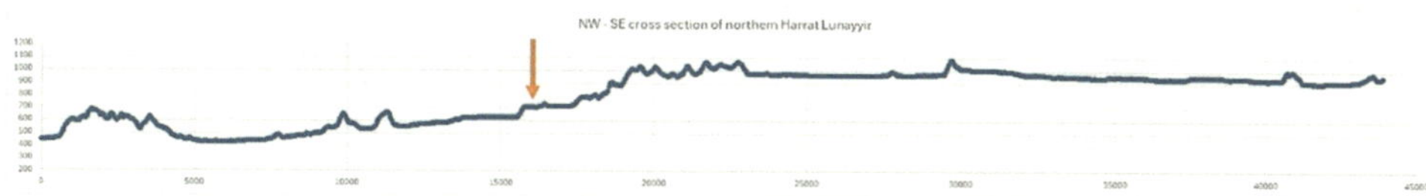

Figure 7.
A representative NW-SE cross section showing the major morphological step along one of the youngest volcanoes erupted within the Harrat Lunayyir (red arrow). Elevations (y-axis) and distances (x-axis) are in meters.

On this map the region has a systematic rectangular valley pattern and adjoining ridge setting all inferred to be highlighting the basement Precambrian rocks structural architecture. Applying various parameters to derive geomorphon maps can be used to locate key geomorphological scales and areas to aid geosite recognition and geodiversity estimates. Convergence index map (**Figure 10**) has also been generated to show the terrain ridge and valley pattern as this method clearly showing the landscape slope changes such as divergent when surrounding pixels getting lower values (less elevation) or convergent if they are forming local heights (more elevation) [46–48].

This method is very sensitive to identifying major structural patterns of the region. Creating an aspect ratio map for the broad regions and color coding revealed the same pattern with a different visualization technique (**Figure 11**). Similar colors show the slopes facing a specific direction segment. Very colorful regions indicate strong variations within the landscape while uniform colors are indicative for uniform morphological settings such as existence of fault planes (**Figure 11**).

The area is characteristically showing a roughly NNW-SSE and a near-perpendicular valley and ridge pattern. This pattern has been obscured in the central part of the volcanic field where most of the vents are located and lava sheets filling the intramountain basins. Testing the viability of satellite imagery usage, it is very clearly

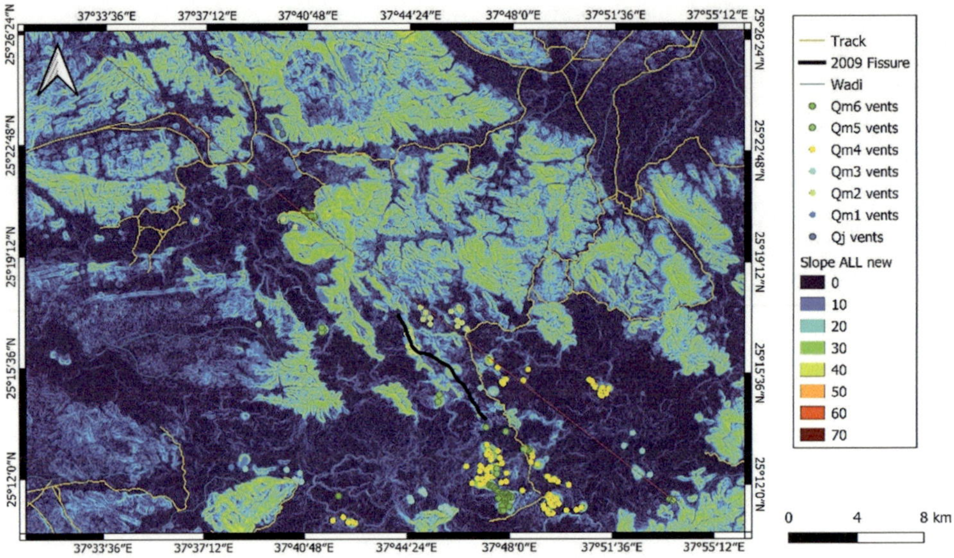

Figure 8.
Slope angle map (in degrees) showing the lava filled flat bottom valleys (blue) and the rugged, steep sloped uplifted basement horst as a contrasting landscape element of the region.

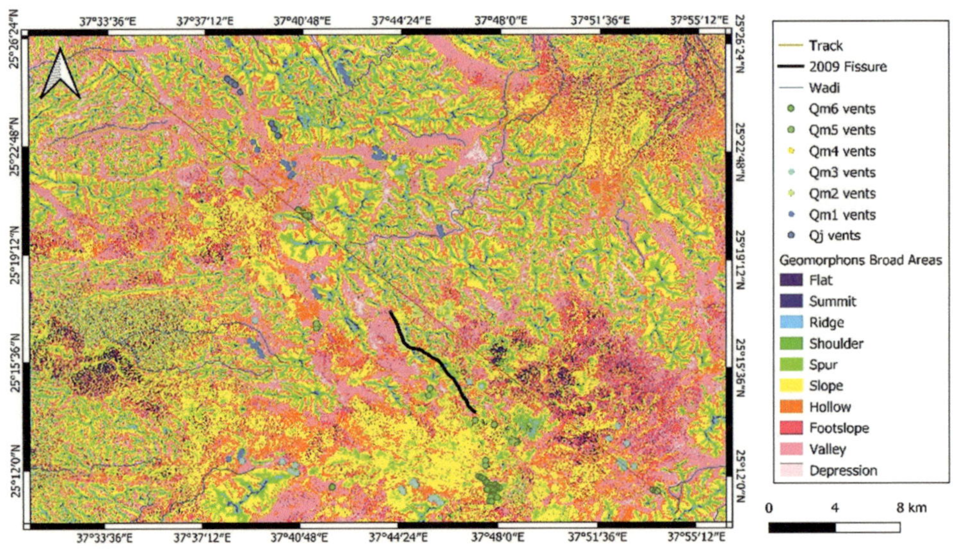

Figure 9.
Geomorphon map showing a dendritic morpho-texture of the basement horsts that are in similar orientation as the valley network filled with Quaternary lava.

visible on such imagery that young lava flows are filling broad basin like features present before the volcanic eruptions modified the landscape. This can be clearly visible on the various Sentinel imagery such as the highlight optimized versions (**Figure 12**). Sentinel moisture index map however shows well that lava flows are likely act as zones where water can be hidden beneath (**Figure 13**).

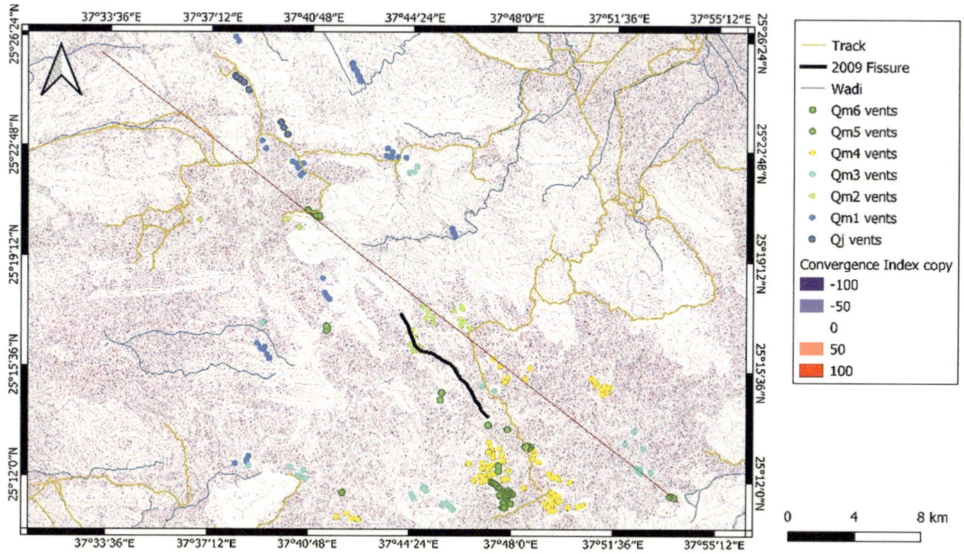

Figure 10.
Convergence index map showing the distribution of divergent (100 = red) - like ridge lines – and convergent (−100 = blue) – like valley axis structures outlining well crater rims or volcanic cone bases.

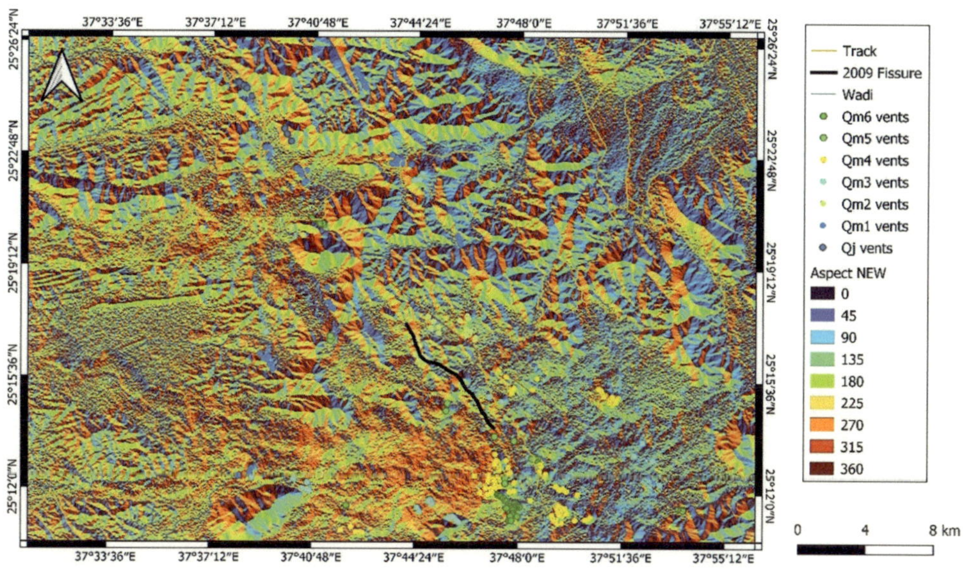

Figure 11.
Aspect ratio map showing planar features facing to the same direction. Similar colors over large, connected areas represent planar features cutting the landscape such as fault plains.

On October 11, 2007, a swarm of earthquakes occurred on the eastern side of the lava fields of Harrat Lunayyir that is relatively young being less than 600,000 years old [41] with the last eruption about estimated to be occurring as young as 1000 years ago [49, 50]. Since early 2009 tens of thousands of earthquakes attaining a magnitude up to MW 5.7 (May 19, 2009) have been observed in the area including19 earthquakes of magnitude (M) 4.0 or greater occurred at depths less than 20 kilometers during

Figure 12.
Highlight optimized Sentinel image of the northernmost part of Harrat Lunayyir (2024-03-04-00_00_2024-03-04-23_59_Sentinel-2_L2A_Highlight_Optimized_Natural_Color). Note the faded colors of older lava surfaces while young flow fields are bright black.

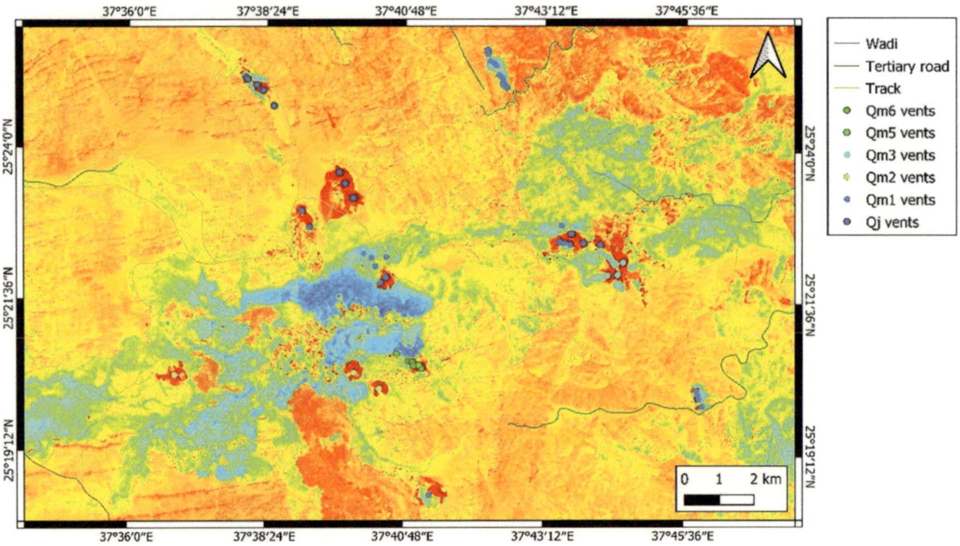

Figure 13.
Moisture index map on a Sentinel satellite image (2024-03-04-00_00_2024-03-04-23_59_Sentinel-2_L2A_Moisture_index) showing bright blue those zones that are holding moist within. It seems lava flow fields, and potentially their tube network and/or lava foot and top breccias are places where relative wetness can be achieved.

the period from May 15, 2009, to September 30, 2010. The May 19 earthquake caused minor damage to structures in the town of Al-Ays located approximately 30 km away the swarm area (**Figure 14**).

The local geologic setting and characteristics of this seismic activity is analyzed in several publications [19, 50–57]. The magnetic data indicate that the recent fracturing and seismicity is controlled by the intersections of old NE faults with one of the major

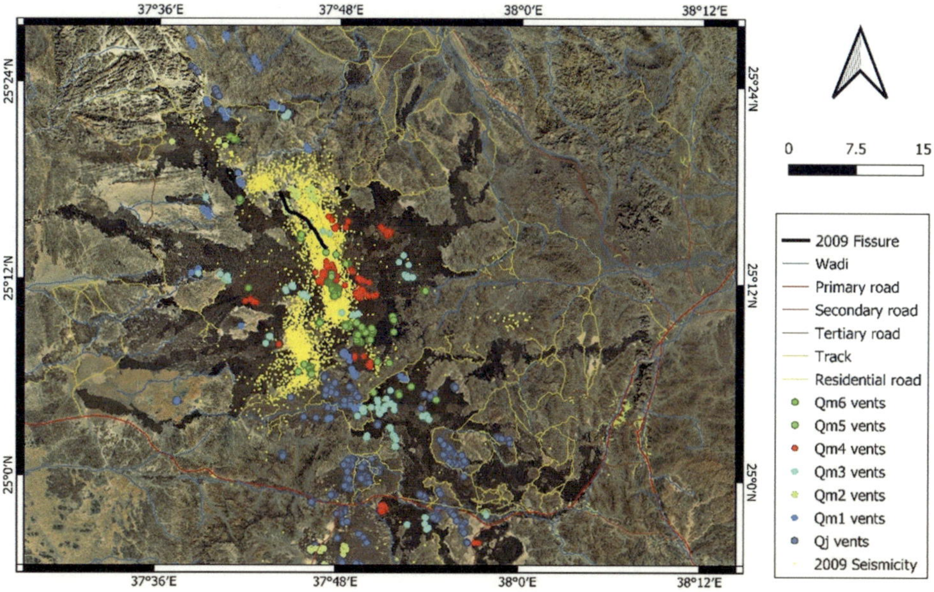

Figure 14.
12,547 earthquake registered by the Saudi Geological Survey Seismic Network in 2009, marking a failed eruption when a dyke stalled about few hundred meters below surface but opening a fissure (marked black). Nearby settlement of Al Ays in the eastern side of the field marked by green diamond.

Red Sea coastal dikes [56]. The intersection may represent the zones of weakness though which magma can ascend. Zahran and El-Hady [58] noted that the location of the swarm corresponds to a distinct linear tectonic element, the dike that is a part of Red Sea Coast Plain Fault system [59, 60]. This suggests that there has been local reactivation of the dike, although the dike system appears to be largely inactive along most of its length (at least 1,500 km). As noted in the previous section, this large system of nearly vertical dikes, typically 200–300 m wide, runs parallel to the Red Sea coast. Based on available geological information summarized by Al-Amri et al. [56], Zahran and El-Hady [58] suggested that at least four lava eruptions have occurred during the last 4500 years. The eruptions may have been accompanied by strong earthquakes or earthquake swarms. The recent earthquake activity in the area has shown that seismic events may also occur without lava eruption; for example, owing to dike intrusions like the 2009 event. The present swarm activity is concentrated in the northern part of Harrat Lunayyir, however the areas around other intersections may also be considered as the zones where future intrusions and corresponding earthquakes may occur. The Harrat Lunayyir area was identified as a seismic source zone in recent studies of seismic source zonation in northwestern Saudi Arabia [49, 50, 61, 62]. Zahran [62] inferred from seismic tomography studies that long-term volcanic and seismic activity in the area may be caused by the existence of a crustal magma conduit that transports magma from the mantle into the crust. The great abundance of monogenetic volcanoes and volcanic field in western Saudi Arabia suggesting fundamental melt generation and transportation processes controlled by various lithosphere structure. The Arabian plate's lithosphere has been commonly modeled from inversion of P- and S-wave receiver functions [53, 63, 64]. These studies suggest an average crustal thickness of ~40 km for the Neoproterozoic Arabian Shield. Toward the Red Sea and the Gulf of Aqaba, the crust thins rather abruptly to about 23–25 km [63, 65]. In the northwestern part of the Arabian Shield including the area of Harrat Uwayrid, just north of the study area of Harrat Lunayyir, the crustal thickness varies from 33 to 37 km, whereas it is much thicker (41–48 km) in the eastern part of the Arabian Platform in the region where Harrat Hutaymah formed [64].

4. Lunayyir volcanic field

The Harrat Lunayyir (HL) [also known as Harrat Shaqa] is one of the small and young volcanic fields in the SW Saudi Arabia, having an area of nearly 1750 km^2, with a generalized structural trend of ~N300 W, located on the western edge of the Great Escarpment and within ~100 km of the Red Sea margin [8]. The HL is composed of Cenozoic basaltic flows, without silicic magma (that makes it different from Harrat Khaybar, Kishb or Rahat), and emplaced along the arc-related plutonic rocks of diorite to tonalite composition with imprints of several cycles of metamorphism and tectonism during Tertiary–Quaternary time. The basement of Harrat Lunayyir comprises of amalgamated belts of sedimentary and metamorphic rock units that are delimited by the regional Precambrian Najd Fault System and intervened by numerous basic dykes [27, 66]. The basaltic lava field has two (main) subdivisions, Quaternary and Holocene basaltic flows [56, 67]. Using remote sensing techniques, however, at least three generations of lava flows has been identified in HL [68]. The HL is a continental volcanic field where composition of lava flows, and scoria cones varies from basanite to alkali olivine basalt to trachybasalt. In this volcanic field, at least 100 volcanic edifices (most of them are overlapping compound or aligned volcanoes with multiple vents are mapped that are aligned) along mostly a NW-SE axis following the orientations of normal faults [56, 69, 70]. Such volcanic cones are served by deep fissures cutting through the crustal rocks of diorite and tonalitic composition. Two sets of fault system which act as feeders are identified: (i) NE-SW trending set, controlled mainly by older Precambrian fault system and reactivated during the Tertiary period, and (ii) other set with NW-SE direction both somehow related to rifting and opening of the Red Sea [67]. Classification of volcanic eruptions is commonly based on characteristics such as magma composition, magma/water mass ratio, volcanic edifice size and geometry, tephra dispersal, dominant grain-size of pyroclasts and (usually eye-witnessed) column height. If the ascending melt or batches of melts reach the near-surface or surface region, it will either behave explosively or intrusively/effusively. Explosive magma fragmentation is triggered either by the dissolved magmatic volatile-content or by the thermal energy to kinetic energy conversion and expansion during magma/water interactions commonly named as phreatomagmatic or hydrovolcanic [71], producing distinctive eruption styles. These eruption styles can be classified based on the dominance of internal or external processes [2]. This distinction is particularly a valid and useful separation of volcano types associated with small-volume magma involving monogenetic volcanism. Such volcanism will be always very sensitive to small changes in the magmatic system, the near surface external or environmental conditions and the variations of all these over the total duration of the evolution of a monogenetic volcanic field over long time (thousands to millions of years). In many senses, external or environmental changes could switch the appearance of a volcanic field dramatically different over short period of time despite no changes exists within their magmatic systems (e.g., in variations of magma flux, chemistry or source regions). Preliminary investigations across the Saudi harrat showed that those regions lived long and appeared to form during pluvial periods more prone to have volcano types more distinct for magma and water explosive interactions such as the region near Hail, at the Harrat Hutaymah [72]. In Harrat Lunayyir so far, no systematic characterization of vents has been done. The existing scientific literature is rather fragmented in this subject and lacks clarity. In some papers commonly mentioned as tuff cones in the central part of the volcanic field that is a distinctly wrong description as there is no indication of such volcanic geoforms within the Harrat Lunayyir. The elevated nature, the arid conditions and the overwhelmingly effusive appearance of the volcanic fields is unlikely the area where explosive magma and abundant external (normally surface) water interaction could have been occurred, or at least not in a way that significant amount of such volcanic geoforms would have been left behind. Harrat Lunayyir, instead is a home of a great variety of small volume mildly explosive to violently explosive magmatic volatiles driven explosion controlled volcanic landforms such as Hawaiian style lava spatter cones (**Figure 15**) or scoria (cinder) cones (**Figure 16**).

Harrat Lunayyir volcanoes has lava spatter producing phases and accumulated the whole spectrum of agglutinate, spatter, agglomerate pyroclasts, rheomorphic lava flows and show high temperature heavily oxidized appearance of pyroclast piles (**Figure 17**). These welded and agglutinated cones are erosion resistant and produce rock-fall-dominated volcanic edifices over time that are apparent on satellite imagery.

Harrat Lunayyir is also abundant with apparently large volume of scoria cones, most of them are dominated with coarse grained ash and lapilli beds typical for high magma flux moderately explosive Strombolian style eruptions. Harrat Lunayyir is also a place where unusually high volume of fine ash preserved over large distances from their source vents, making typical "black sand" across the landscape. While the nature of these deposits has never been reported other than

Figure 15.
Small spatter cone front of a welded core of a scoria cone (left background) [25°22′7.14"N, 37°40′25.26"E].

Figure 16.
One of the youngest scoria cones [25°20′33.55″N, 37°41′5.92″E] within the Lunayyir Volcanic Field.

potential economic raw materials as natural resources, the exact link of them to potential eruption scenarios has not been mentioned yet in the existing scientific literature. On the preliminary remote sensing, DEM and literature data accompanied with non-specific field information indicates however, that these deposits are more extensive (**Figure 18**) and play major role on pyroclast resedimentation (**Figure 19**) and landscape evolution within the region. Their geodiversity and geoheritage value are high and they are also very characteristic for potential volcanic hazard scenario builder approaches. The target scoria cone shows clear evidence of its violent Strombolian style eruptions.

Lava flows has been mapped in 1 to 100 k scale geology mapping programs and scientific information about their nature is very general. From the available records it seems that the flow morphotypes are like other, more studied lava fields such as Harrat Rahat or Khaybar for instance. Based on the lava flow morphotype classification from Harrat Rahat [73], a full spectrum of lava flow surface texture types has been identified that can be compared with the lava flows within Harrat Lunayyir (**Figure 20**). To understand the lava flow surface textures, map them and draft a potential lava flow inundation model will provide key information to create several lava flow inundation simulations [74–78] similarly to a recent work trialed some method for the Harrat Khaybar [79] or Harrat Rahat [80, 81].

Long lava flows in other harrat such as Harrat Rahat, are commonly dammed where accumulation of lava triggered inflation and sudden outbreak showing complex surface textural diversity of typical transitional pāhoehoe lava flows from shelly pāhoehoe in proximal areas to slabby pāhoehoe to transitional ʻaʻā lava types (**Figure 21**).

Figure 17.
Typical coarse lapilli-dominated flank of a scoria cone (foreground) and strongly welded core (background) are erosion resistant landforms within Lunayyir Volcanic Field [25°22′19.31"N, 37°40′17.09"E]. Coarse ash – lapilli unit is about 20 cm thick.

Figure 18.
Extensive tephra cover from one of the youngest vents of Lunayyir Volcanic Field accumulated ash on exposed steep cliffs of the Neoproterozoic monzogranitic horst. This ash partially remobilized and form extensive volcaniclastic fans in the foothill areas of basement blocks [25°20′31.60"N, 37°40′47.86"E].

Figure 19.
Volcaniclastic fan formed on the steep slope of the Neoproterozoic basement covered by ash fall. Note the well-developed channel with side levees marking the original ash infill level of the valley [25°21′25.27″N, 37°41′20.48″E]. Large blocks in mid-view are about 1.5 m across.

Figure 20.
Pile of pāhoehoe slabs in the margin of a medial position lava flow [25°21′20.93″N, 37°41′4.86″E]. Pāhoehoe slabs are about 1 m across.

Figure 21.
Transitional lava flow type toward ʻaʻā surface textures. Note the tube that formed in this highly viscous flow that are contorted the cooling crust and formed wrinkled toothpaste type squeeze out [25°21′9.36"N, 37°41′5.97"E]. The tube is about 1.7 m high.

Figure 22.
Typical rubble pāhoehoe lava surface texture in the medial to distal sections of thick lava flows [25° 4′55.68"N, 37°31′15.35"E]. Pāhoehoe crusts in front about 1 m across.

Long lava flows in their distal regions became more like rubble pāhoehoe where the original lava crusts are heavily disaggregated and form isometric blocks of various parts of former slabs (**Figure 22**).

Such long lava flows in distal areas are also commonly disaggregated by post emplacement superficial processes as water during intense rainfall period tend to follow lava tube network and gradually disaggregate the flow main bodies. Preliminary desktop research revealed that Harrat Lunayyir has such geological pattern, and those regions are critical to understand and draft a workable eruption (lava flow emplacement) scenario model.

5. Volcanic hazard types identified in the northern Harrat Lunayyir

Volcanic hazard types within the study area have been identified based on geological mapping of the volcanic terrain accompanied with spot field observations

Volcanic Hazard Scenario	Impact	Intensity	Duration	Mitigation
Pāhoehoe lava flows	Well-defined zones of smooth lava flow propagation (in main flow channels flow speed can reach over 10 m/s speed). Hot surface areas, lava tube formation (it can be 20+ km^2 area, main lava tubes can reach 100 s of metres long forming lava feeding arteries) Lava tube collapse, skylight opening (on top of flow field) Secondary lava squeeze (meters within the flow margin). In proximal areas flow thickness estimated to be around 10 but potentially even reaching 50 m in the study area.	The intensity of this event is moderate to low. It can be closely watched (within tens of metres)	Total flow emplacement estimated to be in the weeks range. Once lava flow starts, expect an event to last for weeks.	When lava flow effusion starts, control access to active flow fronts and margins. Observe localized pond formation, in real time move objects from regions where marginal squeeze out can take place. When lava appear unexpectedly far from source, expect lava tube formation. Localize lava tube and try to break it to release heat and diver flow. Generally, move out potential flow path.
Transitional lava flows	Common channel axis, and medial to distal flow type, where pāhoehoe crusts are break up and transported on surface of active flow. Such flow moves as a combination of active tube network in its interior, and caterpillar-style in the broad expanded regions and margin. Such flow can travel over 10 km from their source and following topography, but capable climb over obstacles way bigger than the flow thickness.	Flow movement considered to be low intensity but can impact tens of km^2 areas.	Hours to month. In the studied area flow areas and estimated volumes indicates that such flows likely to be active over a month	Observe the lava flow field evolution. Generate real-time flow movement prediction and if flow tend to runover important sites, generate mechanical barriers to divert it. Remove objects from pathways

Volcanic Hazard Scenario	Impact	Intensity	Duration	Mitigation
Lava flow ponding	Lava flow speed can change over terrains and facilitate temporal slowdown of stoppage when ponding can generate inflated zones. In ponded zones flow thickness can be dramatically more than elsewhere.	Low intensity event	Can take hours to weeks to form an inflated ponded flow but if the pond became too large can breakout within minutes and can generate destructive cascading flows.	Locate and observe potential ponding zones. Get ready for deflation event and maintain zones where flow may breakout with care.
Lava spatter	Magmatic volatile pressure can drive lava up in the air (water hose effect). According to the magma flux, lava can nearly continuously spray out, then accumulated in spatter fashion around the vent.	Low intensity.	Spatter cones can be active over weeks to month, but they also can change their activity abruptly. Spatter cone rarely grow over 50-m, and spatter deposits can reach about few hundreds of metres from the active vent.	Locate the active vent and maintain distance from active spattering. Consider vent geometry changes that can quickly change the main thrust direction of the spattering.
Scoria (cinder) cone growth	Normally lava spattering and effusion initiate scoria cone growth when the vent localization along dike-driven fissures completed, normally within days of the onset of the eruption. Scoria cones erupt lapilli size highly vesicular pyroclasts (scoria or cinder) that are normally lifted to about 1–2 km height and deposited around the active vent. Scoria cones rarely grow over 100 tall. The area scoria cones are average by their size parameters.	Low to medium intensity events	Normally scoria cones are active over weeks to month, rarely years. Their impact zone gradually growing, but the changes of the scenario can be easily monitored in real time. The growth of the scoria cone changes the landscape locally.	Monitor the active vents and its regular Strombolian-style explosive eruptions. Note eruption style changes and regularity. Maintain distance from the cone that is normally in a km range exclusion zone around the cone is sufficient.
Ballistic impact	During scoria cone growth the volcanic conduit time to time can be chocked by rolled back debris, accumulate lava ponds where degassing can produce denser lava pods or have a colder and denser top on the active lava pod within the crater. Gas bubble outburst can disrupt those materials and initiate ballistic bomb showers around the cone. The ballistic field location heavily depends on the vent geometry and can change rapidly.	Low to medium intensity event	Ballistic impact can be destructive and deadly as heavy bomb can hit object directly. Normally ballistic bombs cannot travel further than few kilometers from their source vents. At the study area ballistic bomb fields have been identified about a km from the inferred sources.	Observing the cone growth and the vent geometry changes can predict the potential direction of ballistic shower initiation. Keep out from those zones that is normally within few kilometers from the vent. Hard hat can help to reduce impact effect if moving in to hazard zone required.

Volcanic Hazard Scenario	Impact	Intensity	Duration	Mitigation
Cone collapse, rafting	Due to the mechanical and erosional effect of an ongoing scoria cone eruption, part of the cone can collapse and rafter away on top of active lava outflows. In the study area clear evidence showing at least 5 minor collapse event that left a crater opening toward the west at the youngest volcano of the region named as Target Volcano.	Locally intensive and destructive event.	Scoria cones are commonly collapsing due to the explosive activity and the lava outflow that erode one side of the cone (downflow side). Cone collapse can take place rapidly but the situation to reach tipping points can take days to month.	Pay more attention on the downslope side of the eruption surface where potential collapse can take place.
Sub-Plinian explosive eruption	If magma flux, rise rate and the chemical conditions of the fresh magma permit, magma fragmentation can reach a level that can provide pyroclast and upward gas thrust movement above the vent lifting pyroclasts up to the range of 5 to 15 km height. Fine ash can be drifted away and fall back over the land generating typical ash plains and tephra blanket like what observed within Option 4 area.	Moderate intensity but high impact event over large areas.	Ash fall can take place for hours and can repeat several times. From the region it is evident that sub-Plinian stage has been reached in an eruptive episode that produces potentially at least 12 individual ash fall event without significant time break between them. The area impacted can reach 100 km^2. In the study area ash must have covered the entire region that has been stripped away recently due to rainfall over steep terrain, Ash particles however can be recognized in the entire area of the Option 4 region.	When ash fall observed, use face mask, detach aircon filters, or water collector filters and move out. If you drove in ash fall replace air filter in car. Weight of ash fall can be substantial, in proximal areas collapse of roofs need to be considered.
Volcanic gas (vog)	Surface degassing of erupted lava flow that release CO_2, SO_2, CO, H_2O and F as most important substances.	Low intensity, high impact event	The region lava flows chemistry is suitable to have lava flows with substantial vog production. Vog can be carried away from the active region over 100 km.	Use mask, observe gas concentration, and reduce exposure. After event check fittings and electricity connections as acid accumulation can cause rapid corrosion.

Table 1.
Eruption scenario table for a typical expected volcanic eruption for the Option 4 target area showing all the recognized hazards established across the whole volcanic field of Lunayyir.

in critical locations identified during the mapping. The volcanic features are clearly the result of dispersed volcanism representing at least 600 ka eruptive history that yielded about 50 vents to be formed within the northern region of the Lunayyir Volcanic Field that is about 5 to 10% of the total vent number of Harrat Lunayyir. In the study area about 20 volcanic edifices has been recognized as one volcanic edifice (volcano) commonly associated with fissure aligned chain of vents. Here we focus on the eruption types and styles as all these information will serve a good base for developing realistic eruption scenarios without considering their probability to occur other than recognizing them as most likely and less likely. The identified volcanic hazard types are either effusive (lava effusion) or light to moderate intensity explosive eruptions where relatively low viscosity basaltic magma erupted feeding free flowing lava flows that gradually became stickier and generating complex lava fields. Explosive eruptions are typically magmatic volatile (gas) driven eruptions with no sign of phreatomagmatic (magma and water) explosive interaction riven events. Explosive eruptions range from very mild to moderate explosivity types (Volcanic Explosivity Index: VEI = 0–2) with a maximum explosivity inferred to be capable to form sustained eruption column and expanding stabile eruption collum up to 10 km above the vent reaching VEI of 3–4 [82].

Identification of volcanic eruption types is based on mapping eruptive products, observing surface texture characteristics of lava flows, and macro and microtextures of pyroclasts [82]. Recognition of specific eruption types then can be scaled through desktop works by analogies to extract common parameters that can be used for determining regions the identified volcanic eruption type may impact, define exclusion, or buffer zones.

Eruption scenarios are short realistic narratives that summarize the potential type of eruptions, their timelines and impact. Eruption scenario descriptions are very effective information transfer tools as they clearly outline the potential events and their impact [83, 84]. Eruption scenario documentation, or eruption scenario builder is a geology-based method that is widely used in volcanology, especially in volcanic fields where probabilistic methods are difficult to implement or simple too complicated to translate to plain language [85]. The eruption scenarios are drafted based on the collated direct geological observations and the desktop data analysis of the entire field (**Table 1**). In this method, we find two types of volcanic eruption scenario such as well-localized eruption point related and fissure eruptions. Both eruption types have examples in the past 600 ka history of Lunayyir. Somehow the fissure aligned eruption types however more common, and probably can be viewed as a likely eruption scenario future eruption expected to follow. We selected 2 locations where we placed a potential future eruption either point or fissure controlled following the parameters we obtained from the geological mapping. We run a QGIS-based Q-LAVHA lava inundation simulation [75] from a point source placed just in the NW flank of the youngest vent of the northern Lunayyir region (**Figure 23**). In addition, we also run a simulation from a fissure (about 1.5 km long) following the NW-SE strike in the same location (**Figure 24**). We run lava inundation simulation from another point (**Figure 25**) and fissure (**Figure 26**) model from an older vent location that is within a narrow gorge to see how lava would fill the gorge. In each model we used generous lava flow thickness of 25 m (about twice as the value of the DEM resolution of 12.5 m) with a buffer of 50 m to let the calculation able to move lava over of landscape of inaccurate elevation data.

Figure 23.
Lava flow inundation simulation from a likely point source (vent) showing areas where lava likely could flow. Q-LAVHA model was using 50 km flow length, Manhattan simulation with lava maximum thickness of 25 m allowing up to 50 m obstacles to cross. These parameters are likely overestimates, but these values allow the simulation not stopping in morphological irregularities likely to be an artifact of the coarse resolution of the digital elevation model. Note that the lava with these parameters would cover most of the areas currently lava field in a basin-like morphology.

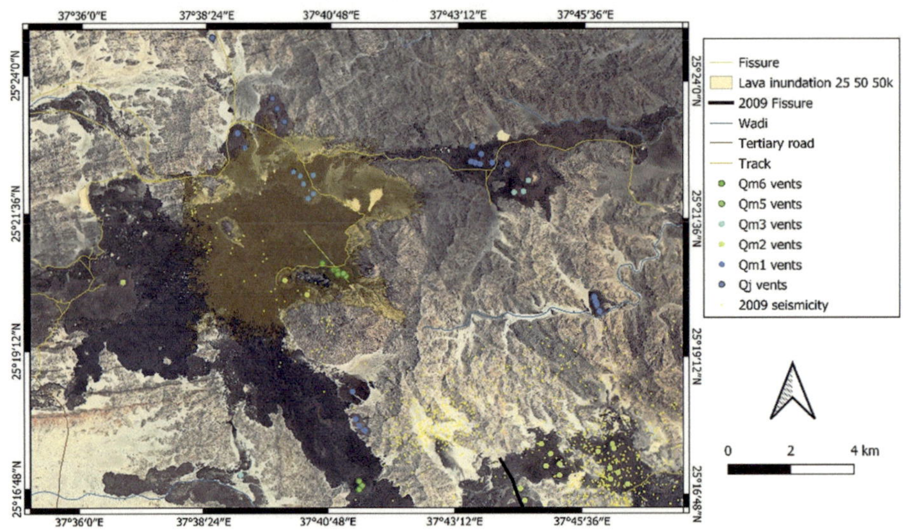

Figure 24.
Lava flow inundation simulation from a fissure source (fissure) showing areas where lava likely could flow. Q-LAVHA model was using 50 km flow length, Manhattan simulation with lava maximum thickness of 25 m allowing up to 50 m obstacles to cross. Fissure length is about 3 km and oriented in the same NW-SE direction than other structural elements in the region. These parameters are likely overestimates, but these values allow the simulation not stopping in morphological irregularities likely to be an artifact of the coarse resolution of the digital elevation model. Note that the lava with these parameters would cover most of the areas currently lava field in a basin-like morphology. Compare this model with the simulation results from a point source. With a fissure model the lava inundation area can increase significantly.

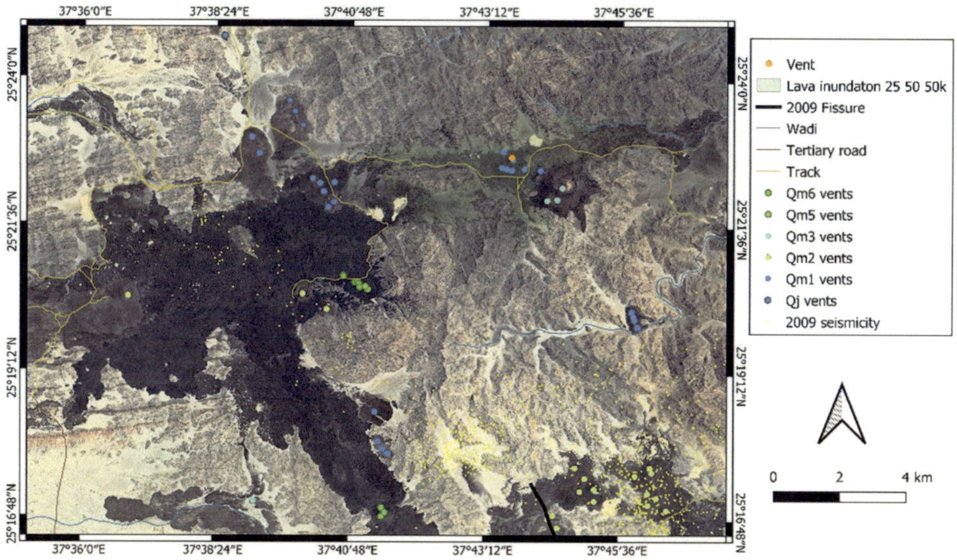

Figure 25.
Lava inundation simulation from a vent located in an older vent cluster area in northern Harrat Lunayyir. Not the capability of the lava flow able to be captured in the narrow gorge and move both toward the west and east.

Figure 26.
Lava inundation simulation from a kilometer long fissure placed in an area of older vent clusters in northern Harrat Lunayyir. Note that in such scenario lava were able to fill perfectly the East-West valley network.

6. Volcanic hazard scenarios for Harrat Lunayyir

In **Table 1** we provided a summary of the volcanic hazard types identified from the geological record of the Pleistocene to Holocene Lunayyir Volcanic Field. To visualize the specific hazards with some key parameters a graphic representation is

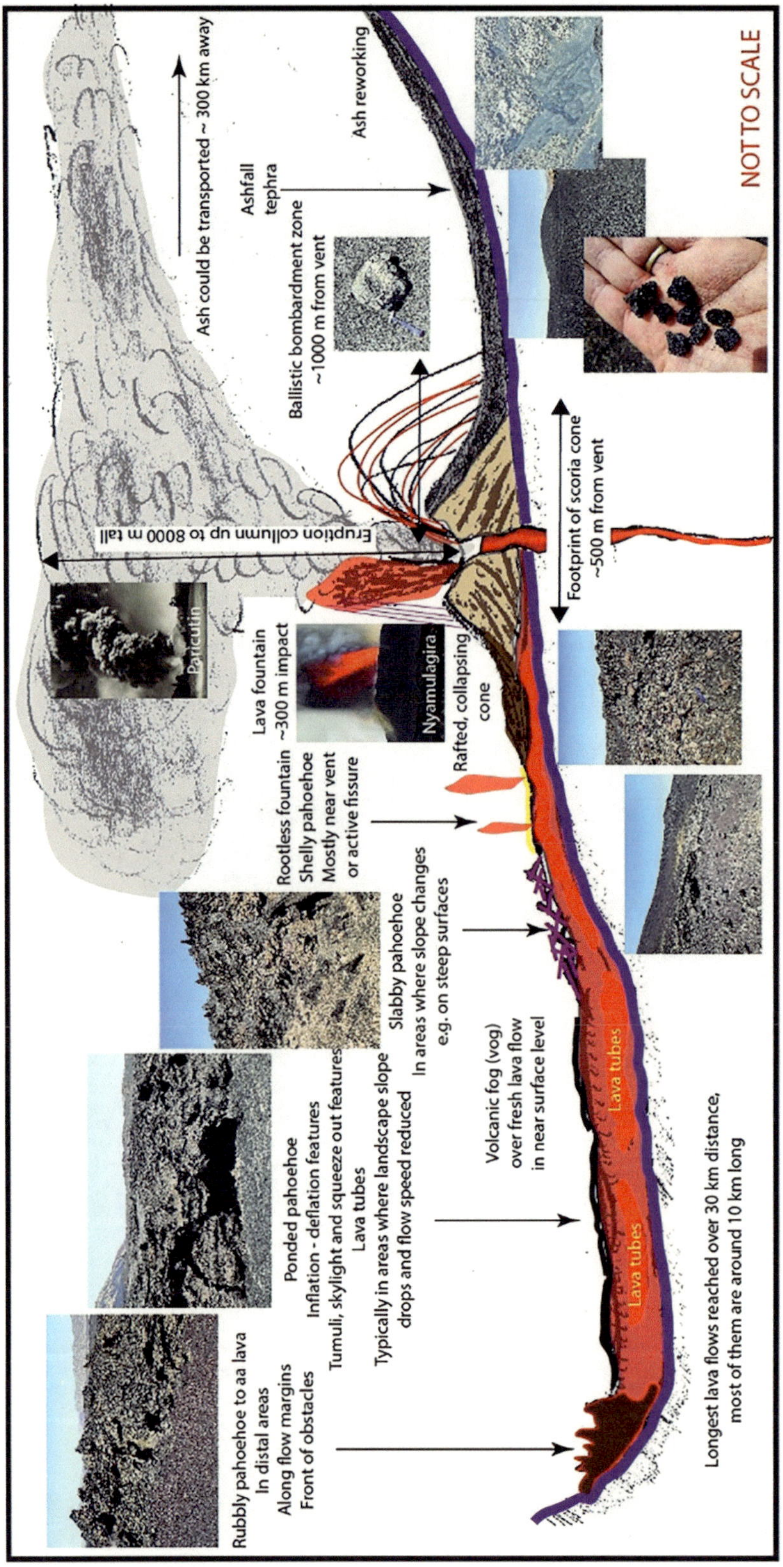

Figure 27.
Summary of the identified volcanic hazard scenarios from the northern Harrat Rahat. Note that the diagram is not to scale.

given in **Figure 27** based on the field observations and limited geological mapping. While this spectrum of identified volcanic hazards is specific to the northern Harrat Lunayyir, preliminary field work confirms that an identical scenario set is likely to apply for the whole volcanic field. The most common volcanic hazard scenarios are the lava spatter and scoria cone building mild explosive eruptions dominated by Hawaiian style explosive eruptions with a potential eruption column may reach about 1 km height and directly impacting an area about 500 meters away from their source vents. Such lava fountain-dominated eruptions are likely form fissures at least 300 m length with a potential to generate lava curtains along several kilometer long fissures. As the 2009 eruption was labeled as a failed eruption, their nearly 20 km long fissure could set an upper limit of a fissure length that could be a row of active vents. However, this scenario cannot be ruled out, its likelihood is rather low to occur. More common and high impact scenario when a fissure opening with active vents along 500 meters to about 3 km length. From field mapping, vent alignments in such length are commonly documented and likely to expected to form in future eruptions. Vent localization is a gradual process associated with initial fissure eruption may reach mature stage and establishment of 1 to few distinct vents that may also go through eruption style changes and governed by more like Strombolian style explosive eruptions. Volcanic cones mapped in the field are the testament for the existence of such process that ultimately led to the growth of scoria cones. Such scoria cones can pose significant hazard by ballistic projectiles to even ballistic curtain deposits potentially within a kilometer from their vent sources. Large cones may reach eruptions styles operate by sustained eruption columns may be able to push pyroclasts up to 6 km above the crater and achieve ash dispersal over 250 km if sustain magma supply is given. Such eruptions are the likely sources of thick (up to meters) tephra cover in proximal settings that can still be in few dm thickness over 3 km from their source vents.

7. Conclusion

In this chapter we provided a comprehensive summary of the volcanic eruption scenarios extracted from the geological record of the 600 ka old, still active Lunayyir Volcanic Field. It is a monogenetic volcanic field with dominantly magmatic effusive and explosive eruption styles that built fissure aligned vents of spatter and scoria cones following the main NW-SE structural pattern of the Neoproterozoic basement rocks. The northern Harrat Lunayyir particularly an interesting place as isolated intermountain basins are filled with transitional type lava flows erupted from local fissure aligned vents. Narrow gorges also captured lava flows and along morphological steps acceleration of lava flows generated slabby pāhoehoes while in more distal areas rubbly pāhoehoe textures. Along major obstacles or far from the main flow axis, lava crystallinity increased generated high viscosity to promote ʻaʻā lava texture formation. Most of the identified volcanic eruption scenarios and hazards are low intensity and manageable by careful planning for future eruption, however especially the lava flows transitional style make an effusive event a potentially high impact event that cannot be controlled or mitigated effectively. One of the most unique eruption scenario and volcanic hazard is the potential sub-Plinian or violent Strombolian style of eruptions. The field observations confirmed the existence of widespread thick tephra blanket nearly in every region for the greater Harrat Lunayyir, suggesting that such violent potentially VEI3 to 4 explosive events were common in the eruption course of the Harrat Lunayyir. Ash falls with the mapped and observed thickness and conditions inferred to be capable to be transported over 250 km, and in the proximal areas accumulate several decimeter thick ash plains.

Acknowledgements

Support on this research by the Saudi Geological Survey is greatly appreciated especially to Mr. Turki Sehli and Mr. Khaled Yousef at the National Center of Earthquakes and Volcanoes at the Saudi Geological Survey. Reviewers' comments are greatly appreciated as it has elevated the quality of this work.

This manuscript is benefited greatly from a volcanic hazard assessment project conducted for the Red Sea Global tourism development projects.

Conflict of interest

The authors declare no conflict of interest.

Author details

Károly Németh[1*], Mostafa Toni[2], Vladimir Sokolov[3], Abdulrahman Sowaigh[3], Mahmoud Ashor[3] and Fawaz Moqeem[3]

1 Saudi Geological Survey, Jeddah, Kingdom of Saudi Arabia, Institute of Earth Physics and Space Science, Sopron, Hungary, Volcanic Risk Solutions, Massey University, Palmerston North, New Zealand

2 Saudi Geological Survey, Jeddah, Kingdom of Saudi Arabia, Geology Department, Faculty of Science, Helwan University, Cairo, Egypt

3 Saudi Geological Survey, Jeddah, Kingdom of Saudi Arabia

*Address all correspondence to: nemeth.k@sgs.gov.sa

References

[1] Smith IEM, Németh K. Source to surface model of monogenetic volcanism; a critical review. Special Publication—Geological Society of London. 2017;**446**(1):1-28. DOI: 10.1144/SP446.14

[2] Nemeth K, Kereszturi G. Monogenetic volcanism: Personal views and discussion. International Journal of Earth Sciences. 2015;**104**(8):2131-2146. DOI: 10.1007/s00531-015-1243-6

[3] Kereszturi G, Németh K. Monogenetic basaltic volcanoes: Genetic classification, growth, geomorphology and degradation. In: Németh K, editor. Updates in Volcanology - New Advances in Understanding Volcanic Systems. Rijeka, Croatia: IntechOpen; 2012. pp. 3-88. DOI: 10.5772/51387

[4] Le Corvec N, Menand T, Rowland JV. Interaction of ascending magma with pre-existing crustal structures: Insights from analogue modeling. In: Eos Trans AGU. 2010;American Geophysical Union, Fall Meeting 2010, Fall Meet. Suppl.,:abstract #V51F-08.

[5] Le Corvec N, Bebbington MS, Lindsay JM, McGee LE. Age, distance, and geochemical evolution within a monogenetic volcanic field : Analyzing patterns in the Auckland volcanic field eruption sequence. Geochemistry Geophysics Geosystems. 2013;**14**(9):3648-3665. DOI: 10.1002/ggge.20223

[6] Le Corvec N, Sporli KB, Rowland J, Lindsay J. Spatial distribution and alignments of volcanic centers: Clues to the formation of monogenetic volcanic fields. Earth-Science Reviews. 2013;**124**:96-114. DOI: 10.1016/j.earscirev.2013.05.005

[7] Moufti MR, Németh K. In: Eder W, Bobrowsky PT, Martínez-Frías J, editors. Geoheritage of Volcanic Harrats in Saudi Arabia. Heidelberg: Springer; 2016. pp. 1-194

[8] Coleman RC, Gregory R, Brown GF. Cenozoic Volcanic Rocks of Saudi Arabia. Open-File Report 83-788. Reston, VA: U.S. Geological Survey; 1983. pp. 1-82. DOI: 10.3133/ofr83788

[9] Camp VE, Roobol MJ. Upwelling asthenosphere beneath Western Arabia and its regional implications. Journal of Geophysical Research-Solid Earth. 1992;**97**(B11):15255-15271. DOI: 10.1029/92JB00943

[10] Camp VE, Roobol MJ, Hooper PR. The Arabian continental alkali Basalt Province 2. Evolution of Harrats Khaybar, Ithnayn, and Kura, Kingdom of Saudi-Arabia. Geological Society of America Bulletin. 1991;**103**(3):363-391. DOI: 10.1130/0016-7606(1991)103<0363:TACABP>2.3.CO;2

[11] Runge MG, Bebbington MS, Cronin SJ, Lindsay JM, Moufti MR. Integrating geological and geophysical data to improve probabilistic hazard forecasting of Arabian shield volcanism. Journal of Volcanology and Geothermal Research. 2016;**311**:41-59. DOI: 10.1016/j.jvolgeores.2016.01.007

[12] Runge MG, Bebbington MS, Cronin SJ, Lindsay JM, Moufti MR. Sensitivity to volcanic field boundary. Journal of Applied Volcanology. 2015;**4**(1):22. DOI: 10.1186/s13617-015-0040-z

[13] Runge MG, Bebbington MS, Cronin SJ, Lindsay JM, Kenedi CL, Moufti MRH. Vents to events:

Determining an eruption event record from volcanic vent structures for the Harrat Rahat, Saudi Arabia. Bulletin of Volcanology. 2014;**76**(3):804. DOI: 10.1007/s00445-014-0804-z

[14] Coleman RG. Geologic evolution of the Red Sea. In: Oxford Monographs on Geology and Geophysics. Vol. 24. New York: Oxford University Press; 1993. pp. 1-186

[15] Brown GF, Schmidt DL, Huffan AC Jr. Geology of the Arabian peninsula, shield area of western Saudi Arabia. US Geological Survey professional paper. 1989;**560-A**:1-188. DOI: 10.3133/pp560A

[16] Abdelwahed MF, Alqahtani F, El-Masry NN, Aboud E, El-Hady S, Faried A, et al. The role of the Red Sea rift and the inherited geological structures in the seismo-volcanic activity along the rift flanks. Journal of Asian Earth Sciences. 2024;**260**:105964. DOI: 10.1016/j.jseaes.2023.105964

[17] Abdelwahed MF, Alqahtani FA, El-Masry NN, El-Hady SM. Insights into the relationship between the Red Sea rift-related structures and the seismo-volcanic activity in Harrat Lunayyir, Saudi Arabia: A seismic tomography study. Journal of Asian Earth Sciences. 2023;**241**:105484. DOI: 10.1016/j.jseaes.2022.105484

[18] Li B, Németh K, Palmer J, Palmer A, Wu J, Procter J, et al. Basic volcanic elements of the Arxan-Chaihe volcanic field, Inner Mongolia, NE China. In: Németh K, editor. Updates in Volcanology - Transdisciplinary Nature of Volcano Science. Rijeka, Croatio, London, UK: IntechOpen; 2021. Available from: https://www.intechopen.com/chapters/73823

[19] Duncan RA, Al-Amri AM. Timing and composition of volcanic activity at Harrat Lunayyir, western Saudi Arabia. Journal of Volcanology and Geothermal Research. 2013;**260**:103-116. DOI: 10.1016/j.jvolgeores.2013.05.006

[20] Adams RD, Barazangi M. Seismotectonics and seismology in the Arab region: A brief summary and future plans. Bulletin of the Seismological Society of America. 1984;**74**(3):1011-1030. DOI: 10.1785/BSSA0740031011

[21] Al-Amri AM. Seismotectonics and seismogenic source zones of the Arabian platform. Frontiers in Earth Science. 2013;**5**:295-316. DOI: 10.1007/978-3-642-30609-9_15

[22] Al-Husseini MI. Origin of the Arabian plate structures: Amar collision and Najd rift. GeoArabia. 2000;**5**(4):527-542. DOI: 10.2113/geoarabia0504527

[23] Bohannon RG. Tectonic configuration of the Western Arabian continental margin, southern Red Sea. Tectonics. 1986;**5**(4):477-499. DOI: 10.1029/TC005i004p00477

[24] Girdler RW. The Afro-Arabian rift system-an overview. Tectonophysics. 1991;**197**(2-4):139-153. DOI: 10.1016/0040-1951(91)90038-T

[25] Greenwood WR, Anderson RE, Fleck RJ, Roberts RJ. Precambrian geologic history and plate tectonic evolution of the Arabian shield. Saudi Arabian Directorate General of Mineral Resources, Bulletin. 1980;**24**:1-35

[26] Le Pichon X, Gaulier JM. The rotation of Arabia and the Levant fault system. Tectonophysics. 1988;**153**(1-4):271-294. DOI: 10.1016/0040-1951(88)90020-0

[27] Stern RJ, Johnson P. Continental lithosphere of the Arabian plate: A geologic, petrologic, and geophysical

synthesis. Earth-Science Reviews. 2010;**101**(1-2):29-67. DOI: 10.1016/j.earscirev.2010.01.002

[28] Vita-Finzi C. Neotectonics at the Arabian plate margins. Journal of Structural Geology. 2001;**23**(2-3):521-530. DOI: 10.1016/S0191-8141(00)00117-6

[29] Vita-Finzi C. Neotectonics on the Arabian Sea coasts. Geological Society Special Publication. 2002;**195**:87-96. DOI: 10.1144/GSL.SP.2002.195.01.06

[30] Johnson PR. The Arabian–Nubian shield, an introduction: Historic overview, concepts, interpretations, and future issues. In: Hamimi Z, Fowler A-R, Liégeois J-P, Collins A, Abdelsalam MG, Abd Ei-Wahed M, editors. The Geology of the Arabian-Nubian Shield. Cham: Springer International Publishing; 2021. pp. 1-38. DOI: 10.1007/978-3-030-72995-0_1

[31] Bosworth W, Huchon P, McClay K. The Red Sea and Gulf of Aden basins. Journal of African Earth Sciences. 2005;**43**(1-3):334-378. DOI: 10.1016/j.jafrearsci.2005.07.020

[32] Roobol MJ, Camp VE. Geologic map of the Cenozoic lava field of Harrat Kishb, Kingdom of Saudi Arabia. Saudi Arabia Directorate Gen Min Res, Map GM-132, 1:250,000 Geol map and text. Jeddah: Saudi Geological Survey; 1991

[33] Roobol MJ, Camp VE. Structural control of young basaltic fissure eruptions in the plateau basalt area of the Arabian plate, Northeast Jordan - Comment. Journal of Volcanology and Geothermal Research. 1990;**43**(1-4):365-366

[34] Camp VE, Roobol MJ. The Arabian continental alkali Basalt Province. 1. Evolution of Harrat-Rahat, Rahat, Kingdom-of-Saudi-Arabia. Geological Society of America Bulletin. 1989;**101**(1):71-95. DOI: 10.1130/0016-7606(1989)101<0071:TACABP>2.3.CO;2

[35] Camp VE, Hooper PR, Roobol MJ, White DL. The Madinah eruption, Saudi Arabia: Magma mixing and simultaneous extrusion of three basaltic chemical types. Bulletin of Volcanology. 1987;**49**(2):489-508. DOI: 10.1007/BF01245475

[36] Ebinger CJ, Sleep NH. Cenozoic magmatism throughout East Africa resulting from impact of a single plume. Nature. 1998;**395**(6704):788-791. DOI: 10.1038/27417

[37] Mukhopadhyay M, Mukhopadhyay B, Mogren S, Nandi BK, Ibrahim E. Regional significance of crustal and sub-crustal rheological heterogeneities beneath the Harrat Lunayyir and their continuity into the neighboring harrats, Western Saudi Arabia – Perspectives of the Afar plume activity. Journal of African Earth Sciences. 2022;**186**:104432. DOI: 10.1016/j.jafrearsci.2021.104432

[38] Stelten ME, Downs DT, Dietterich HR, Mahood GA, Calvert AT, Sisson TW, et al. The duration and characteristics of magmatic differentiation from basalt to trachyte within the Matan volcanic Center, northern Harrat Rahat, Kingdom of Saudi Arabia. In: US Geological Survey Professional Paper. Vol. 2023. Reston, VA: U.S. Geological Survey; 2023. pp. 1-56. DOI: 10.3133/pp1862F

[39] Stelten ME, Downs DT, Champion DE, Dietterich HR, Calvert AT, Sisson TW, et al. Eruptive history of northern Harrat Rahat—Volume, timing, and composition of

volcanism over the past 1.2 million years. In: Professional Paper. Reston, VA: U.S. Geological Survey; 2023. p. 46. DOI: 10.3133/pp1862D

[40] Robinson JE, Downs DT. Overview of the Cenozoic geology of the northern Harrat Rahat volcanic field, Kingdom of Saudi Arabia. Professional Paper. U.S. Geological Survey; 2023. p. 20. DOI: 10.3133/pp1862R

[41] Pallister JS, McCausland WA, Jonsson S, Lu Z, Zahran HM, El Hadidy S, et al. Broad accommodation of rift-related extension recorded by dyke intrusion in Saudi Arabia. Nature Geoscience. 2010;**3**(10):705-712. DOI: 10.1038/ngeo966

[42] Jonsson S, Pallister J, McCausland W, El-Hadidy S. Dyke intrusion and arrest in Harrat Lunayyir, western Saudi Arabia, in April-July 2009. Geophysical Research Abstracts. 2010;**12**:EGU2010-7704

[43] Al-Amri AM, Abdelrahman K, Mellors R, Harris D. Geothermal potential of Harrat Rahat, northern Arabian shield: Geological constraints. Arabian Journal of Geosciences. 2020;**13**(6):268. DOI: 10.1007/s12517-020-5254-7

[44] Mellors RJ, Camp V, Harris D, Al-Amri A, Investigating Potential Geothermal Resources in Western Saudi Arabia: Transactions—Geothermal Resources Council; 2015. Available from: https://alamrigeo.com/input/docs/technical-reports/Geothermal-Resources.pdf [Accessed: December 1, 2024]

[45] Jasiewicz J, Stepinski TF. Geomorphons-A pattern recognition approach to classification and mapping of landforms. Geomorphology. 2013;**182**:147-156. DOI: 10.1016/j.geomorph.2012.11.005

[46] Fidelus-Orzechowska J, Wronska-Walach D, Cebulski J, Zelazny M. Effect of the construction of ski runs on changes in relief in a mountain catchment (inner Carpathians, southern Poland). Science of the Total Environment. 2018;**630**:1298-1308. DOI: 10.1016/j.scitotenv.2018.02.305

[47] Kholia N, Kotlia BS, Joshi N, Kandregula RS, Kothyari GC, Dumka RK. Quantitative analysis of Khajjiar and Rewalsar lakes and their surroundings, Himachal Pradesh (India): Remote sensing and GIS-based approaches. Journal of Applied Geophysics. 2023;**211**:104976. DOI: 10.1016/j.jappgeo.2023.104976

[48] Płaczkowska E, Cebulski J, Bryndza M, Mostowik K, Murawska M, Rzonca B, et al. Morphometric analysis of the channel heads based on different LiDAR resolutions. Geomorphology. 2021;**375**:107546. DOI: 10.1016/j.geomorph.2020.107546

[49] Zahran HM, El-Hady SM, Sokolov V. Integrated Studies for Assessment of Seismic Hazard in Harrat Lunayyir: Geophysical and Seismological Studies and Seismic Hazard Assessment for Cenozoic Lava Field in. Western Saudi Arabia: LAP LAMBERT Academic Publishing; 2018

[50] Zahran HM, El-Hady SM, Abuelnaga HS. Aeromagnetic data over Harrat Lunayyir and surrounding areas, western Saudi Arabia. Arabian Journal of Geosciences. 2017;**10**(3):63. DOI: 10.1007/s12517-017-2849-8

[51] Abdelfattah AK, Mogren S, Mukhopadhyay M. Mapping b-value for 2009 Harrat Lunayyir earthquake swarm, western Saudi Arabia and coulomb stress for its mainshock. Journal of Volcanology and Geothermal Research. 2017;**330**:14-23. DOI: 10.1016/j.jvolgeores.2016.12.001

[52] Abdelfattah AK, Almadani S, Fnais M, Alfaifi HJ, Al-Arifi N, Al-amri A, et al. Rupture characteristics of a small-sized earthquake (M W 4.2), onshore the South Red Sea, Saudi Arabia. Journal of African Earth Sciences. 2019;**151**:315-323. DOI: 10.1016/j.jafrearsci.2018.12.014

[53] Abdelfattah AK, Al-amri A, Sami Soliman M, Zaidi FK, Qaysi S, Fnais M, et al. An analysis of a moderate earthquake, eastern flank of the Red Sea, Saudi Arabia. Earth, Planets and Space. 2020;**72**(1):34. DOI: 10.1186/s40623-020-01159-5

[54] Abdelfattah AK, Jallouli C, Fnais M, Qaysi S, Alzahrani H, Mogren S. The key role of conjugate fault system in importing earthquakes into the eastern flank of the Red Sea. Earth, Planets and Space. 2021;**73**(1):178. DOI: 10.1186/s40623-021-01513-1

[55] Abdelfattah AK, Al-amri A, Alzahrani H, Abuamarah BA. Ambient noise tomography in the upper crust of north Harrat Rahat, Saudi Arabia. Journal of King Saud University—Science. 2023;**35**(2):102523. DOI: 10.1016/j.jksus.2022.102523

[56] Al-Amri A, Fnais M, Abdelrahman K, Mogren S, Al-Dabbagh ME. Geochronological dating and stratigraphic sequences of Harrat Lunayyir, NW Saudi Arabia. International Journal of Physical Sciences. 2012;7:2791-2805. DOI: 10.5897/IJPS12.178

[57] Saibi H, Mogren S, Mukhopadhyay M, Ibrahim E. Subsurface imaging of the Harrat Lunayyir 2007-2009 earthquake swarm zone, western Saudi Arabia, using potential field methods. Journal of Asian Earth Sciences. 2019;**169**:79-92. DOI: 10.1016/j.jseaes.2018.07.024

[58] Zahran HM, El-Hady SM. Seismic hazard assessment for Harrat Lunayyir—A lava field in western Saudi Arabia. Soil Dynamics and Earthquake Engineering. 2017;**100**:428-444. DOI: 10.1016/j.soildyn.2017.06.009

[59] Roobol MJ, Kadi KA. Cenozoic Faulting in the Rabigh Area, Central West Saudi Arabia (Including the Sites of King Abdullah Economic City and King Abdullah University for Science and Technology). Saudi Geological Survey Technical Report; 2008 SGS-TR-2008-6, 1250,000 map:1-12

[60] Roobol MJ, Stewart ICF. Cenozoic faults and recent seismicity in northwest Saudi Arabia and the Gulf of Aqaba region. Saudi Geological Survey Technical Report. 2009;**SGS-TR-2008-7**:1-35

[61] Al-Arifi NS, Fat-Helbary RE, Khalil AR, Lashin AA. A new evaluation of seismic hazard for the northwestern part of Saudi Arabia. Natural Hazards. 2013;**69**(3):1435-1457. DOI: 10.1007/s11069-013-0756-1

[62] Zahran HM. Seismic Hazard Assessment in Harrat Lunayyir, Western Saudi Arabia. Jeddah, Saudi Arabia. Jeddah, KAU: King Abdulaziz University; 2017

[63] Al-Damegh K, Sandvol E, Barazangi M. Crustal structure of the Arabian plate: New constraints from the analysis of teleseismic receiver functions. Earth and Planetary Science Letters. 2005;**231**(3):177-196. DOI: 10.1016/j.epsl.2004.12.020

[64] Al Amri A, Abdelrahman K, Andreae MO, Al-Dabbagh M. Crustal and upper mantle structures beneath the Arabian shield and Red Sea. In: Roure F, Amin A, Khomsi S, Al Garni M, editors. Lithosphere Dynamics and Sedimentary Basins of the Arabian Plate and Surrounding Areas. Frontiers in

Earth Sciences. Cham: Springer; 2017. DOI: 10.1007/978-3-319-44726-1_1

[65] Almalki KA, Betts PG, Ailleres L. The Red Sea—50years of geological and geophysical research. Earth-Science Reviews. 2015;**147**:109-140. DOI: 10.1016/j.earscirev.2015.05.002

[66] Blasband B, White S, Brooijmans P, De Boorder H, Visser W. Late Proterozoic extensional collapse in the Arabian-Nubian shield. Journal of the Geological Society. 2000;**157**:615-628. DOI: 10.1144/jgs.157.3.615

[67] Mukhopadhyay B, Mogren S, Mukhopadhyay M, Dasgupta S. Incipient status of dyke intrusion in top crust—Evidences from the Al-Ays 2009 earthquake swarm, Harrat Lunayyir, SW Saudi Arabia. Geomatics Natural Hazards and Risk. 2013;**4**(1):30-48. DOI: 10.1080/19475705.2012.663794

[68] Mogren S, Saibi H, Mukh-opadhyay M, Gottsmann J, Ibrahim E-KH. Analyze the spatial distribution of lava flows in Al-Ays volcanic area, Saudi Arabia, using remote sensing. Arabian Journal of Geosciences. 2017;**10**(6):133. DOI: 10.1007/s12517-017-2889-0

[69] Baer G, Hamiel Y. Form and growth of an embryonic continental rift; InSAR observations and modelling of the 2009 western Arabia rifting episode. Geophysical Journal International. 2010;**182**(1):155-167. DOI: 10.1111/j.1365-246X.2010.04627.x

[70] Trippanera D, Ruch J, Passone L, Jónsson S. Structural mapping of dike-induced faulting in Harrat Lunayyir (Saudi Arabia) by using high resolution drone imagery. Frontiers in Earth Science-Switz. 2019;7:168. DOI: 10.3389/feart.2019.00168

[71] Németh K, Kósik S. Review of explosive hydrovolcanism. Geosciences. 2020;**10**(2):44. DOI: 10.3390/geosciences10020044

[72] Moufti MR, Nemeth K, El-Masry N, Qaddah A. Volcanic geotopes and their geosites preserved in an arid climate related to landscape and climate changes since the neogene in northern Saudi Arabia: Harrat Hutaymah (Hai'il region). Geoheritage. 2015;**7**(2):103-118. DOI: 10.1007/s12371-014-0110-3

[73] Murcia H, Németh K, Moufti MR, Lindsay J, El-Masry N, Cronin S. Late Holocene lava flow morphotypes of northern Harrat Rahat, Kingdom of Saudi Arabia: Implications for the description of continental lava fields. Journal of Asian Earth Sciences. 2014:**84**;131-145. DOI: 10.1016/j.jseaes.2013.10.002

[74] Martín-Raya N, Díaz-Pacheco J, López-Díez A, Dorta Antequera P, Cabrera A. A lava flow simulation experience oriented to disaster risk reduction, early warning systems and response during the 2021 volcanic eruption in cumbre Vieja La Palma. Natural Hazards. 2023;**117**(3):3331-3351. DOI: 10.1007/s11069-023-05989-w

[75] Mossoux S, Saey M, Bartolini S, Poppe S, Canters F, Kervyn M. Q-LAVHA: A flexible GIS plugin to simulate lava flows. Computers & Geosciences. 2016;**97**:98-109. DOI: 10.1016/j.cageo.2016.09.003

[76] Becerril L, Larrea P, Salinas S, Mossoux S, Ferres D, Widom E, et al. The historical case of Paricutin volcano (Michoacan, Mexico): Challenges of simulating lava flows on a gentle slope during a long-lasting eruption. Natural Hazards. 2021;**107**(1):809-829. DOI: 10.1007/s11069-021-04607-x

[77] Prieto-Torrell C, Rodriguez-Gonzalez A, Aulinas M, Fernandez-

Turiel JL, Cabrera MC, Criado C, et al. Modelling and simulation of a lava flow affecting a shore platform: A case study of Montana de Aguarijo eruption, El Hierro (Canary Islands, Spain). Journal of Maps. 2021;**17**(2):502-511. DOI: 10.1080/17445647.2021.1972853

[78] Vilches M, Ureta G, Grosse P, Nemeth K, Aguilera F, Aguilera M. Effusion rate estimation based on solidified lava flows: Implications for volcanic hazard assessment in the Negros de Aras monogenetic volcanic field, northern Chile. Journal of Volcanology and Geothermal Research. 2022:**422**;422. DOI: 10.1016/j.jvolgeores.2021.107454

[79] Németh K, Moufti MR. Lava flow hazard and its implication in geopark development for the active Harrat Khaybar intracontinental monogenetic volcanic field, Saudi Arabia. Land. 2023;**12**(3):705. DOI: 10.3390/land12030705

[80] Dietterich HR, Downs DT, Stelten ME, Zahran H. Reconstructing lava flow emplacement histories with rheological and morphological analyses: The Harrat Rahat volcanic field, Kingdom of Saudi Arabia. Bulletin of Volcanology. 2018;**80**(12):85. DOI: 10.1007/s00445-018-1259-4

[81] Kereszturi G, Németh K, Moufti MR, Cappello A, Murcia H, Ganci G, et al. Emplacement conditions of the 1256 AD Al-Madinah lava flow field in Harrat Rahat, Kingdom of Saudi Arabia—Insights from surface morphology and lava flow simulations. Journal of Volcanology and Geothermal Research. 2016;**309**:14-30. DOI: 10.1016/j.jvolgeores.2015.11.002

[82] Giordano G, Cas R, Wright JV. Explosive eruption styles, columns, and pyroclastic fallout deposits. In: Cas R, Giordano G, Wright JV, editors. Volcanology: Processes, Deposits, Geology and Resources. Cham: Springer International Publishing; 2024. pp. 519-691. DOI: 10.1007/978-3-319-66613-6_9

[83] Hayes JL, Tsang SW, Fitzgerald RH, Blake DM, Deligne NI, Doherty A, et al. The DEVORA Scenarios; Multi-Hazard Eruption Scenarios for the Auckland Volcanic Field. GNS Science Report; 2018. p. 138. DOI: 10.21420/G20652

[84] Hayes JL, Wilson TM, Deligne NI, Lindsay JM, Leonard GS, Tsang SWR, et al. Developing a suite of multi-hazard volcanic eruption scenarios using an interdisciplinary approach. Journal of Volcanology and Geothermal Research. 2020;**392**:106763. DOI: 10.1016/j.jvolgeores.2019.106763

[85] Ang PS, Bebbington MS, Lindsay JM, Jenkins SF. From eruption scenarios to probabilistic volcanic hazard analysis: An example of the Auckland volcanic field, New Zealand. Journal of Volcanology and Geothermal Research. 2020;**397**:106871. DOI: 10.1016/j.jvolgeores.2020.106871